ACS SYMPOSIUM SERIES **710**

Structure and Properties of Glassy Polymers

Martin R. Tant, EDITOR
Eastman Chemical Company

Anita J. Hill, EDITOR
Commonwealth Scientific and Industrial Research Organisation

Developed from a symposium sponsored by the
Division of Polymeric Materials: Science and Engineering
at the 213th National Meeting
of the American Chemical Society,
San Francisco, California,
April 13–17, 1997

American Chemical Society, Washington, DC

Library of Congress Cataloging-in-Publication Data

Structure and properties of glassy polymers / Martin R. Tant, editor, Anita J. Hill, editor.

p. cm.—(ACS symposium series, ISSN 0097–6156 ; 710)

"Developed from a symposium sponsored by the Division of Polymeric Materials: Science and Engineering at the 213th National Meeting of the American Chemical Society, San Francisco, California, April 13–17, 1997."

Includes bibliographic references and index.

ISBN 0–8412–3588–0

1. Polymers. 2. Amorphous substances.

I. Tant, Martin R., 1953– . II. Hill, Anita J. III. American Chemical Society. Division of Polymeric Materials: Science and Engineering. IV. American Chemical Society. Meeting (213th : 1997 : San Francisco, Calif.) V. Series.

TA455.P58S77444   1998
620.1′92—dc21   98–26922
                                                                                      CIP

Foreword

THE ACS SYMPOSIUM SERIES was first published in 1974 to provide a mechanism for publishing symposia quickly in book form. The purpose of the series is to publish timely, comprehensive books developed from ACS sponsored symposia based on current scientific research. Occasionally, books are developed from symposia sponsored by other organizations when the topic is of keen interest to the chemistry audience.

Before agreeing to publish a book, the proposed table of contents is reviewed for appropriate and comprehensive coverage and for interest to the audience. Some papers may be excluded in order to better focus the book; others may be added to provide comprehensiveness. When appropriate, overview or introductory chapters are added. Drafts of chapters are peer-reviewed prior to final acceptance or rejection, and manuscripts are prepared in camera-ready format.

As a rule, only original research papers and original review papers are included in the volumes. Verbatim reproductions of previously published papers are not accepted.

ACS BOOKS DEPARTMENT

Contents

PHYSICAL AGING

MECHANICAL PROPERTIES

TRANSPORT PROPERTIES

INDEXES

Preface

This volume was developed from selected contributions to the Symposium on Structure and Properties of Glassy Polymers, sponsored by the Division of Polymeric Materials: Science and Engineering (PMSE) and held during the 1997 Spring ACS Meeting in San Francisco. This three-day symposium provided a forum for lively discussions on diverse topics ranging from ionic transport in polymer glasses to glass formation of spatially confined molecules. The discussions continuously evoked the principles of glassy polymer physics to interpret new or recent data on physical, mechanical, electrical, and transport properties. Application of recent and developing mechanistic models (in place of phenomenological models) to such newly emerging data showed promise for developing unified interpretations of glassy behavior. It became clear during the symposium that only by obtaining a fundamental understanding of the underlying physics of polymeric glasses will polymer scientists and engineers be able to develop the ability to tailor the structure and state of the polymer glass to achieve the properties desired for specific applications.

This book, like the symposium, is organized into five sections: Physics of Glassy Polymers, Molecular Mobility and Relaxations in the Glassy State, Physical Aging, Mechanical Properties, and Transport Properties and includes an introductory chapter written by the editors. We know of no other book that covers such a diverse range of topics concerning the structure and properties of glassy polymers in such an up-to-date fashion. Some chapters could have been placed in any of several sections. Thus, readers are encouraged to peruse the entire book for only then will a complete picture of glassy polymer physics emerge. Note that the physics of polymer melts is not covered except where an understanding is needed of, for example, the approach toward the glass transition or where comparing glassy behavior to that of the liquid.

We express our deep gratitude to the sponsors of the symposium: Eastman Chemical Company, Commonwealth Scientific and Industrial Research Organization (CSIRO) Division of Manufacturing Science and Technology, EG&G Ortec, the Petroleum Research Fund of the American Chemical Society, and the ACS Division of Polymeric Materials: Science and Engineering, Inc. We are also grateful for the expert assistance of Anne Wilson of the ACS Books

Department. In addition, the editors acknowledge Georgina Roderick of CSIRO for her assistance with the cover design. Contributors to the symposium and to the volume are thanked for sharing their work and opinions with the scientific community.

MARTIN R. TANT
Research Laboratories
Eastman Chemical Company
Kingsport, TN 37662

ANITA J. HILL
CSIRO
Manufacturing Science and Technology
Clayton, Victoria 3168 Australia

Chapter 1

The Structure and Properties of Glassy Polymers

An Overview

Anita J. Hill[1] and Martin R. Tant[2]

[1]CSIRO Manufacturing Science and Technology, Private Bag 33, South Clayton MDC, Clayton, Victoria 3169, Australia
[2] Research Laboratories, Eastman Chemical Company, P.O. Box 1972, Kingsport, TN 37662

An overview of the physics of glassy polymers and the relationships between molecular mechanisms and macroscopic physical, mechanical and transport properties of polymer glasses is presented. The importance of local translational and/or rotational motions of molecular segments in the glass is discussed in terms of the implications for thermodynamic descriptions of the glass (configurational states and energy surfaces) as well as history dependent properties such as expansivity, refractive index, gas permeability, and viscoelastic mechanical behaviour.

Glassy polymers are technologically important across the gamut of materials applications from structural (hyperbaric windows) to electronic (ionic conductors, surface coatings for printed circuit boards) to environmental (membranes for industrial gas separation). The properties of glassy polymers are widely varying and not easily predictable due to the variety of possible chemical structures that can be synthesized and the range of non-equilibrium states available to the polymer. Early theoretical work in glassy polymers focused on two aspects of their nature: the glass transition and their viscoelastic behaviour (1-7). A formal description and understanding of the glass transition temperature (the temperature below which the supercooled liquid behaves as a glass) is necessary in order to determine the configurational state and hence physical (and possibly mechanical and transport) properties of the glass. The contributions to this book and other symposia compilations (8-12) illustrate the quest for theories and characterization techniques that adequately predict and measure phenomena responsible for the mechanical, electrical, magnetic, physical and transport properties of polymer glasses.

The Glass Transition

Glasses can be easily formed from non-crystallizable polymers and can be formed as well from crystallizable polymers if the cooling rate from the melt to below the glass transition temperature is sufficiently rapid. The glass transition, or glassy solidification, has been described as either a thermodynamic transition or a kinetic phenomenon (*13*). Glassy solidification shows similarities to a formal second-order transition in the pressure dependence of T_g; however, as pointed out by Rehage and Borchard (*14*) internal thermodynamic equilibrium does not exist on both sides of T_g (the Ehrenfest relationships do not hold) making it impossible to describe the solidification process as a formal second order transition. Alternatively, the kinetic argument suggests that our measurement techniques and their characteristic frequencies define the phenomenological glass transition temperature. As the melt is cooled, viscosity increases to such an extent (*ca.* 10^{11} - 10^{13} poise) that the time scale of the experiment becomes similar to or longer than the molecular relaxation times. The temperature dependence of the coefficient of thermal expansion, elastic modulus, specific heat, dielectric constant, chemical shift linewidth, diffusivity, etc. defines the glass transition temperature, but (it is argued) the change in these parameters is due to the frequency of molecular motion with respect to the measurement frequency or time, not due to the existence of a thermodynamic second order transition temperature. DiMarzio suggests in Chapter 2 that there is an intimate connection between kinetic and equilibrium quantities and argues for a thermodynamic transition on the basis of theory and experiment, as do Bohn and Krüger in Chapter 5. For the purposes of this discussion, the transition in behaviour from that of a liquid to a glass will be described by the observation that the large-scale cooperative rearrangements of the molecules that are possible in the liquid are effectively arrested in the glass in relation to experimental time scales for motion; however, whether this limited mobility occurs due to formation of glassy regions which coexist with the liquid until eventual percolation or due to the limited patience of the experimentalist is left to the reader to decide (see arguments presented in this volume by DiMarzio: Chapter 2, Baschnagel: Chapter 4, and Bohn and Krüger: Chapter 5). In either case, the viscosity becomes very large as the glass transition is approached on cooling such that the number of configurational states on an energy landscape is greatly reduced in the glass as compared to the liquid, and the ease of rearrangement from one configurational state (or energy minimum) to another is dependent on the nature of the energy landscape. Angell (*15*) has suggested that the nature of the energy landscape reflects the polymer "personality", and knowledge of this energy landscape gives information regarding the solidification history, the structure of the polymer glass, the population of configurational states available and the energy necessary for configurational rearrangements which lead to relaxations in the glassy state. Angell discusses in Chapter 3 the factors that affect the energy landscape placing emphasis on the α–β bifurcation temperature (or the temperature at which the β relaxation diverges from the main α relaxation). In this case the α relaxation is used to describe the large-scale cooperative rearrangements of the molecules in the region of the glass transition whilst the β relaxation is used to describe slower segmental or sub-T_g relaxations. The persistence of local-scale cooperative, as well as uncoupled segmental, motion of the molecules in the glass is of interest from the polymer physics as well as applied property points of view and will be explored further in this chapter as well as Chapters 8-12. It is thus important to be able to predict the energy landscape of a polymer glass, or if this landscape is infinitely complicated, to be able to predict and measure the main factors that reflect this landscape, namely the relaxation times for the molecular motions.

Figure 1 shows an isobaric specific volume-temperature schematic for glassy, liquid and crystalline states of a polymer. The theoretical 100% crystalline material has a first order thermodynamic transition at T_m, which represents the melt temperature. The first feature of interest for glassy polymers is the departure from equilibrium on cooling. As the polymer is cooled below T_g, nonequilibrium properties (in this case volume, but as shown in Chapter 13, enthalpy as well) are quenched in and will relax to some quasi-equilibrium value with time. The quasi-equilibrium glassy state is shown in Figure 1 as a dotted line extrapolated from the liquid. At much lower temperatures the quasi-equilibrium dotted line displays a change of slope associated with vanishing configurational entropy. As mentioned above, in addition to volume and enthalpy changes observed on cooling through T_g, the entropy of the polymer changes as indicated by a significant increase in relaxation times which signifies a decrease in the number of configurations available to the polymer system. Kauzmann (16) originally noted the thermodynamic inconsistencies associated with the extrapolation of equilibrium data far below T_g by calculating unrealistic quantities such as negative configurational entropy and negative free volume. Gibbs and DiMarzio (17) resolved the paradox of negative configurational entropies by predicting a second order transition at a temperature T_2 much less than T_g. At T_2 the configurational entropy vanishes and remains zero to 0 K. Chapter 2 by DiMarzio and Chapter 3 by Angell (15,18) further discuss Kauzmann's paradox and its resolution via extension of equilibrium liquid theory to low temperatures (DiMarzio) and its implications for strong and fragile behaviour of liquids as well as cooperativity of relaxations (Angell). Relaxation of the volume (and enthalpy) takes place below T_g due to the departure from quasi-equilibrium as shown in Figure 1. The relaxation process is nonlinear and nonexponential and can be modeled using an approach outlined by Hodge (19,20). This approach is discussed in a subsequent section entitled Physical Aging and in more detail by Simon in Chapter 13.

The second feature of interest in Figure 1 for this discussion of glassy polymers is the change of slope (increase in expansivity) as the material is heated through various relaxation regions labeled with Greek symbols. The largest change in linear expansion coefficient occurs at T_g (the α relaxation). The schematic in Figure 1 can be compared to experimental results for poly(methylmethacrylate) shown in Figure 2 (21). Each relaxation can be attributed to a particular molecular rearrangement in the glassy polymer and research to discover the molecular mechanisms responsible for the sub-T_g relaxations in glassy polymers is ongoing. Cartoons in Figure 2 suggest the motions attributed to each relaxation in PMMA. Chapter 10 by Pratt and Smith discusses assignment of sub-T_g dielectric relaxations in polycarbonate and polyester blends. The importance of these relaxational processes to physical, mechanical and transport properties is discussed in a subsequent section of this chapter entitled Links Between Sub-T_g Mobility and Macroscopic Properties. The sub-T_g mobility responsible for these relaxations reflects the energy landscape of the glass and as such, the activation energies for the relaxations and the distribution of relaxation times can give information on the intersegmental motion that affects macroscopic properties.

4

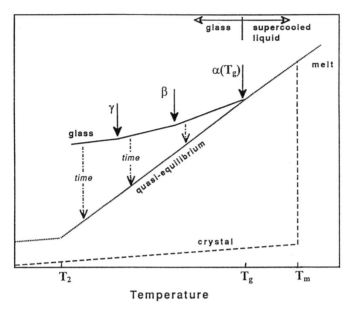

Figure 1. Schematic illustration of volume-temperature behaviour for an amorphous and a crystalline polymer.

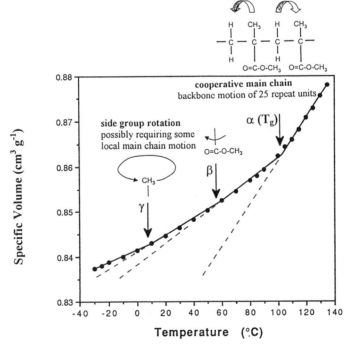

Figure 2. Specific volume-temperature data for poly(methylmethacrylate). (Adapted from ref. 21.)

Relaxation Times

One method of describing non-equilibrium glassy polymer behaviour is through the use of relaxation or retardation times. However, glassy behaviour cannot, in general, be described using a single relaxation or retardation time (or single parameter) model. The local structure of loosely packed polymer chains allows cooperative reorientation processes which can be represented by a sum of single relaxation times (simple process) or a relaxation time distribution (complex process) (22). There are many experimental ways to measure the relaxation times over a wide temperature/frequency range and the relaxation spectrum near T_g as well as the spectra for sub-T_g processes are used to model behaviour (23). Figure 3 shows the volume response of poly(vinylacetate) PVAc on cooling through T_g (24). Cooling from the liquid to a temperature below T_g, where the glass is held isothermally and the relaxation of a property (volume, free volume, refractive index, etc.) is measured, is termed a contraction or physical aging "annealing" experiment. In a homogeneous isotropic glass, density is proportional to refractive index such that the refractive index response mimics the volume relaxation. Also shown in Figure 3 is the volume response for an expansion 'memory' experiment. The expansion experiment consists of quenching a sample at quasi-equilibrium to a temperature T_1 below T_g, storing for some time, and then heating to a testing temperature T still below T_g at which the property of interest is measured as a function of time. Memory behaviour is observed in polymer glasses when the double temperature change of the expansion experiment is performed in a time frame that does not allow the polymer to reach quasi-equilibrium at T_1. The memory effect is illustrated by a volume relaxation peak during the expansion experiment resulting from the distribution of relaxation times associated with molecular mobility. The contraction and expansion isotherms are asymmetric and approach the same quasi-equilibrium volume. Much work has been devoted to the development of models to describe the kinetics of polymers in the non-equilibrium glassy state and to accurately predict the volume and enthalpy response during relaxation experiments. One can anticipate the importance of the volume response, following processing and/or thermal history in service, to vital characteristics such as dimensional stability, residual stress, etc. Further discussion of contraction, expansion, and memory volume relaxation experiments can be found elsewhere (24-28).

Modeling of relaxation behaviour in the glass can be done using either phenomenological models or models that attempt to describe bulk behaviour using thermodynamic or molecular arguments. An example of the former is the transparent mulitparameter model of Kovacs, Aklonis, Hutchinson, and Ramos (29), now commonly referred to as the KAHR model. They used a sum of exponentials and a normalized departure from equilibrium, $\delta = (v-v_\infty)/v_\infty$, where v is the volume at time t and v_∞ is the volume at equilibrium. The distribution of relaxation or retardation times is a function of δ and can be written:

$$\delta(t) = -\Delta\alpha \int_0^z R(z-z') \frac{dT}{dz'} dz' \qquad (1)$$

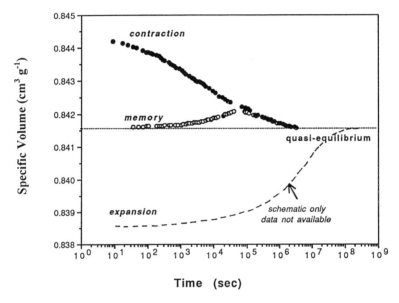

Figure 3. Volume relaxation in poly(vinylacetate) upon cooling from 40°C, through T_g to 30°C (•), *contraction*; upon cooling from 40°C, through T_g to 25°C holding for 90 hr and heating to 30°C (o), *memory*; and upon cooling from 40°C, through T_g to 25°C holding until quasi-equilibrium is reached and heating to 30°C (---); *expansion*. (Adapted from ref. 24.)

where z is a reduced time, $\Delta\alpha$ is the change in coefficient of thermal expansion at the glass transition, T is absolute temperature, and R(t) is a retardation function describing structural recovery. The model successfully describes the various features of glassy behaviour such as asymmetry, nonlinearity and the memory effect.

An approach that can be more closely linked to thermodynamic and molecular arguments is the empirical Kohlrausch-Williams-Watt (KWW) (30) function, in which the relaxing quantity $\varphi(t)$ is given as:

$$\varphi(t) = \exp[-(t/\tau)^\beta] \quad (1 \geq \beta > 0) \tag{2}$$

where the relaxation time is represented by τ and the breadth of the relaxation time distribution is represented by β. Use of the empirical KWW stretched exponential function to describe relaxation behaviour in the glass has been tied to molecular mechanisms using the models of Tool (31) and Narayanaswamy (32). Derivation of τ from the nonlinear form of the Adam-Gibbs (33) equation expresses the relaxation time in terms of structural parameters (20):

$$\tau = A\exp[(x\Delta h^*/RT) + (1-x)\Delta h^*/RT_f] \tag{3}$$

where $\Delta h^*/R$ reflects the energy barrier to intersegmental motion, x is governed by the activation energy for cooperative rearrangement of segments, and T_f is the fictive temperature which represents the structural state of the relaxing glass in terms of the temperature at which the relaxing property (volume, enthalpy, refractive index, etc.) would be the equilibrium value. (Comparing this model with the KAHR model, it is clear that the stretched exponential replaces the distribution of relaxation times and the state is represented by the fictive temperature instead of volumetric departure from equilibrium.) Hodge (19,20) has shown that the nonlinearity and nonexponentiality of the relaxation are contained in the parameters x and β respectively, and that these parameters (Δh^*, x, β) are dependent on the configurational landscape of the polymer glass and the cooperativity of the intersegmental motions involved in the relaxation, thus modeling of the relaxation gives information on the "state" of the glassy polymer. As discussed earlier, the ability to measure relaxation spectra and relate them to the polymer structure is necessary for property prediction. Further discussion of modeling of volume and enthalpy relaxations in the glass and the importance of linking the polymer structure and state to model parameters (as opposed to the use of empirical models) is given by Simon in Chapter 13.

In addition to the physical aging experiment (a quench from above T_g to below, hold and measure isothermal relaxation), glassy state relaxations occur in response to an applied field (stress, electric, magnetic). Figure 4 shows dielectric relaxation spectroscopy (DRS) data and dynamic mechanical thermal analysis (DMTA) data for polycarbonate (34,35). For DRS, the dielectric loss factor ε" or the dielectric loss tangent $\tan\delta_\varepsilon$ ($\tan\delta_\varepsilon = \varepsilon''/\varepsilon'$) are used to follow the molecular rearrangement of polar groups in the glass. For DMTA, the loss modulus E" or the quantity $\tan\delta$ ($\tan\delta = E''/E'$) is used to show the regions of energy dissipation (loss) as a function of temperature at a fixed frequency. Pratt and Smith discuss assignment of the DRS features for polycarbonate in detail in Chapter 10. Dallas et al. report the effect of

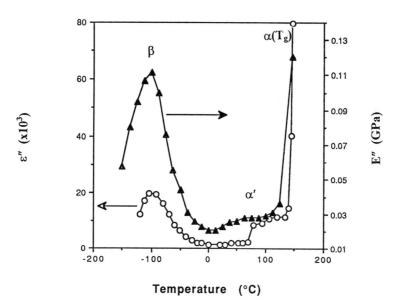

Figure 4. Dielectric loss factor ε" at 10Hz and dynamic mechanical loss modulus E" at 1Hz as functions of temperature for polycarbonate. (Adapted from ref. 34 and 35.)

small molecule penetrants on the relaxation behaviour of polyimide-amide, as measured by DMTA and DRS, in Chapter 11. Other methods of detecting sub-T_g mobility include solid state nuclear magnetic resonance (*36*), ultrasonic (*37*), creep and stress relaxation measurements (*38*). Each of these techniques covers a particular frequency regime and allows measurement of relaxation times and activation energies for a wide range of glassy mobility including main chain, side group, and local mode or cooperative relaxations. The relaxation time distribution is measured and the temperature dependence of the relaxation time is used to calculate the activation energies for the relaxation involving cooperative rearrangement of molecular segments. The relaxation times have been linked to the configurational entropy by Adam and Gibbs (*33*) whilst the relaxation times and configurational entropy have been linked to polymer behaviour as described by the configurational states on an energy landscape via Angell (*15, herein Chapter 3*).

Given that mechanical creep and stress relaxation measurements are sensitive to sub-T_g relaxations, it comes as no surprise that prediction of macroscopic mechanical properties can be made based on knowledge of the relaxation behaviour. The recent work of Matsuoka (*39, 42*) is an example of the continuing development of theoretical approaches to describe molecular motions and their effect on properties at temperatures near and below the glass transition. He has used the concept of conformers to describe molecular relaxation in both the rubbery and glassy states. A conformer is defined simply as the smallest unit of conformational change. For example, an addition polymer typically consists of two conformers per repeat unit. If substituent groups are long enough to contain an additional conformer, then the number of conformers per repeat unit will be greater than two. When a force field (mechanical, electrical, magnetic) is applied, the conformer responds with a change in bond angle which may or may not lead to a change in conformation, e.g. from gauche to trans or vice versa. At very high temperatures, relaxations are not restricted by adjacent conformers and can occur in picoseconds. As temperature is reduced, molecular crowding occurs to the point where adjacent conformers often must relax simultaneously in order for either relaxation to take place. Cooperative relaxation occurs when a number of adjacent conformers relax (or change conformation) simultaneously. Matsuoka defines the domain of cooperativity as the number of cooperative conformers, and this domain size grows as temperature decreases. It is shown that the excess entropy is proportional to the number of domains and that the domain size is inversely proportional. From these basic concepts, predictions can be made for such properties as T_g, rheological shift factors, the distribution of relaxation times, the dynamics of physical aging, and the kinetics and T_g's of thermosets during cure.

Another model which is also closely linked to the physics of glasses (and condensed matter in general) is the coupling model (*40,41*) which presents a response function based on fundamental relaxation modes existing in domains of short range order that are coupled to their complex surroundings. Relaxation involves cooperative adjustments by the fundamental modes within the domains as well as by the complex environment. The coupling model of relaxation has been used to predict the time dependent response of glassy polymers following a perturbation including the influence of variables such as physical aging, molecular additives, molecular weight, and strain history (*42,43*).

The connection of relaxation times (which can be measured) to molecular pair interaction parameters to macroscopic properties has been extensively pursued by Chow (*23*) who has shown that yield behaviour of miscible blends can be predicted from knowledge of the compositional dependent relaxation time and strength of volume interactions. In particular the connectivity of hole motions is used to reflect the relaxation spectrum (*23*). These relationships are of particular interest for modeling of polymer mobility and transport property predictions (*44,45*) as well as application of positron annihilation lifetime spectroscopy (PALS) to measure hole dynamics (*46*). In Chapter 8 Simha discusses extension of the Simha-Somcynsky equation of state below the glass transition and prediction of relaxation in the glass. These predictions are compared to measurements of hole relaxation via PALS. Building the quantitative connection between hole dynamics, molecular relaxation times, molecular mechanisms (chain dynamics), and macroscopic properties in glassy polymers is a topic of much current research. With this background information on the features of the glassy state, it is possible to examine the effects of common variables (residual stress, orientation, physical aging, antiplasticization, etc.) on the relaxation behaviour and macroscopic properties of polymer glasses.

Links Between Sub-T_g Mobility and Macroscopic Properties

Physical Aging. Examination of the effect of a quench from above T_g to an annealing temperature below T_g on the secondary relaxation behaviour shows a suppression of the sub-T_g relaxation processes (*47-50*). Figure 5 illustrates the decrease in intensity of the dielectric loss tanδ for the sub-T_g relaxation of glassy poly(propylene oxide) as a function of aging or "annealing" time. The inset in Figure 5 shows the decrease in the amplitude of the tanδ peak for the β relaxation as a function of aging time. Analogous measurements of mechanical relaxation of glassy polymers (*48,50-52*) show a qualitatively similar decrease in the sub-T_g mechanical loss tanδ on aging. Reduction in the number of segments participating in the β relaxation (i.e. the number of segments able to dissipate energy whether electrical or mechanical) should be reflected in macroscopic properties such as yield strength and impact energy. Fracture of glassy polymers is typified by either shear yielding and ductile failure or craze initiation and brittle fracture (*53*). As shown by Tant et al. in Chapter 17 the yield stress of glassy poly(ethylene terephthlate) increases with aging time while the craze stress remains constant, eventually resulting in a transition from ductile to brittle behaviour. It is thought that the reduction of the number of segments participating in energy dissipating relaxations is due to improved local order (reduction of configurational entropy) which limits mobility and an accompanying decrease in the hole fraction. Simha in Chapter 8 as well as Tant et al. in Chapter 17 explore this link between reduction of free volume due to physical aging (relaxation) in the glassy state. The concept of a free volume fraction that is "frozen in" at T_g and thus constant at all temperatures below T_g (*54,55*) is not supported by the results in Chapters 8 and 17.

In addition to the effect on mechanical and electrical properties, an important aspect of physical aging and volume relaxation is the dimensional change that

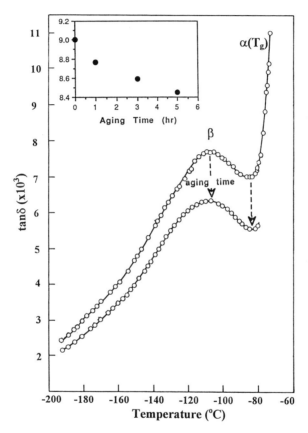

Figure 5. Dielectric loss tanδ as a function of temperature at 1kHz for poly(propylene oxide) in the unaged (o) and aged (•) states. The aged sample was held for 5 hr at 5.3°C below T_g. Inset shows the tanδ peak intensity for the β relaxation as a function of aging time at 10.3°C below T_g. (Adapted from ref. 49.)

accompanies the macroscopic property changes. In order to predict dimensional stability of glassy polymers, one can study the density distribution, or hole fraction distribution, because the density is affected by the applied or residual stress, the state of stress during vitrification, the cooling rate during vitrification and the thermal history post-vitrification (*27,56*). In essence the density distribution can give information on the energy landscape.

Residual Stress and Orientation. Application of stress to a polymer melt causes partial orientation of the molecules in the direction of the applied stress. As the melt is cooled through the glass transition, the orientational relaxation times increase and preferred orientation is quenched in the glassy state (*56,57*). The macroscopic properties of the polymer glass are affected by this orientation which causes macroscopic anisotropy evidenced in thermal conductivity, thermal expansion, elastic modulus, refractive index, etc. Because residual stress as well as flow-induced molecular orientation affect macroscopic properties, prediction of residual molecular orientation in glassy polymers is important. Relaxation of residual molecular orientation in the glass is similar to the physical aging process in that it is dependent on molecular mobility. The molecular mobility is in turn dependent on the level of orientation and residual stress in much the same way as the rate of physical aging is dependent on the departure from equilibrium. Modeling of the relaxation of residual molecular orientation below T_g can be accomplished using the approach of Tool (*31*) and Narayanaswamy (*32*) as discussed in a previous section entitled Relaxation Times. The relaxing quantity in this case is the stress optical coefficient that allows calculation of the stress tensor from the deviatoric refractive index tensor (*56*). Just as the concept of free volume "frozen in" at the glass transition has persisted in the literature (*54,55*) so has the "frozen in" birefringence in the glassy state for prediction of molecular orientation and residual stress in molded parts (*58*). The mobility of the glassy state is much reduced from that of the equilibrium liquid; however, as this overview has illustrated, the rotational and/or translational mobility present in the glass determines the relaxation times which are used to model the macroscopic behaviour, thus presenting a dynamic picture of the glass that is far from "frozen."

Surfaces, Interfaces, and Confinement. The relaxation behaviour of glassy polymers can be affected by the presence of a surface or interface or confined/constrained geometry. Recent computer simulations of glassy polymers (*59-62*) have predicted a variation in the packing and mobility of chains at surfaces and interfaces compared to chains in the bulk. These modeling results lend support to postulations that the variation from bulk polymer properties found in thin films, latex particles, and polymers in confined spaces, are due to conformational differences of polymer chains at surfaces/interfaces. It is postulated that polymer chains in contact with a surface may have a different conformational distribution, and thus energy landscape, than chains in the bulk. It follows from previous discussion that changes in the energy landscape will be reflected in the relaxations in the glassy state, and hence the properties of the glass. The surface may be a free surface or a hard wall; the type of surface and the distance to the next interface strongly influence the preferred or possible conformations and hence the packing and chain entanglements (*63*).

Conformational changes are expected to affect density (packing) and chain dynamics (mobility) such that surface/interface properties vary from that of the bulk glass. The surface effects are evident only within several chain segment diameters of the surface/interface (63). Chapter 6, by Haralampus et al., presents experimental and theoretical results on the effect of confinement on T_g.

Experimental techniques to probe polymer molecules at surfaces and interfaces include X-ray and neutron reflectivity (64), near edge X-ray absorption fine structure (NEXAFS) (64), spectroscopic ellipsometry (65), Brillouin light scattering (66), PALS using a variable energy positron beam (67), and chromophore-based techniques such as second harmonic generation and fluorescence nonradiative energy transfer (68). Surface/interface properties are critical for polymers used in thin film applications or constrained/confined geometry systems such as resists, interlayer dielectrics in microelectronics fabrication (69), alignment layers for liquid crystal displays (64), latex particles and films from latex (70), gas separation membranes (71), and interfaces in copolymers and blends (63,72). As such, the chain dynamics, the distribution of relaxation times, the polymer chain topology, and the physical, mechanical and transport properties of polymer chains at surfaces/interfaces and in confined spaces are the topics of much current research.

Antiplasticization. By examining the effect of plasticizer or antiplasticizer on sub-T_g relaxations, one can assess the effects on macroscopic properties (73,74). Figure 6 shows suppression of the sub-T_g β relaxation in glassy polycarbonate as antiplasticization occurs. The β relaxation is detected via solid state NMR and has been presented here (Figure 6) in a manner similar to the loss peaks observed by DMTA and DLS. The addition of 5 wt % phthalate diluent results in a slight increase in elastic modulus and a suppression of the β loss peak. As discussed by Lui et al. (36), the low temperature part of the peak is due to phenylene groups that are suppressed by addition of the antiplasticizer (5 wt % phthalate diluent). Removal of this low temperature component of the distribution of relaxation times shifts the β relaxation peak to higher temperatures. The increase in relaxation times as measured by NMR (75) or DMTA (74) reflects an increase in the activation energy of the relaxation. The restricted mobility results in an increase in elastic modulus similar to that observed due to physical aging. In addition a decrease in gas transport properties is observed (76). The change in macroscopic physical (increase in T_g), mechanical (increase in yield strength) and transport (decrease in gas permeability) properties of glassy polymers due to antiplasticization has been attributed to the reduction of configurational entropy reflected in the increase in relaxations times and the reduction in free volume due to improved packing.

Transport Properties. Sorption and transport properties are highly dependent on the post-vitrification history of glassy polymers (77); hence one would expect parameters such as physical aging, antiplasticization and amorphous orientation to affect transport properties. The reduction in diffusivity and permeability due to aging, orientation, and antiplasticization can be modeled via entropy or free volume arguments (77). In addition, diffusive jumps of penetrant molecules in glassy polymers can be affected by (facilitated by) the segmental mobility that is manifested in sub-T_g relaxations (78).

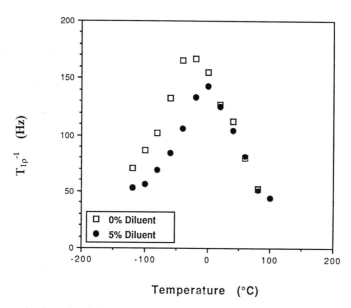

Figure 6. Inverse of the proton spin lattice relaxation time $(T_{1\rho})^{-1}$ as a function of temperature for polycarbonate as-extruded () and on addition of antiplasticizer (•). (Adapted from ref. 36.)

Molecular dynamics simulations have shown that penetrant jumps between free volume elements require cooperativity (44). Molecular mechanics and molecular dynamics modeling of the size, distribution, and dynamics of the free volume elements has been used to predict penetrant transport (44,45,79-82). In Chapter 21 Freeman and Hill present transport data for glassy polymers modeled using free volume arguments. Nakagawa discusses in Chapter 22 the exceptionally high permeability of poly(1-trimethylsilyl-1-propyne) PMSP or PTMSP and illustrates that aging in glassy PTMSP is responsible for the time dependent reduction in gas permeability following casting of PTMSP membranes. Antiplasticization, which causes a decrease in the β relaxation (see Figure 6) and an increase in relaxation times and activation energy, has also been shown to decrease permeability in glassy polymers (76). The correlation between sub-T_g mobility (and the relaxation times) and macroscopic transport properties is clear.

The apparent diffusivity for carbon dioxide of glassy poly(vinyl acetate) is shown as a function of aging time upon cooling from above T_g in Figure 7 (83). The decay of diffusion coefficient with aging time is similar to the relaxation-induced decay (or saturation growth) shown by properties such as yield stress, volume, enthalpy, and refractive index.

Mechanical Properties. Crazing and shear band formation have been shown to require significant changes in chain conformation, again suggesting the importance of segmental mobility. Indeed it has been suggested (53) that crazes and shear bands may propagate via highly cooperative segmental rotation of isomers. Hence the ability to calculate (and confirm by measurement) the energy barriers (20) which control the intersegmental motion is an important step in mechanical property prediction. Matsuoka's model (39,84) for intermolecular cooperativity among conformers leads to a yield criterion for glassy polymers based on the activation energy for conformational relaxation and intermolecular segmental cohesive energy. Figure 8 shows the stress-strain data and theory for polycarbonate at three different strain rates. The fits are superb, and as Matsuoka (84) points out, since the stress strain curves are modeled based on relaxation times and the relaxation times reflect the aging time, the stress-strain curves can be scaled to predict mechanical response as a function of physical aging time.

Boyer (85) argued in 1968 that polymer scientists must understand mechanical properties of polymers in terms of the molecular structure and molecular motions in order to be able to design better plastics. The same may be said of understanding physical, electrical, and transport properties of glassy polymers, and the chapters contained in this book present current research efforts toward that end.

Concluding Comments

Scientists and engineers pursue development and application of theories for the physics of glassy polymers in order to understand vitrification and the glassy state in molecular solids. This understanding is coupled with the investigation of glassy state phenomena and properties so that new materials and new applications are continuously developed to improve the quality of life. Examples of this approach are found in the following chapters and an illustrative example can be found in Chapter 25 by Forsyth

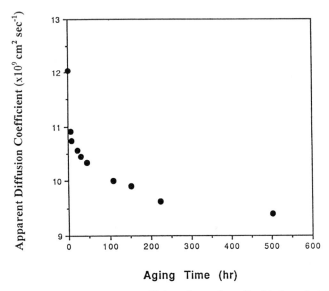

Figure 7. Apparent diffusion coefficient for carbon dioxide in poly(vinyl acetate) as a function of aging time after cooling through T_g to 10°C. (Adapted from ref. 83.)

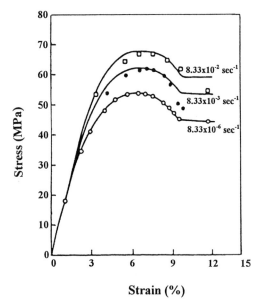

Figure 8. Stress-strain data and theoretical prediction (solid lines) for glassy polycarbonate at ambient temperature. (Adapted from ref. 84.)

et al. These authors discuss the modeling of ionic conductivity at ambient temperatures for glassy polymers and show the development of novel glassy ionic conductors with conductivities equivalent to or exceeding those of rubber electrolytes (which should have the advantage of much higher molecular mobility, and hence conductivity, at ambient temperatures). For equivalent ionic conductivities between glass and rubber electrolytes, the advantages of glassy polymer electrolytes can be found in their transparency and rigidity. Possible applications include electrochromic windows, photoelectrochemical solar cells, and low power capacitors.

The importance of being able to measure and predict the relaxation spectra for glassy polymers was illustrated in a number of examples. The glassy state relaxation behaviour (sub-T_g segmental mobility: α', β, γ processes, as well as physical aging) was shown to affect macroscopic mechanical and transport properties. The development of sensitive techniques and the ability to synthesize systematic series of polymers, copolymers, and blends enables researchers to test theory and experiment and to determine unifying molecular mechanisms for macroscopic properties of the glassy state. The importance of chain conformation dynamics and the ability of statistical mechanical theories to describe macroscopic behaviour emphasize the dynamic nature of polymer glasses and the inadequacy of models that rely on properties "frozen in" at the glass transition during cooling. The picture of the polymeric glassy state that emerges in the following chapters is, rather, that of a dynamic, constantly changing system of long chain molecules whose motions directly control the physical, mechanical, and transport properties of the bulk material.

Literature Cited

1. Russell, J. and Van Kerpel, R. G. *J. Appl. Phys.* **1958**, *29*, 1438.
2. Dannis, M. L. *J. Appl. Polym. Sci.*, **1959**, *1*, 121.
3. McCrum, N. G. *J. Polym. Sci.* **1961**, *54*, 561.
4. Read, B. E. and Williams, G. *Polymer,* **1961**, *2*, 239.
5. Baccaredda, M. and Butta, E. *J. Polym. Sci.*,**1961**, *51*, S-39.
6. Read, B. E. *Polymer,* **1962**, *3*, 529.
7. Natta, G. *SPE Trans.*, **1963**, *3*, 99.
8. Goldstein, M., and Simha, R. Eds. *Annals New York Academy of Science*, Vol. 279, *The Glass Transition and Nature of the Glassy State*, NY Academy of Sciences: New York, **1976**.
9. O'Reilly, J. M and Goldstein, M. Eds. *Annals New York Academy of Science*, Vol. 371, *Structure and Mobility in Molecular and Atomic Glasses*, NY Academy of Sciences: New York, **1981**.
10. Angell, C. A. And Goldstein, M. Eds. *Annals New York Academy of Science*, Vol. 484, *Dynamic Aspects of Structural Change in Liquids and Glasses*, NY Academy of Sciences: New York, **1986**.
11. Dorfmuller, Th. and Williams, G., Eds. *Lecture Notes in Physics*, Vol. 277, *Molecular Dynamics and Relaxation Phenomena in Glasses*, Springer-Verlag: New York, **1987**.
12. Roe, R. J., and O'Reilly, J. M., Eds. *MRS Symp. Proc.* Vol. 215, *Structure, Relaxation, and Physical Aging of Glassy Polymers*, MRS: Pittsburg, **1991**.

13. Saito, N., Okano, K., Iwayanogi, S., and Hideshima, T. in *Solid State Physics* Vol 14, Seitz, F. and Turnbull, D. Eds., Academic Press: New York, **1963**, 344.
14. Rehage, G. and Borchard, W. in *The Physics of Glassy Polymers*, Haward, R. N. Ed., Applied Science Publ. Ltd.: London, **1973**, 54.
15. Angell, C. A., Monnerie, L., and Torell, L. M. in *MRS Symp. Proc.* Vol. 215, *Structure, Relaxation, and Physical Aging of Glassy Polymers*, MRS: Pittsburg, **1991**, 3.
16. Kauzmann, W. *Chem. Rev.*, **1948**, *43*, 219.
17. Gibbs, J. H., and DiMarzio, E. A. *J. Chem. Phys.*, **1958**, *28*, 373.
18. Angell, C. A. *J. Res. NIST*, **1997**, *102*, 171.
19. Hodge, I. M. *Macromolecules*, **1986**, *19*, 936.
20. Hodge, I. M. in *MRS Symp. Proc.* Vol. 215, *Structure, Relaxation, and Physical Aging of Glassy Polymers*, MRS: Pittsburg, **1991**, 11.
21. Wittmann, J. C. and Kovacs, A. J. *J. Polym. Sci. C*, **1969**, *16*, 4443.
22. Starkweather, H. W. *Macromolecules*, **1981**, *14*, 1277.
23. Chow, T. S. *Advances in Polm. Sci.*, **1991**, *103*, 152.
24. Kovacs, A. J. *Fortschr. Hochpolym.-Forsch.* **1963**, *3*, 394.
25. Greener, J., O'Reilly, J. M. and Ng, K. C. in *MRS Symp. Proc.* Vol. 215, *Structure, Relaxation, and Physical Aging of Glassy Polymers*, MRS: Pittsburg, **1991**, 99.
26. Tant, M. R. and Wilkes, G. L. *Polym. Engr. Sci.* **1981**, *21*, 874.
27. Struik, L. C. E. *Physical Aging in Amorphous Polymers and Other Materials*, Elsevier: Amsterdam, **1978**.
28. McKenna, G. B. in *Comprehensive Polymer Science*, Vol. 2, *Polymer Properties*, Booth C. and Price, C. Eds., Pergamon: Oxford, **1989**, 311.
29. Kovacs, A. J., Aklonis, J. J., Hutchinson, J. M., and Ramos, A. R. *J. Polym. Sci. B*, **1979**, *17*, 1097.
30. Williams, G. and Watts, D. C. *Trans. Farad. Soc.* **1970**, *66*, 80.
31. Tool, A. Q. *J. Amer. Ceram. Soc.*, **1946**, *29*,240.
32. Narayanaswamy, O. S. *J. Amer. Ceram. Soc.*, **1970**, *53*, 380.
33. Adam, G. and Gibbs, J. H. *J. Chem. Phys.*, **1965**, 43, 139.
34. Watts, D. C. and Perry, E. P. *Polymer* **1978**, *19*, 248.
35. Shelby, M. D. *PhD Dissertation* **1996**, Virginia Polytechnic Institute and State University.
36. Lui, Y., Roy, A. K., Jones, A. A., Inglefield, P. T., and Ogden, P. *Macromolecules*, **1990**, *23*, 968.
37. Phillips, D. W., North, A. M., and Pethrick, R. A. *J. Appl. Polym. Sci.*, **1977**, *21*, 1859.
38. Locati, G. and Tobolsky, A. V. *Advances in Molecular Relaxation Processes*, **1970**, *1*, 375.
39. Matsuoka, S. and Hale, A. *J. Appl. Polym. Sci.*, **1997**, *64*, 77.
40. Ngai, K. L. *Comments Solid State Phys.* **1979**, *9*, 127.
41. Rendell, R. W. and Ngai, K. L. in *Relaxations in Complex Systems*, Ngai, K. L. and Wright, K. B. Eds., US Govt. Printing Office: Washington, D.C., **1985**, 309.
42. Ngai, K. L., Rendell, R. W., Rajagopal, A. K., and Teitler, S. *Annals New York Academy of Sciences* **1986**, *484*, 150.

43. Rendell, R. W., Ngai, K. L. and Yee, A. F. in *MRS Symp. Proc.* Vol. 79, *Scattering, Deformation and Fracture in Polymers*, MRS: Pittsburg, **1987,** 311.
44. Greenfield, M. L. and Theodorou, D. N. *Polym. Mater. Sci. and Engr.*, **1997**, *76*, 429.
45. Arizzi, S., Mott, P. H., and Suter, U. W. *J. Polym. Sci. B*, **1992**, *30*,415.
46. Higuchi, H., Jamieson, A. M, and Simha, R. *J. Polym. Sci. B*, **1996**, *34*, 1432.
47. Mininni, R. M, Moore, R. S., Flick, J. R., and Petrie, S. E. B. *J. Macromol. Sci.-Phys.*, **1973**, *B8*, 343.
48. Petrie, S. E. B. *J. Macromol. Sci. Phys.*, **1976**, *12*, 225.
49. Johari, G. P. *J. Chem. Phys.* **1982**, *77*, 4619.
50. Allen, G., Morley, D. C. W., and Williams, T. *J. Mater. Sci.*, **1973**, *8*, 1449.
51. Frank, M. and Stuart, H. A. *Kolloid Z.* **1968**, *225*, 1.
52. Flick, J. R. And Petrie, S. E. B. *Studies in Physical and Theoretical Chemistry* **1978**, *10*, 145.
53. Sternstein, S. S. in *Treatise on Materials Science and Technology*, Vol 10, Schultz J. M. Ed., Academic Press: New York, **1977**, 541.
54. Fox, T. G. and Flory, P. J. *J. Appl. Phys.*, **1950**, *21*, 581.
55. Shen, M. C. and Eisenberg, A. *Progress in Solid State Chem.* **1966**, *3*, 407.
56. Wimberger-Friedl, R. *Prog. Polym. Sci.*, **1995**, *20*, 369.
57. Struik, L. C. E. *Internal Stresses, Dimensional Instabilities, and Molecular Orientations in Plastics,* Wiley: New York, **1990**.
58. Flaman, A. A. M. *Polym. Engr. Sci.*, **1993**, *33*, 193.
59. Mansfield, K. F. and Theodorou, D. N. *Macromolecules*, **1990**, *23*, 4430.
60. Mansfield, K. F. and Theodorou, D. N. *Macromolecules*, **1991**, *24*, 4295.
61. Mansfield, K. F. and Theodorou, D. N. *Macromolecules*, **1991**, *24*, 6283.
62. Baschnagel, J. and Binder, K. *Macromolecules*, **1995**, *28*, 6808.
63. Brown, H. R. and Russell, T. P. *Macromolecules*, **1996**, *29*, 798.
64. Russell, T. P. *MRS Bull.*, **1996**, *21*, 49.
65. Beaucage, G., Composto, R., and Stein, R. S. *J. Polym. Sci. B*, **1993**, *31*, 319.
66. Forrest, J. A., Rowat, A. C., Dalnoki-Veress, K., Stevens, J. R., and Dutcher, J. R. *J. Polym. Sci. B,* **1996**, *34*, 3009.
67. DeMaggio, G. B., Frieze, W. E., Gidley, D. W., Zhu, M., Hristov, H. A. and Yee, A. F. *Phys. Rev. Lett.*, **1997**, *78*, 1524.
68. Hall, D. B., Miller, R. D., and Torkelson, J. M. *J. Polym. Sci. B*, **1997**, *35*, 2795.
69. Frank, C. W., Rao, V., Despotopoulou, M. M., Pease, R. F. W., Hinsberg, W. D., Miller, R. D., Rabolt, J. F. *Science*, **1996**, *273*, 912.
70. Parker, H-Y. *Polym. Mater. Sci. and Engr.*, **1997**, *76*, 536.
71. Pfromm, P. H. and Koros, W. J. *Polymer*, **1995**, *36*, 2379.
72. Russell, T. P. *Curr. Opin. Colloid Interface Sci.*, **1996**, *1*, 107.
73. Petrie, S. E. B., Flick, J. R., Garfield, L. J., Marshall, A. S., and Papanu, V. D. *J. Polym. Prepr.*, **1973**, *14*, 1254.
74. Wyzgoski, M. G. and Yeh, G. S. Y. *J. Macromol. Sci.-Phys.*, **1974**, *B10*, 441.
75. Parker, A. A., Shieh, Y. T., and Ritchey, W. M. in *MRS Symp. Proc.* Vol. 215, *Structure, Relaxation, and Physical Aging of Glassy Polymers*, MRS: Pittsburg, **1991,** 119.
76. Maeda, Y. and Paul, D. R. *J. Polym. Sci. B*, **1987**, *25*, 1005.

77. Vieth, W. R. *Diffusion in and Through Polymers*, Oxford University Press: New York, **1991**.
78. Light, R. R. and Seymour, R. W. *Polym. Engr. Sci.*, **1982**, *22*, 857.
79. Vrentas, J. S., Duda, J. L., and Ling, H.-C. *Macromolecules*, **1988**, *21*, 1470.
80. Cohen, M. H. and Turnbull, D. *J. Chem. Phys.*, **1959**, *31*, 1164.
81. Gusev, A. A., Muller-Plathe, F., van Gunsteren, W. F., Suter, U. W. *Adv. Polym. Sci.*, **1994**, *116*, 207.
82. Gray-Weale, A. A., Henchman, R. H., Gilbert, R. G., Greenfield, M. L., and Theodorou, D. *Macromolecules*, **1997**, *30*, 7296.
83. Toi, K., Ito, T., and Ikemoto, I. *J. Polym. Sci. B*, **1985**, *23*, 525.
84. Matsuoka, S. in *MRS Symp. Proc.* Vol. 215, *Structure, Relaxation, and Physical Aging of Glassy Polymers*, MRS: Pittsburg, **1991,** 71.
85. Boyer, R. F. *Polym. Engr. Sci.*, **1968**, *8*, 161.

PHYSICS OF GLASSY POLYMERS

Chapter 2

The Use of Configurational Entropy To Derive the Kinetic Properties of Polymer Glasses

Edmund A. Di Marzio

National Institute of Standards and Technology, Gaithersburg, MD 20899

A logical chain of reasoning containing 8 links is described which leads to formulas for the frequency and temperature dependent viscosity $\eta(\omega.T)$, diffusion coefficient $D(\omega,T)$, and dielectric response $\epsilon(\omega,T)$. Our tentative result derived from our minimal model for the zero frequency viscosity is $Log\eta = B - AF_c/kT$ where F_c is the thermodynamic configurational free energy. Thus, contrary to the Vogel-Fulcher equation or the Adam-Gibbs equation, there is no discontinuity of the viscosity at the glass transition even though the glass transition occurs when the configurational entropy S_c approaches zero. An alternate model which is a nesting of minimal models results in $Log\eta = B - JS_c + KU/kT$ where the linear combination of configurational entropy and energy is not the free energy.

We have recently attempted to derive the frequency dependent viscosity $\eta(\omega,T)$, diffusion coefficient $D(\omega,T)$ and dielectric response $\epsilon(\omega,T)$ as a logical consequence of the principles of statistical mechanics (*1*). In this paper we wish to outline the logic of our development, describe an extension of our method to a hierarchical nesting of minimal models and discuss some of the results.

We know that there is **no undiscovered deep law of physics** on which the existence of glasses depends. So we are confident in asserting that there is a chain of reasoning using the known laws of physics which will lead to an understanding of glasses and glass behavior. The only question is...can we forge the links of this chain... Can we list a series of steps which lead to an understanding of glasses...and can each step be built securely? Can the links be forged without flaws? In our view an understanding of glasses is accomplished in the following 8 steps

[1] We first show that the crystal phase is not ubiquitous (*2,3*). There exist certain molecules that cannot possibly have a crystal phase. For these materials the liquid phase which everyone admits exists at high temperatures extends to all low temperatures. Thus, certain materials must of necessity have the amorphous phase as the low temperature equilibrium phase. See the Appendix. One is thus forced to face squarely the problem of determining the equilibrium low temperature properties of those materials that form glasses.

22

[2] A simple way (actually the only way) to determine this low temperature amorphous phase is to use a theory that correctly predicts the behavior of the liquid and extend it to low temperatures. One obviously should use the most realistic existing equilibrium theory to obtain the low temperature phase. The predictions are the two equations of state, S-V-T and P-V-T which are each derived from the Helmholtz free energy F which is in turn obtained from the partition function ($F = -kTLnQ$). In obtaining the S-V-T equation of state it is discovered that the configurational entropy S_c defined as the total entropy minus the vibrational entropy, approaches zero at a finite temperature (4). This vanishing of S_c is taken as the thermodynamic criterion of glass formation (5,6).

[3] One then tests these equilibrium predictions of the theory to see if they make sense. Since we have obtained hard results (quantitative predictions) before proceeding further we must show that indeed every prediction is consistent with experiment. Were this not so we would have a flawed link and the chain would be broken. Are all quantitative predictions of thermodynamics of glasses correct? If not stop here. We think that we have agreement with experiment. See below.

[4] Are all quantitative predictions of the thermodynamics of liquid crystals correct. If not stop here. The reason for this step is that the theory (Flory-Huggins lattice model) also predicts the occurrence of the isotropic to nematic phase transition in liquid crystals. If the theory had predicted correctly the properties of glasses but had failed for liquid crystals we would have had to abandon it, especially since in both cases the cause of the transition is ascribed to the vanishing of the configurational entropy. Alternatively the correctness of the prediction for liquid crystals argues for the correctness of the prediction for glasses. Since we have not been stopped by steps 3 and 4 we proceed to step 5.

[5] Having established a proper equilibrium foundation we look to derive the kinetics. How do we connect these two classes of phenomena? It is already known that as the configurational entropy approaches zero the viscosity gets very high. This is because the elementary flow event is a jumping from one allowed configuration to another. As S_c decreases the allowed configurations become fewer and therefore get further and further apart in configuration space so it becomes more and more difficult to jump from one allowed configuration to another. Also, we know from the fluctuation-dissipation theorem (7) that a deep interrelationship exists between kinetic and equilibrium quantities. We also know from experience that it is very difficult to implement the connection, so we use the principle of detailed balance (8) to relate the thermodynamics to the kinetics. If N_j is the number of systems in state j and if α_{ji} is the rate of jumping from state j to state i then the principle of detailed balance states that $N_j\alpha_{ji} = N_i\alpha_{ij}$.

[6] We now propose a minimal model which is a trapping model in configuration space. The potential energy surface in configuration space is highly convoluted with deep wells interconnected along ridges and saddle points (9). The principle of detailed balance can be used to determine the rate constants for jumping into and out of the wells and the number of wells of a

given depth is known from the structure of the potential energy surface. The (master) equations of the minimal model are then written down and solved.

[7] In principle, knowing how the phase point moves from well to well allows us to extract the viscosity $\eta(\omega,T)$, diffusion constant $D(\omega,T)$ and dielectric response $\epsilon(\omega,T)$. We solve this problem and obtain formulas for each of these quantities.

[8] These quantities are then are compared to experiment. And agreement with experiment is noted.

The logic of our "eight-fold way" is unassailable. The more important question is...Have we implemented it properly? Have we forged the links in the chain so that they are without serious flaws? So, we now discuss each of these links in more detail.

Discussion

Link 1. The crystal phase is not ubiquitous: We believe that this was proved for polymers (when the molecules are not compact) in two previous papers (2,3). The gist of the argument is as follows. Consider a polymer molecule whose lowest energy shape appears to be a random walk with linear dimensions of $n^{1/2}$, n being the number of monomers. If they are to space-fill on a lattice the distance between the centers of mass of neighboring molecules must vary as $n^{1/3}$. There are then approximately $n^{1/2}$ molecules within the volume spanned by any one of them. Can each of these molecules have the same shape without double occupancy of lattice sites? In order for the crystal phase to be ubiquitous it must be possible for every given polymer shape to space-fill the lattice without double overlap at even one site. An estimate is made of the number of polymer molecules that have a shape which can space-fill the lattice of coordination number z without double occupancy. For large n the number zero is obtained. Coupling this with the fact that ubiquity of the crystal phase implies that every shape must be able to space-fill, i.e. the number estimated (10) should have been on the order of z^n, we conclude that the crystal phase is not universal. So, we are certain of a well-forged first link.

Link 2. Extending a liquid-like theory to low temperatures: This is just what we need to resolve Kauzmann's paradox (11). What better way is there to extrapolate equilibrium behavior than by devising an equilibrium theory which describes the high temperature region correctly and then seeing what it predicts for low temperature? This was our original approach and we discovered that the configurational entropy was approaching zero at a finite temperature (5). The configurational entropy S_c which is defined for all materials as the total entropy S_t minus the vibrational entropy S_v is easily calculated for polymers by using Flory-Huggins lattice statistics. This model was adapted by us to polymer glasses by using empty lattice sites to allow for volume changes within the bulk polymer and by making the molecules semiflexible (5,6). The volume fraction of holes V_0 is controlled by the hole energy E_h and the fraction of bonds f flexed out of the lowest energy wells (isomeric state model with nearest neighbor energies) is controlled by the flex energy $\Delta\epsilon$ which is the energy difference between the lowest well and the upper wells on rotation about a bond. The total number of configurations W can be expressed as a function of the volume fraction of holes V_0 and the amount of flexed polymer bonds f. The glass temperature occurs when the configurational entropy $(S \equiv kTLnW(V_0,f))$ approaches zero.

$$S_c(V_0,f) \rightarrow 0 \tag{1a}$$

V_0 is obtained from the P-V-T equation of state through the relations $F_c = -kTLnQ$, and $P = -\partial F/\partial V$, while f is given by $f=(z-2)exp(-\Delta\epsilon/kT)/(1+(z-2)exp(-\Delta\epsilon/kT))$, where z is the coordination number of the lattice. This enables us to express the configurational entropy as a function of T and P. At the glass transition we can write

$$S_c(T_2,P) = 0 \tag{1b}$$

Eqs. 1b divides T-P space into two regions. The high temperature region for which $S_c(T,P) > 0$ and the low temperature region (12) for which $S_c(T,P) = 0$. For a given pressure the values of V_0 and f in the low temperature region are independent of T- they have the same values as at the glass temperature T_2. V_0 for temperatures lower than T_2 is however dependent (13) on P. One immediate consequence of Eq. 1a is that the constant free volume assumption, usually stated as $V_0=.025$, must be abandoned. Not only is V_0 a function of P, it is also dependent on E_h, $\Delta\epsilon$ and other details of polymer architecture such as molecular weight for linear polymers. In the past, because it was much easier to obtain, both experimentally and theoretically, the P-V-T equation of state rather than the S-V-T equation of state, it was natural for the science community to give undue emphasis to volume criteria rather than entropy criteria. It's the fallacy of looking for a lost article under the lamppost because that's where the light is.

Of course our modified FH theory which is 40 years old can obviously be improved. Any theory that does a better job of predicting the properties of bulk liquid polymer is an obvious candidate. The theories of Milchev (14), Gujrati (15), Freed (16) and the composite analytical-computer intensive theories of Binder (17) et al all need to be examined to see which one models liquids more perfectly, and is therefore a better candidate for extrapolation to the glassy region.

Another improvement in theory would be to treat f, V_0 as local rather than global order parameters (18). This leads to the idea that as one cools a polymer (because of spatial-temporal fluctuations in f and V_0) pockets of material become glassy while other regions stay liquid. The glass transition then occurs when the glassy pockets connect up so as to span the space occupied by polymer. The configurational entropy at which percolation first occurs is now slightly greater than zero. This view of glasses has the nice feature that liquid-like pockets exist below the glass transition, thus allowing for both the α and β phases observed in bulk polymers. Reference 1 discusses the local order parameter approach in somewhat more detail.

More fundamental fluctuation approaches that do not rely on the order parameters f and V_0 are possible. Such treatments however, do need to incorporate the concept of chain stiffness if they are to explain polymer glasses, since it is only through chain stiffness that the configurational entropy can approach zero at finite temperature. See also the discussion under Link 4.

Link 3. Comparison to Experiment for glasses: This theory has been successful in locating the glass transition in temperature-pressure space as a function of molecular weight (*2,5*), polymer architecture (linear molecules (*2,5*), rings (*19,20*), cross-linking (*21*)), stretch ratio λ in a rubber (*21*), plasticizer content (*22,23*), blend composition (*24*) and copolymer content (*25*) as well as predicting the specific heat break at the glass temperature (*26*). In all these cases we have what we believe is quantitative concord with experiment.

One must be careful here because one can prove too much. The entropy theory of glasses, also called the Gibbs-Di Marzio (GD) theory, is a theory of equilibrium thermodynamic quantities only, it is not a theory of the kinetics of glasses. Polymer viscosities do in fact get so large at the glass transition that the relaxation times in the material equal and exceed the time scale of the experiment. At such temperatures one should not expect to have a perfect prediction of the various thermodynamic quantities. It is sensible to suppose that our predictions should not accord perfectly with experiment in these high viscosity regions.

Nevertheless, certain of the predictions are clear evidence of the underlying thermodynamic character of the transition. First, the depression of glass temperature by diluents is predicted to depend only on the mole fraction of diluent and on no other quantity provided only that the diluent molecules are small (*22,23*). The dependence on mole fraction is a colligative property and the independence of T_2 on the energy of the diluent-polymer interaction is clear evidence that entropy controls the location of T_2. Second, the glass temperature of ring polymers in bulk is predicted to increase as the ring molecular weight is lowered while the glass temperature of linear polymers is predicted to decrease(*19,20*). Both predictions agree with experiment. This is ascribed to the fact that the configurational entropy of the ring system **decreases** as we lower the ring molecular weight (*27*). Thus one does not have to cool as far for low molecular weight rings to reach the $S_c=0$ condition. For a system of linear polymers lowering the molecular weight raises the configurational entropy, and this means that we must cool further to reach the $S_c=0$ condition. Third, any deformation of a rubber lowers the configurational entropy; this means that the glass temperature of a rubber should increase with increasing deformation of any kind. This seems to be the case (*21*). Fourth, the specific heat break at the glass transition has been predicted to within 10% for 11 different polymers without the use of any parameters (*26*).

Link 4. Comparison to Experiment for Liquid Crystals: Liquid crystals and polymer glasses are really two faces of the same coin. In both cases the transition occurs because of a geometrical frustration arising from the stiffness of the chain.

Liquid crystals consisting of rigid rods with asymmetry ratio x are easiest to understand so we discuss them first. If the concentration of rigid rods is small a particular rigid rod chosen at random can achieve every orientation and every location. It is not frustrated from doing so. However, as the concentration of rods is increased, whether a particular rigid rod chosen at random can achieve a particular orientation depends on its neighbors. In fact interferences prevent the rod from freely choosing each orientation. Although the mathematics of this is well established a proper intuitive appreciation of the geometrical frustration involved can be obtained by attempting to pack pencils at random on a table top (2-d problem). One sees that frustration occurs at relatively large empty volumes. The 3-dimensional phenomenon of geometrical frustration is neatly observed when trimming branches from shrubs or trees. On throwing them onto a pile one observes, provided the twigs are stiff, relatively large empty spaces within the pile. By cutting the branches in half (and half again etc.) the volume of the pile can be reduced considerably. If the branches are straight one can obtain unit density by laying the branches parallel.

Flory gives for rigid rods the formula (28) $xV_x = 8$ while we give (29) $xV_x = 4$. Here V_x is the volume fraction of rigid rods and x the asymmetry ratio. If xV_x is less than the critical value the chains can pack at random but when the critical value is exceeded the chains begin to order and a two phase system is formed.

The first one to calculate the interferences of a collection of isotropically ordered rigid rods and compare them to the interferences of an anisotropic collection of rigid rods was Onsager (30). He showed quantitatively that as the concentration of isotropically ordered rigid rods was increased it became preferable for the rigid rods to align rather than for them to remain isotropic. However, he restricted his analysis to dilute solution and did not examine what happened if the concentration of the isotropically ordered rigid rods became high. Flory (31), using the lattice model which though approximate could treat all concentrations did show that the configurational entropy became zero at a finite concentration, thereby implying through the relation $S_c = kTLnW$ that the number of configurations became 1 at this concentration. At higher concentrations the rods simply could not pack together in an isotropic fashion. Anisotropic (more ordered) collections of rigid rods have the larger entropy! Later, Zwanzig (32) improved on the Onsager calculation by going out to the 7th virial coefficient and Di Marzio (33) improved on the Flory calculation by deriving formulas which reduced to the exact result for perfectly aligned rods and to the Huggins result for an isotropic orientation. The net result of these calculations was that the approach of the configurational entropy towards zero at a finite concentration as the rod concentration increased was understood as geometric frustration.

To apply this concept of geometric or packing frustration to polymer glasses imagine that, starting at high temperatures, we lower the temperature. Each molecule has a lowest energy shape which it would like to obtain at zero K, but at high temperatures every shape is equally likely. For concreteness imagine an isomeric state model with rotations about each bond having one lowest energy well and two higher energy wells. As we lower the temperature the system of molecules goes from having many ways to pack together to having fewer. Let us now model the polymer as linearly connected rigid rods pivoting freely about their connections. At high temperatures the rods are one monomer long and there are many pivots. As we lower the temperature, because of the increasing chain stiffness the number of pivot points is fewer and fewer until at T=0K we have only one rigid rod. Obviously this is the same problem as the rigid rod problem except that **for liquid crystals we keep the rod length the same and increase concentration while for glasses we keep concentration constant but increase the effective rod length**. In both cases we have an entropy catastrophe caused by the geometrical frustration of rods trying to pack together in a space too small to accommodate them.

Notice that as the temperature is lowered the molecules are getting stiffer and stiffer. If they are also getting straighter and straighter, as is the case for polyethylene, the option of orienting the chains presents itself and the packing difficulty can be avoided by the system becoming crystalline or liquid crystalline. However if the lowest energy configuration is stiff but not straight then the system does not have the option of crystallizing because the energy penalty for straightening is simply too great, Rather the system simply gets stuck in one of its few low temperature configurations. This is the glassy phase.

It is a quantitative result of the theory that at a finite temperature T_2 the number of configurations is so small that the associated packing or configurational entropy is zero (5). Above this temperature the polymer configurations occur with Boltzmann probability (5); if the energy of shape j is E_j then the probability of shape j is proportional to $\exp(-E_j/kT)$. Below this temperature the chain configurations occur with the Boltzmann probabilities appropriate to T_2; the probability of shape j is proportional to $\exp(-E_j/kT_2)$. Thus, geometric frustration prevents the molecules from assuming their Boltzmann distribution of shapes at the glass temperature and below.

We note that the concept of geometrical frustration (4,5,30) predates the notion of frustration in spin glasses (34) by many years.

Link 5. Principle of Detailed Balance Connects Equilibrium to Kinetic Considerations: Are our equilibrium results then unrelated to the kinetics of glasses? It is easy to show that there is indeed an intimate connection between kinetic and equilibrium quantities. An experimental proof is in Kauzmann's paradox[11]. If we did our experiments infinitely slowly we would, according to extrapolation of experimental curves for entropy and volume, reach negative entropies and volumes less than crystal volumes. Yet, kinetics always intervenes to save the day for thermodynamics. Since this always happens no matter what the experimental system we are forced to conclude on experimental grounds alone that there is a fundamental connection between viscosity and entropy such that whenever the entropy is approaching zero the viscosity is becoming very large.

Fundamental theory also tells us there is an intimate connection. The fluctuation-dissipation (FD) theorem (7) relates dissipative quantities such as viscosity to the ever present fluctuations in equilibrium systems, such as correlations in local velocities. A significant statement about the viscosity can now be made. The dissipative quantity has the same kind of temperature and pressure discontinuities as does the correlation function. Thus, if we have a first-order thermodynamic transition then we expect the viscosity to display first-order character in its T-P behavior. This is the case. Since we predict an underlying second-order transition for polymer glasses we expect to have second-order transition behavior in the viscosity. This is contrary to the Vogel-Fulcher equation which predicts an infinite value for the viscosity at a finite critical temperature.

The principle of detailed balance is another connection (8). This principle relates the rate constants, α_{ij}, α_{ji} for the system jumping from state i to j to the occupation numbers N_i, N_j.

$$N_j/N_i = \alpha_{ij}/\alpha_{ji} = \exp(-[E_j-E_i]/kT) \qquad (2)$$

where E_i is the energy of state i. One notes that equation 2 plus another as yet unspecified equation will define the kinetics completely in terms of equilibrium quantities. We can obtain such an equation by deciding how to apportion the energy into the forward and backward reactions. We have argued previously (35) that the energy should appear entirely as a barrier to escape from the wells rather than as an energy sucking particles into wells. Thus the α_{ji} are known except for a multiplicative constant and we can now proceed to develop a kinetic model in which the α_{ij} play the fundamental role

Link 6. Theoretical Approach; A Minimal Model for the Kinetics: Our approach examines the topology of the potential energy surface in configuration space and the motion of the phase (configuration) point in this space (1). This potential energy surface is a mix of deep and shallow wells separated by energy barriers. We use a trapping model in configuration space because at low temperatures the configuration point escapes the deep wells only rarely. Trapping models have been used previously to describe glasses (36,37).

The principle of detailed balance discussed above allows us to determine kinetic properties from equilibrium properties. We infer that the escape from a deep well is exponential in time with a relaxation time that is exponential in the well depth E_j. The GD configurational entropy theory can be used to estimate $Q(E_j)$, the number of wells of depth E_j.

The problem then reduces to that of solving the master equation for diagrams that describe the essentials of the connectivity of the potential energy minima.

Our minimal model consists in replacing the potential energy surface in configuration space by a diagram for which we know the transition rates α_{ij} between points (wells). Our minimal model diagram is (1)

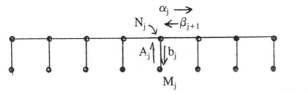

The Master Equations describing the minimal model of the diagram are given by the simple set of equations

$$dN_1/dt = -(\alpha_1 N_1 - \beta_2 N_2) - b_1 N_1 + A_1 M_1 \qquad (3)$$

....

$$dN_j/dt = (\alpha_{j-1}N_{j-1} - \beta_j N_j) - (\alpha_j N_j - \beta_{j+1}N_{j+1}) - b_j N_j + A_j M_j \qquad ......$$
$$dM_j/dt = +b_j N_j - A_j M_j$$

where the Greek rate constants denote stepping to the right. α_j, or left, β_j, and the Roman rate constants denote stepping down, b_j, or up, A_j.

This minimal model displays the feature that there is no transport of the configuration point when it is in one of the deep wells. Transport from well to well is represented by the rate constants for horizontal motion α_j and β_j. This diagram can be generalized by imagining that the horizontal lines are themselves replaced by figures with the same topology as the above figure but with a different set of rate constants. In this way we have a hierarchy of structures or a nesting of nests.

Link 7. Theoretical Approach; Obtaining the Viscosity, $\eta(\omega,T)$: By solving these equations we should in principle be able to obtain the frequency and temperature dependent viscosity, $\eta(\omega,T)$, and diffusion coefficient $D(\omega,T)$. This is a very difficult problem and we are not sure it was done right. The idea we used was that in deep wells there is no motion. Motion only occurs when the configuration point is roaming the configurational sea of shallow wells that connect to the deep wells. Additionally, in real space there must be a critical number of particles per unit volume residing in shallow wells before flow can occur. Please consult reference 1 for our treatment.

However, once the complex viscosity is known the dielectric response is easily obtained through a generalization of the Debye equation, viz.

$$[\epsilon(\omega,T)-\epsilon(\infty,T)]/[\epsilon(0,T)-\epsilon(\infty,T)] = 1/(1+G^*(\omega,T)A) \qquad (4)$$

where $G^*(\omega,T) = i\omega\eta^*(\omega,T)$ is the complex shear modulous and η^* is the complex shear viscosity. The Debye model assumes that each electric dipole is imbedded in a rigid sphere which can rotate in a viscous liquid. The spheres are imagined to be sufficiently far apart that they interact neither electrically nor through the fluid. We showed that under these assumptions the above equation is the correct one (*38*). If these assumptions fail then we expect it will be because of the mechanical interaction. This is because the Green's function (Oseen Tensor) for the Navier-Stokes equation varies inversely with the distance while the dipole-dipole interaction varies as inversely as the cube of the distance. We would then opt for a Havriliak-Negami (HN) form (*39*) with the frequency dependent complex viscosity again replacing the zero frequency viscosity. An interesting question is whether the two parameters of the HN equation have reduced variation from material to material when the complex viscosity is used instead of the zero frequency viscosity.

The Fourier transforms of both $G^*(\omega,T)$ and $\eta^*(\omega,T)$ display "KWW-like" behavior.

Link 8. Some Results: We began our quest fully expecting to derive the Vogel-Fulcher (VF) equation (*40*) along the lines of the Adam-Gibbs (AG) approach (*41*) and then extend the results to frequencies other than zero. Instead, much to our surprise we found that our logical extension of the equilibrium theory (what we believe to be a logical extension) results in a much different functional form for the zero frequency viscosity. Whereas the VF equation

$$Log\eta = B + A/(T-T_c)$$ (5)

and the AG equation

$$Log\eta = B + A/TS_c$$ (6)

both show divergences, one at T_c, the other at the underlying second-order transition temperature T_2. Our equation

$$log\eta = B - AF_c/kT$$ (7)

shows neither a discontinuity, nor even a discontinuity in slope at T_2. Here F_c is the thermodynamic free energy appropriate to the system (G_c for constant pressure-temperature, F_c for constant volume-temperature). We are aware of only one other work that stressed there should be no discontinuity in η versus T plots (*42*).

We note that although the location of the glass transition in temperature-pressure space is determined solely by the configurational entropy S_c (equation 1b) the zero frequency viscosity is determined solely by the configurational free energy F_c (equation 7).

It is remarkable that the viscosity should depend on a purely thermodynamic quantity. However for frequencies other than zero this feature is lost[1].

To test equation 7 one needs to know F_c either from experiment or from theory. If we use experiment there will be a two-parameter fit rather than a three-parameter fit as in the VF equation. There is a general consensus that the specific heat above the glass transition, $C_{p,c}$, varies approximately inversely with temperature (*43*). We therefore use as an approximation to the experimental situation the form

$$C_{p,c} = \alpha/T \rightarrow S_c = \alpha(1/T_2 - 1/T) \rightarrow$$ (8)

$$G_c = -K - \alpha(T/T_2 - 1) + \alpha Ln(T/T_2), \quad T_2 \leq T$$ (9a)
$$G_c = -K, \qquad\qquad\qquad\qquad T \leq T_2$$ (9b)

where the constant of integration K is the energy of activation.

To obtain these equations we integrated $C_{p,c} = T\partial S_c/\partial T$, $S_c = -\partial G_c/\partial T$ and ignored any pressure dependence. Below the transition temperature T_2 the configurational entropy is zero according to the simple version of the GD theory so that we have only energy of activation while above T_2 the free energy is consistent with the inverse temperature dependence of the specific heat.

Using equation 7 we can eliminate B by choosing a reference temperature T^* for which the viscosity equals 10^{12} Pas (10^{13} poise). A little algebra results in

$$Log\eta = 12 + \zeta xLnx + (1-x)(\zeta[1 + Ln(T^*/T_2)] - \theta)$$ (10a)
$$T_2 \leq T^* \leq T, \ T_2 \leq T \leq T^*$$

$$Log\eta = 12 + \theta(x-1) - \zeta[(T^*/T_2 - 1) + Ln(T^*/T_2)]$$ (10b)
$$T \leq T_2 \leq T^*$$

$$Log\eta = 12 + \theta(x-1) + \zeta[(T^*/T_2 - x) + xLn(xT_2/T^*)]$$ (10c)
$$T^* \leq T_2 \leq T$$

$$\text{Log}\eta = 12 + \theta(x-1) \qquad (10d)$$
$$T \leq T^* \leq T_2, \ T^* \leq T \leq T_2$$

where $\theta = KA/kT^*$, $\zeta = \alpha A/kT^*$, $x = T^*/T$. T^* is the temperature for which $\eta = 10^{12}$ Pas. If we had picked 10^y as the reference viscosity then the above equations would be the same with y replacing 12 and T^* being the temperature at which the viscosity is 10^y Pas.

Equations 10a, 10c give a positive curvature (curve is concave up) and the curvature is greater the larger the specific heat discontinuity at T_g. Also, as the value of T^*/T decreases the curvature is larger. These features are also features of Angell's classification of glasses into strong and fragile varieties (44). Below the glass temperature we predict pure Ahrennius behavior. An interesting prediction is that if $T^*/T_2 = 1$ then the initial slope at $T^*/T = 1$ is independent of specific heat. It does however depend on the constant K.

We tested these predictions for polymers using data for Polydimethylsiloxane of varying molecular weight. Roland and Ngai (45) using dielectric relaxation data of Kirst et al (46) and specific heat data of Bershstein and Egorov (47) created fragility plots of the logarithm of relaxation time versus T_g/T where T_g was defined as the temperature for which the relaxation time was one second. Roland and Ngai observed, 1) that the slope of the curves at $T^*/T = 1$ are independent of specific heat. We predict this; 2) The curvature is larger the smaller the value of T_g/T. We predict this; 3) The curves flare out for low T_g/T with the higher specific heat (low M.W.) material flaring up and the low specific heat (high M.W.) material flaring down. We predict this. See figure 7 of reference 1.

In the future we hope to make a more rigorous test of equation 7. First, assuming equations 10 we can check to see if there is agreement with the WLF form. Preliminary results show that provided $.85 < T_g/T < 1$ there is a good fit to data (48). Second, F_c can be obtained from lattice model statistics. We can view that the constant A, B of equation 7 are adjusted to fit the high and low values of the viscosity in the temperature range of interest. Then all that is required of the theory is that it give the correct curvature. Finally, there is simply no substitute for good experimental data. This will allow us to examine the dependencies of viscosity on such quantities as specific heat, compressibility and thermal expansion coefficients, and to make a decisive test of equation 7.

Finally, we are aware that the use of different diagrams for our minimal model gives different results. A hierarchical nesting of the diagram as described in the text results in

$$\log\eta = B - JS_c + KU/kT \qquad (11)$$

which is a linear combination of energy and entropy, but not that combination that results in free energy. It has an additional parameter. This formula implies that $\eta(T)$ is continuous through the glass transition but that $d\eta(T)/dT$ is discontinuous at T_2.

It is obvious that if a system of molecules cannot possibly crystallize then at low temperatures the system must per force be amorphous and of necessity this amorphous system has low temperature equilibrium properties. In this appendix we prove that such systems exist. We classify all systems into crystallizable (Class I) and non-crystallizable (Class II).

Class IIA consists of many different kinds of molecules each of which can take on many different shapes. Atactic polymer molecules constitute an example within this class.

Class IIB consists of many kinds of molecules each of which can take on only one shape. Such systems are the subject of tiling theory (*49*). Penrose has a system of several (the minimum being two) kinds of 2-dimensional molecules that can completely cover a surface without any repetition of pattern (*50*).

Class IIC consists of one kind of molecule that can take on many shapes. Many polymer and biopolymer molecules are of this type. We can perform a gedankin experiment which shows that in principle at least, this class contains non-crystallizable systems. let us imagine ourselves to be master chemists and then create a polymer molecule so that so that each shape has the same energy as every other shape. The equilibrium state of a system of such molecules will be an amorphous phase with each shape occurring, even at a temperature of absolute zero. This is our first example. A second more realistic example occurs when we allow one of the shapes to be the lowest energy shape and all other shapes to be of the same higher energy. We choose this lowest energy shape to be one which cannot pack in regular array in a space-filling manner. We shall prove later that such shapes exist. At high temperatures each of the shapes exist with equal probability. As we lower the temperature more and more molecules fall into their lowest energy shape. As we approach absolute zero all of the molecules want to go into their lowest energy shape. However, because this cannot happen only two other options remain. (1) Molecules can always be placed in regular array if we are willing to place them far apart from each other. However, being master chemists we have made our molecules with strong intermolecular interaction energies so that the molecules must pack tightly in space. This leaves only option (2) A certain fraction of the molecules fall into their lowest energy shape and the remainder are free to sample continuously their vast number of higher energy state. Such an equilibrium state is obviously not a crystal.

To complete the proof we must now show that certain polymer shapes cannot pack together densely in regular array (i.e. cannot crystallize). Let us pick as the lowest energy shape one of the many shapes that has linear dimensions on the order of $n^{1/2}$ where n is the molecular weight. For simplicity it is useful to imagine polymer chains on a lattice but a lattice is not necessary to the argument. Imagine also that the shape we have chosen is roughly spherical in shape, being one of the many such random-walk shapes available to a polymer molecule. Now, because of close packing (no voids) the distance between adjacent molecule centers varies as $n^{1/3}$. However, the size of the molecule varies as $n^{1/2}$. Thus there is large overlap and extensive interpenetration of neighboring molecules. The volume occupied by one molecule, $n^{3/2}$, is enough to accommodate $n^{1/2}-1$ other molecules. In order for the system to be crystalline each of the $n^{1/2}$ molecules within the sphere of volume $n^{3/2}$ must have the same shape. This is clearly impossible.

To make an estimate of the probability p that a shape chosen at random can close-pack with each of the other chains having the same shape we proceed as follows. The first chain defines the volume into which we place the subsequent $n^{1/2}-1$ chains. The first segment of the $(j+1)$th chain can be placed in $(n^{3/2}-jn)$ ways. Each subsequent segment of the $(j+1)$th chain has a probability $j\ n^{-1/2}$ of suffering interference and a probability $1-jn^{-1/2}$ of not suffering interference. The probability that the $(j+1)$th chain sufferers no interference on being placed in the volume circumscribed by the first chain is $(n^{3/2}-jn)(1-jn^{-1/2})^{n-1}$. Thus,

$$p = \prod_{j=0}^{n^{1/2}-1}[(n^{3/2}-jn)(1-jn^{-1/2})^{n-1}] \qquad (A1)$$

This problem is identical to that of calculating the configurational entropy for the case of perfectly stiff (stiff but not straight) chains and has been given previously (4-6). A little algebra shows that the $\log_e(p)$ varies as

$$\log_e(p) \sim n^{1/2}\log_e(n/e) - n^{3/2} \qquad (A2)$$

which means that p decreases exponentially with $n^{3/2}$. p is an average value and since it is non-zero there is a small chance that a chain configuration chosen at random can pack in crystalline array. However, for there to **always** be a an equilibrium crystalline state **every** possible shape must be able to pack in a regular array. But this possibility is clearly in contradiction to equation A1. Equation A1 shows that the vast majority of molecular shapes allowed to a polymer molecule are shapes that cannot pack in regular array in a space filling manner.

Class IID consists of one kind of molecule, one kind of shape. The open structures we considered above have a Hausdorff dimension of $D=d\log M/d\log r = 2$ and we have shown that they indeed do not pack together in a space-filling manner. However, even for the case of compact molecules there are many shapes that do not form crystals. Hoare (51) has discussed the problem of densely packing compact molecules that locally at least abhor regular arrangements. A prototype of this kind of molecule is a pentagon which does not pack nicely because because of the non-existence of five-fold crystal symmetry. Higher-order regular polygons with numbers of sides that are not multiples of 2 or 3 retain this amorphous nature and do not pack well locally. Such objects and their analogues in dimensions other than 2 are called amorphons or vitrons (52). Monte Carlo calculations on such objects always result in amorphous liquids (51). Though suggestive these calculations do not constitute a proof because one must be open to the possibility that small clusters of amorphons (say three pentagons) can pack in regular array.

References

[1] Di Marzio, E. A.; Yang, A. J.-M. *J. Research NIST*, **1997**, *102* 135.

[2] Di Marzio, E. A. *Annals N. Y. Acad. Sci.* **1981**, *371* 1.

[3] Di Marzio, E. A. *The Nature of the Glass Transition*, In *Relaxation in Complex Systems*, K L. Ngai and G. B. Wright, Eds.; Blacksburg, VA, 1984,

pp 43-52, Available from NTIS, USDOC, 5285 Port Royal Rd., Springfield, VA 22161.

[4] Gibbs, J. H. *J. Chem. Phys.* **1956**, *25*, 185.

[5] Gibbs, J. H.; Di Marzio, E. A. *J. Chem. Phys.* **1958**, *28*, 373; Di Marzio, E. A.; Gibbs, J. H. *J. Chem. Phys.* **1958**, *28*, 807.

[6] Di Marzio, E. A.; Gibbs, J. H.; Fleming, P. D.; Sanchez, I. C. *Macromolecules* **1976**, *9*, 763.

[7] Callen, H. B.; Welton, T. A. *Phys. Rev.* **1951**, *83*, 34; Callen, H. B.; Greene, R. F. *Phys. Rev.* **1952**, *86*, 702; **1953**, *88*, 1387.

[8] The principle of detailed balance is proved by relating it to microscopic reversibility in a micrococonical ensemble. See Cox, R. T. *Statistical Mechanics of Irreversible Change*, Johns Hopkins Press: Baltimore, MD, 1955.

[9] Stillinger, F. H.; Weber, T. A. *Science*, **1984**, *225*, 983; Stillinger, F. H. *Phys. Rev. B*, **1985**, *32*, 3134.

[10] deGennes, P. G. In *Scaling Concepts in Polymer Physics*, Cornell Univ. Press: Ithaca, NY, 1979, gives the estimate $4.68^n n^{1/6}$ for a 3-dimensional cubic lattice with coordination number $z=6$. Thus, the main effect of self excluded volume on thermodynamics is to reduce the effective value of the coordination number.

[11] Kauzmann, W. *Chem. Revs.*, **1948**, *43*, 219.

[12] Although the entropy is zero below the glass transition this does not mean that the number of configurations is only 1 as might supposed by using the relation $S_c = kTLnW$. In the canonical partition function approach every state that is available above the glass transition is also available below the transition. However their contributions to the partition function appear differently above and below the glass transition. This was explained in Reference 5.

[13] According to the Gibbs-Di Marzio theory of glasses, viscous flow is permitted below the glass transition. A change in the pressure on the glassy state then, means that the system will attain, albeit slowly, the same free volume V_0 which the glass would have had if one had cooled through the glass transition at the new pressure.

[14] Milchev, A. I. *C. R. Acad. Bulg. Sci.* **1983**, *36*, 1413.

[15] Gujrati, P. D.; Goldstein, M. *J. Chem. Phys.* **1981**, *74*, 2596.

[16] Foreman, K. W.; Freed, K. F. *J. Chem. Phys.* **1997**, *106*, 7422; Ibid, In press.

[17] Binder, K.; Paul, W. *J. Poly Sci. Pt. B* **1997**, *35*, 1; Wolfgardt, M.; Bashnagel, J.; Paul, W.; Binder, K. *Phys. Rev. E* **1996**, *54*, 1535.

[18] Baumgartner, A. *J. Phys.* **1984**, *A17*, L975 has, through computer modeling, shown strong spatial variation in the orientation order parameter for rigid rod systems. His *J. Physique Lett.* **1985** *46*, L659 paper strongly implies that global order parameters will not give a correct view of liquid crystals.

[19] Di Marzio, E. A.; Guttman, C. M. *Macromolecules* **1987**, *20*, 1405.

[20] Yang, A. J.-M.; Di Marzio, E. A. *Macromolecules* **1991**, *24*, 6012.

[21] Di Marzio, E. A. *J. Research NBS* **1964**, *A68*, 611.

[22] Di Marzio, E. A.; Gibbs, J. H. *J. Poly. SCi. A* **1963**, *1*, 1417.

[23] Di Marzio, E. A.; Castellano, C.;Yang, A. J.-M. *J. Poly Sci., Part B* **1996**, *34*, 535.

36

[24] Di Marzio, E. A.; Gibbs, J. H. *J. Poly Sci.* **1959**, *40*, 121.

[25] Di Marzio, E. A. *Polymer*, **1990**, *31*, 2294; Schneider, H. A.; Di Marzio, E. A. *Polymer* **1992**, *33*, 3453.

[26] Di Marzio, E. A.; Dowell, F. *J. Applied Phys.* **1979**, *50*, 5061.

[27] If a ring of x monomer units is made into two rings of x/2 monomer units each the delocalization entropy increases but the ring entropy decreases. This latter statement is proved by using the formula w $\propto z^x/x^{3/2}$ for the number of configurations of a ring. When these terms are put into a FH lattice model calculation it is seen that the ring entropy decreases more than the delocalization entropy increases (*19,20*). Thus the glass temperature of rings increases as we lower the ring molecular weight. At very low molecular weight there is a reversal of the effect.

[28] Flory, P. J. *Advances in Polymer Sci.* **1984**, *59*, 1.

[29] Di Marzio, E. A.; Yang, A. J.-M.; Glotzer, S. C. *J. Research NIST* **1995**, *100*, 173.

[30] Onsager, L. *Ann. N. Y. Acad. Sci.* **1949**, *51*, 627.

[31] Flory, P. J. *Proc. Roy. Acad. Sci.* **1956**, *A234*, 73.

[32] Zwanzig, R. *J. Chem. Phys.* **1963**, *39*, 1714).

[33] Di Marzio, E. A. *J. Chem. Phys.* **1961** *35*, 658.

[34] Toulouse, G. *Commun. Phys.* **1977**, *2*, 115.

[35] Sanchez, I. C.; Di Marzio, E. A. *J. Chem. Phys.* **1971**, *55*, 893.

[36] Di Marzio, E. A.; Sanchez, I. C. In *Transport and Relaxation in Random Materials*; Klafter, J.; Rubin, R. J.; Shlesinger, M., Eds. World Scientific: 1986; Di Marzio, E. A. *Computational Matl. Sci.* **1995**, *4*, 317.

[37] Odagaki, T.; Hiwatari, Y. *Phys. Rev.* **1990**, *A41*, 929; Hiwatari, Y; Miyagawa, H.; Odagaki, T. *Solid State Ionics* **1991**, *47*, 179.

[38] Di Marzio, E. A.; Bishop, M. *J. Chem. Phys.* **1974**, *60*, 3802.

[39] Havriliak, S.; Negami, S. *J. Poly. Sci.* **1996**, *C14*; *Polymer*, **1967**, *8*, 161.

[40] Vogel, H. *Phys. Zeitschrift*, **1921**, *22*, 645; Fulcher, G. S. *J. Amer. Cer. Soc.* **1925**, *8*, 339, 789.

[41] Adam, G.; Gibbs, J. H. *J. Chem. Phys.* **1965**, *43*, 139.

[42] Stillinger, F. H. *J. Chem. Phys.* **1988**, *88*, 7818.

[43] Angell, C. A. *Science* **1995**, *267*, 1924.

[44] Angell, C. A. *J. Non-Crystalline Solids* **1991**, *131-133*, 13.

[45] Roland, C. M.; Ngai, K. L. *Macromolecules* **1996**, *29*, 5747.

[46] Kirst, K. U.; Kremer, F.; Pakula, T.: Hollingshurst, J. *Colloid and Poly. Sci.* **1994**, *272*, 1420.

[47] Bershstein, V. A.; Egorov, V. M. *Differential Scanning Calorimetry of Polymers*; Horwood: NY, 1994, Chapter 2.

[48] Colucci, D. M.; Di Marzio, E. A.; McKenna, G. B. *Connection of the Relaxation Response in Polymers to Thermodynamic Quantities: Comparison of Experiment to Theory*. In Preparation.

[49] Gardner, M. *Extraordinary nonperiodic tiling that enriches the theory of tiles* In *Sci. Am.* **January 1977**, p. 10.

[50] Penrose, R. *J. Inst. Math. and its Application*, **1974**, *7-8*, 266.

[51] Hoare, M. *Ann. N. Y. Acad. Sci.* **1976**, *271*, 186.

[52] Tilton, L. W. *J. Res. Nat. Bur. Stnds.* **1957**, *59*, 139.

Chapter 3

Entropy, Landscapes, and Fragility in Liquids and Polymers, and the ΔCp Problem

C. A. Angell

Department of Chemistry, Arizona State University, Tempe, AZ 85287

We examine the relation of the mode coupling theory of glassforming liquid dynamics to the Gibbs-Goldstein landscape picture of relaxation, and identify, using both scaling relationships and thermodynamic calculations, where the crossover between the two domains occurs. This "landscape" approach is then applied to chain polymers, using a neglected relation between the WLF C_1 parameter and the high frequency limit for relaxation, to establish the characteristic temperature of the landscape ground state and the appropriate fragility parameters. A problem with the relation between thermodynamic to relaxational assessments of the landscape "height" is used to focus attention on a problem with the heat capacity changes at T_g in the case of chain polymer melts and their glasses.

The problem of viscous liquids and the glass transition is currently enjoying one of its characteristic thirty-year cycles of high activity, stimulated jointly by the review of Anderson in 1979 (1), and the advent of Götze and colleagues' mode coupling theory of the glass transition in 1984 (2) The latter gave, as its central features, the existence of a two-step response to a perturbation of the equilibrium state about which detailed predictions were made, and the prediction of a dynamical "jamming," i.e. vitrification at finite temperature due to power law divergences in the response times with decreasing temperature. The first of these has been extensively verified for liquids and polymers whereas the second has not.

The failure to jam in the predicted manner has been attributed to the crossover in relaxation mechanism at some temperature between normal liquid (high fluidity, short relaxation times) temperatures and the glass transition temperature, an occurrence which was anticipated by Goldstein in a 1969 paper which is now a classic.

Here we will (1) give a brief review of what might be called the Gibbs-Goldstein picture of the phenomenology, thermodynamic and dynamic, of glassforming liquids, (2) add to this a classification of liquid types ranging from the highly non-Arrhenius glassformers (to which the mode coupling description was expected to apply) to the almost Arrhenius liquids (to which it was not expected to apply but, according to recent studies, does), and then (3) to examine the extent to which the phenomenology of polymer liquids and glasses can be interpreted in terms of (1) and (2). This will lead us to recognize a problem in the polymer case for which we will then supply a partial resolution and a proposal for further work. In the process, we will encounter not only a variety of insights into the utility of the "landscape paradigm" for discussions of glassformers but also both thermodynamic and dynamic markers for the crossover temperature between "landscape-dominated" and "free diffusion" domains.

The Gibbs-Goldstein Picture

The fact that glasses are brittle solids at temperatures below their glass transition temperatures implies that the arrangement of particles taken up as a liquid cools below T_g can be described by a point in configuration space near the bottom of a potential energy minimum in this space (3,4). If this were not so, the system would move in the direction dictated by the collective unbalanced force acting on it, and some sort of flow would occur. On the other hand, the existence of the annealing phenomenon, in which the density and energy of a glass formed during steady cooling can change with time on holding at a temperature below but close to the "glass transition temperature" means that there is more than one such mechanically stable minimum available to the system. Indeed, there would appear to be a huge number, of order e^N, where N is the number of particles in the system (5, 6, 7). The minima, or "basins of probability" (5) obviously are distributed over a wide range of energies, usually scaling with density. However, there are also many ways of organizing the same collection of particles into minima which differ negligibly in energy from one another. Each minimum is a configurational microstate, or "configuron," of the system. Each laboratory "glass" is thus a palpable configuron, though there is evidence that the distinguishing properties of a glass, viz. its structure, heat capacity, and diffusivity are determined within very small distances, so that the properties of a glassformer can be evaluated by consideration of only a tiny subset of its particles.

The fact that annealing ("aging" for polymers) proceeds more slowly the lower the temperature at which the annealing is carried out suggests that the process of finding deeper minima becomes more difficult statistically as the temperature is decreased. One arrives at the notion of an interconnected series of minima on a landscape of inconceivable complexity, in which increasing depth is associated with decreasing configuron population. The important implication of Kauzmann's paradox *(8)* is that for each system, at least for each "fragile" system, there must exist a statistically small number of minima at energies still well above that representing the crystal and that these must set an absolute limit on the energy lowering achievable by annealing the amorphous system. It is into one of these last few minima that the system is tending to settle at the temperature where the excess entropy is tending to vanish. The energy of these lowest minima $k_B T_K$ defines the ground state for the amorphous system, to be reached at the Kauzmann temperature T_K on indefinitely slow cooling. Hence at this temperature, the excess entropy would vanish.

The Adam-Gibbs equation for viscous liquid relaxation asserts that the time scale for re-equilibration after some perturbation is related to the excess entropy of liquid over crystal, according to

$$\tau = \tau_0 \exp C\Delta\mu/TS_c \tag{1}$$

where τ_0 is a quasi-lattice vibration time, 10^{-14} s, $\Delta\mu$ is a free energy barrier per particle to cooperative rearrangements, and S_c is the excess (configurational) entropy. Evaluating S_c as the entropy generated above T_K according to

$$S_c = \int_{T_K}^{T} \frac{\Delta C_p}{T} dT . \tag{2}$$

leads to the well-known Vogel-Fulcher-Tamann (VFT) equation as an identity or as a good approximation depending on how the excess heat capacity vs. T relation is approximated *(9)*. Although the VFT equation is itself only a rough description of the behavior of supercooling liquids over the 16 orders of magnitude of relaxation times which can now be measured *(10)*, a fitting of data to that equation under the constraint that τ_0 have a physical value 10^{-14} s, yields an agreement between the VFT relaxation time divergence temperature T_0 and the thermodynamically determined T_K, within a variance of ~2% for 40 liquids for which the T_K data are available with T_g ranging from 50 to 1000 K *(11)*.

The departure from Arrhenius behavior of the relaxation time arises, in the Adam-Gibbs theory, from the temperature dependence of the S_c term in equation (1), which itself is a consequence of the excess heat capacity ΔC_p of equation (2). For constant $\Delta\mu$ in equation (2), the degree of non-Arrhenius character, now called the fragility, is determined by the magnitude of ΔC_p. A general but incomplete accord seems to exist between the fragility and ΔC_p and exceptions, like the alcohols, can be rationalized by the presence of unusual $\Delta\mu$ terms.

Thus, in broad brush, the Adam-Gibbs approach within the landscape paradigm, which is summarized pictorially in Figure 1, provides a good basic understanding of key features of the behavior of glassforming liquids and polymers in their ergodic states above T_g. Furthermore, by setting S_c in equation (1) at a value fixed by cooling rate or annealing time, the Adam-Gibbs equation goes far towards a description of behavior in the non-ergodic regime below T_g.

The viscosity of polymers above T_g can be incorporated in this picture by inclusion of a molecular weight-dependent pre-exponent in equation 1, so as to obtain a local viscosity relaxation

$$\tau_{\eta,loc} = \frac{\eta_{loc}}{G_\infty} = \eta/M^x G_\infty = \tau_0 \exp\left(\frac{C\Delta\mu(\eta)}{TS_c}\right) \qquad (3)$$

time, where $x = 1$ for pre-entanglement conditions, $x = 3.2$ for the entanglement range of M, and τ_0 again has values of ~10^{-14} s. The exponent in equation (3) becomes identical with that of the Vogel-Fulcher-Tammann equation ($\exp(B/[T-T_0])$) with $B = DT_0$, when *(9)* S_c is developed using the hyperbolic temperature dependence of the excess heat capacity ΔC_p found so frequently for molecular liquids. D is an inverse measurement of the "fragility" of the liquid or polymer, to which more reference will be made below.

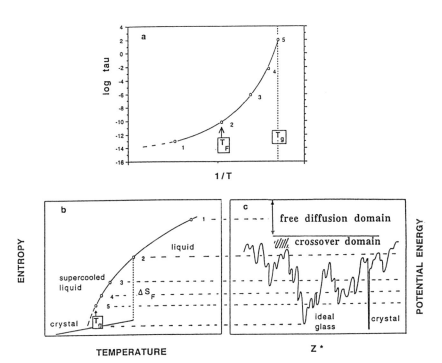

Figure 1. Summary of phenomenology of glassformers showing diverging relaxation times (in part (a)) being related, by points 1, 2, 3. ... on the plot, to vanishing excess entropy (in part (b)) and finally, in part (c), to the level of energy minima on the potential energy hypersurface (or "landscape"). The temperature T_K corresponds with the energy of the lowest minimum on the amorphous phase hypersurface. Many vertical spikes, corresponding to configurations in which particle core coordinates overlap, are excluded from this diagram for clarity.

The number of minima on the surface is of order exp(N), where N is the number of particles. The height of the landscape relative to $k_B T_K$ is a measure of the "strength" of the liquid and seems to be about 1.4-$1.6 T_K$ for fragile liquids. Near the top of the landscape is a crossover region to a domain in which only disconnected remnants of the low temperature "structure" remain to provide the "caging" effects responsible for the slow relaxation described by mode coupling theory. At higher temperatures, $>2T_K$, (above the melting points of glassforming liquids, but still below the melting points of non-glassformers) the cages are dismantled, relaxation becomes exponential, and diffusion becomes "free." Boiling typically occurs near 3.5-$4 T_K$.

Scaling Schemes, and "Height" of the Landscape

In a recent scaling proposal, that of Rössler and Sokolov *(12,13)* , the α-glass transition temperature used by the author in developing the fragility concept from scaled Arrhenius plotting *(14,15)*, is retained as a scaling parameter and a second characteristic temperature, T_c, is introduced in order to collapse all liquid viscosities on to the same VTF curve. This second characteristic temperature is higher relative to the first (T_g) for the stronger liquids, and indeed lies close to the crossover temperature of mode coupling theory *(12)*. In the following argument, we show that, if this second characteristic temperature is a measure of T_c, then both are measures of T_X, the crossover temperature where landscape domination of the relaxation dynamics takes over. This is because, as we suggested some time ago *(14)* but now have effectively proved *(16,17)*, T_X is near the point where the system encounters the "top" of the landscape. We "prove" it by establishing the nature of ΔC_p in terms of landscape exploration and then integrating ΔC_p over a temperature interval sufficient to exhaust all the configurational states (configurons) of the system. This upper limit of the integration, T_u, is found *(16,17)* not only to relate closely to T_X ($\approx T_c$), but also with the α-β bifurcation temperature, and the Stickel temperature, T_B.

The "height" of the landscape $k_B T_u$ can be estimated *(17)* by accepting, after Speedy and Debenedetti *(7)* and Stillinger *(5)* that there are, to good approximation, e^N configurons per mole of heavy atoms, and then calculating the temperature T_u in the expression,

$$S_c = k_B \ln W = \sim k_B \ln (e^N) = R/\text{mole} = \int_{T_K}^{T_u} \frac{\Delta C_p}{T} dT. \qquad (4)$$

Here ΔC_p is the heat capacity increment associated with exploration of the landscape (*i.e.* the jump in C_p as $T > T_g$). Of course the relative height T_u/T_K is then greater for liquids with small ΔC_p (*i.e.* strong liquids), in accord with the higher T_c/T_g found by Rössler *et al (12,13)* for the stronger liquids. The value of T_u turns out to depend only weakly on the functional form assumed for ΔC_p. The two simple choices,

$$\Delta C_p = \text{constant}, \quad \text{and} \quad \Delta C_p = \text{constant}/T$$

yield, respectively from equation (3),

$$T_u = T_K \exp(R/\Delta C_p) \qquad (5a)$$

and

$$T_u = T_K/[1 - R/\Delta C_p(T_K)] \qquad (5b)$$

For simple laboratory glassformers like S_2Cl_2 *(16)*, and also for Lennard-Jones argon (18) ΔC_p at T_g is found to be about 17J/K per mole of heavy atoms, and T_u is then $1.59T_K$ (which is about $1.27T_g$) for each of Eqs. (5a) and (5b). Interestingly enough, this is almost the same as the value of T_c/T_g found by Rössler and co-workers (we prefer T_x/T_g) for the ratio of upper scaling temperature T_c to T_g, which means it is the same as the ratio T_c/T_g where T_c is the mode coupling theory T_c. The identification of T_c (T_x) with the temperature characterizing the top of the landscape is consistent with the long-standing idea that MCT fails at low temperature because of the crossover to landscape-dominated dynamics (which must be the real meaning of the term "hopping" used in many MCT papers *(2,19,20)*.

Our estimate of T_u is probably a minimum value because a part of the ΔC_p at T_g (of unknown magnitude, but thought to be small in the general case) is vibrational in nature, and there are also other possible non-configurational contributions to the heat capacity difference between liquid and crystal *(21)*. A value of T_u somewhat above T_c would seem appropriate because T_c presumably represents a crossover in *dominant* relaxation mechanism rather than a real end-point in structural character. In some model systems for which data are available, e.g o-terphenyl, T_c is also found to correspond with the bifurcation temperature into distinguishable α- and β-relaxations (22) and with the lower limit temperature for accurate data-fitting by the "high temperature Vogel-Fulcher law" in the Stickel-plot analysis *(10*. The reason that a VTF law (with unphysical parameters) should fit in the free diffusion domain is not at all clear. This is the domain in which a power law fit of the same data yields the T_c found by the other criteria, so the VTF fit may be a trivial consequence of its relation to the power law through the Bardeen singularity *(23)*. If ΔC_p drops to a value like that of $ZnCl_2$, 7.5 J/m.K of heavy atoms *(24)*, then

T_u/T_K = 2.94 (equation 5a) (equation 5b becomes inappropriate because for low ΔC_p substances, the value of ΔC_p is either constant above T_g or increases with increasing temperature). Using T_K = 250 K, we obtain T_u = 736 K from equation (5a). This compares poorly with the value of 580K obtained by Rössler scaling, but this would be improved if the initial increase in ΔC_p observed in the laboratory studies *(24)* were taken into account.

Referring to the relation between entropy, relaxation time, and the position of $k_B T$ on the energy surface summarized in Figure 1, we note the presence of a regime of free diffusion above the highest features of the landscape but below the boiling point. T_c falls somewhat below the highest energy features of the landscape because, as noted above, this is merely the temperature about which the landscape dominance of relaxation, not the landscape itself, terminates.

This raises a problem which has so far not been addressed in our discussions (11,17) of the landscape limits assessed by the above argument. The problem is that no thermodynamic signature of T_u, where the landscape contribution to the excess C_p would be expected to drop out, is seen in the heat capacity vs temperature relation. Since T_u falls at a value where there is still a six decade difference between vibrational and structural relaxation time scales, the separability of degrees of freedom should still be clear. Either the landscape concept must be at fault, the non-landscape contributions to the ΔC_p are seriously underestimated, or the landscape limit is much more tenuously demarcated than a simple endpoint, at T_u, would suggest. Some information on this point may be forthcoming from the discussion of the "ΔC_p problem" for polymers given in the next section. There we will suggest that the presence of the polymer chain has an effect comparable to that of examining ΔC_p in liquids at a very short relaxation time hence in the vicinity of T_u.

The Polymer Case: Fragility and the ΔC_p Problem

To incorporate polymers into the above scheme, we need to recognise that the special effects of polymer chain length on viscosity enter the problem in the **pre**-exponents of Vogel-Fulcher or Adam-Gibbs-like expressions (see equation (2)) and, accordingly, to scale them out. This is achieved by use of the WLF representation. To get the appropriate landscape ground state temperature T_K in the face of the uncrystallizability of many polymer systems, one must turn to the relation between T_o of the Vogel-Fulcher relation and the Kauzmann temperature. T_K can be obtained from the WLF parameter C_2 via $T_o \equiv T_K = T_g - C_2$, **provided** C_1 is set at the number 16 or 17 (depending on the relaxation time at the reference temperature, τ_g). This is because of the little recognised (25) identity

$$C_1 = \log(t_r/t_o) \tag{6}$$

where τ_o is the Vogel Fulcher pre-exponent which must be on the phonon time scale for the other fit parameters to be physically acceptable.

With T_K identified, a convenient representation of the fragility which varies between extremes of 1 and 0, can be obtained from the ratio $T_K/T_g = 1 - C_2/T_g$ according to the commonly recognized relation between WLF and VTF equation parameters, C_2 and T_o ($C_2 = T_g - T_o$). To see if polymer behavior is consistent with the small molecule scenario, a second measure of fragility can be obtained from the upper limit of the landscape using either assessments of T_c from mode coupling theory (e.g., breaks in the quadratic behavior of the Debye Waller factor) , fits to the high temperature relaxation time data, or the $\alpha - \beta$ bifurcation temperature. According to the recent report of Frick and Richter (26) on polybutadiene, a fairly fragile polymer (the D value (D = B/T_o) extracted from the Vogel Fulcher parameters in ref 26 is 10.9) there is indeed a close accord between T_c (215K) and the $\alpha - \beta$ bifurcation temperature (217K). The data are reproduced in Figure 2. Furthermore, T_o from Vogel -Fulcher fits of the monomeric friction coefficient for viscosity yield (26) a T_o of 126K which then gives $T_c/T_o = 1.69$, a little higher than the value for the height of the landscape for fragile liquids given above. In footnote 26 we give evidence that T_o should actually be higher, hence the ratio T_c/T_o lower.

When we look for a *thermodynamic* confirmation of the landscape height, however, we encounter a problem. Instead of finding that the jump in heat capacity at T_g is large in proportion to how fragile the polymer is, we find that it is much the same for all polymers (despite wide variation in fragility (27,28), and usually rather small compared with the polymer glass heat capacity at T_g. If this value is adopted for calculations using equation (3), all

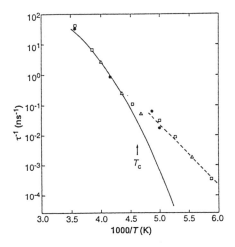

Figure 2. The α-β relaxation bifurcation in the case of the polymer polybutadiene as observed in inelastic neutron scattering(•) and dielectric relaxation (open symbols Δ, ±). Note that the bifurcation temperature accords with the MCT T_c, as in the case of molecular liquids. $T_c/T_{0(VFT)}$ is 1.69 in this case[26], or $1.59T_0$ if T_0 is taken to be T_g-50. (reproduced form ref. 26, by permission).

polymers would have comparable values of T_u/T_K far above their T_c/T_o values. Clearly there is something to be understood about the relation of ΔC_p in polymers to the corresponding quantity in molecular liquids.

Some insight into, if not resolution of, this problem can be obtained by examining the relation between the ΔC_p and the chain length. As the latter decreases and the T_g correspondingly drops quickly, one finds (29,32) that the value of ΔC_p rapidly increases. As with the increase of T_g, most of the effect is obtained with the first few repeat unit additions. The relation is shown in Figure 3. However the chain length of polymers has generally been found not to affect their fragilities (27,28) (though there is some recent conflict from different estimates in the case of polystyrene (31,32)). Clearly, though, the correlation between ΔC_p and fragility seen in the inorganic ionic (33), and covalent (34) network polymers does not hold for carbon chain polymers (in which the in-chain bonds are inviolable).

To look into this further, we show in Fig 4, in part (a), the behavior of the heat capacity of polypropylene, in units of J/K.(mol of $-CH_2-CH_2(CH_3)-$ repeat units) (35,36) in comparison with that of the molecular liquid 3-methyl pentane (37) (divided by 2 to have the same mass basis as the polymer repeat unit) (38). It is seen that the liquid heat capacity of the hexane isomer (x 0.5) falls not much above the natural extrapolation to lower temperatures of the heat capacity per repeat unit of the polymer. This implies that the main effect of polymerization, as far as the change in heat capacity at Tg is concerned, is to postpone the glass transition until a much higher vibrational heat capacity has been excited. This not only reduces the value of ΔC_p but has a disproportionate effect on the ratio $C_{p,l}/C_{p,g}$ at T_g. This happens despite a lower *glassy* heat capacity in the polymer than in the molecular liquid at the same temperature. The latter effect is a direct consequence of the lower Debye temperature (and lower vibrational anharmonicity) at a given temperature for in-chain interactions in the polymer than for *inter*molecular interactions in the same mass of molecules.

A somewhat similar effect is seen in the effect of pressure on the heat capacity of 3-methyl pentane (37) where the origin of the effect is a little clearer. Takara et al (37), in a study which established the correctness of the Adam-Gibbs conclusion that the glass transition should occur at constant TS_c, found that, as pressure acting on the glass state increased, the heat capacity of the glass rose to higher values before the glass transition occurred. This can

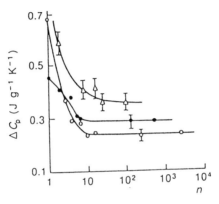

Figure 3. The heat capacity of several polymer glasses and liquids as a function of increasing number or repeat units in the chain, showing the rapid decrease in ΔC_p which accompanies MW increase (and consequent T_g increase). The cases illustrated are PDMS (triangles), polycarbonate (filled circles) and polymethyl siloxane (open circles). (reproduced from ref.29, by permission)

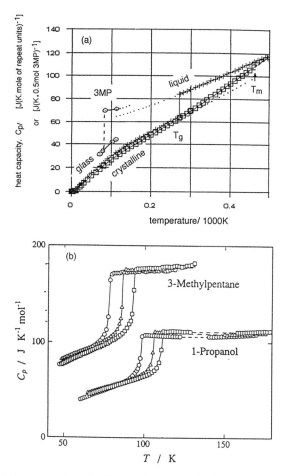

Figure 4. (a) Heat capacity of polypropylene crystal, glass and liquid in
J/K per mole of -CH$_2$-CH$_2$(CH$_3$)-repeat units, from refs. 35 and 36 compared
with the same properties of 3-methyl pentane (1/2 mole) from ref. 37. The
comparison shows how the postponement of the glass transition to the higher
temperature, consequent on the extension of the carbon chain, completely changes
the relation between liquid and glass heat capacities, causing polymers to appear
"strong" by the C$_p$(l)/C$_p$(g) (at T$_g$) criterion. The ΔC$_p$/C$_{p,g}$ value observed for
polymers is characteristic of that of a molecular liquid studied on the nanosecond
time scale.

(b) heat capacity of 3-methyl pentane glass and supercooled liquid at
three different pressures 0.1, 108.4, and 198.6 MPa, from ref. 37, showing how
the effect of increasing pressure is somewhat similar to the effect of increasing
molecular weight in its influence on the magnitude of the vibrational heat capacity
at the temperature of the glass transition. (reproduced from ref.37, by permission)

be understood as a result of increase of pressure, at constant temperature, rendering the vibrational motions more harmonic, since it raises the vibrational frequencies (positive Gruneisen constants dlnv/dV). Consequently if also raises the glass transition temperature (since the glass transition is itself a consequence of anharmonicity (39-41)).

While the effect of pressure in this respect is weaker than the effect of increase of molecular weight, both factors, through their effects on the mean vibrational anharmonicity, permit the glass (i.e. vibrational) heat capacity to build up to larger values before T_g intervenes, while less strongly affecting the liquid heat capacity. This progressively diminishes the observed jump in C_p and consequently, but misleadingly, makes the substance appear "stronger" by the change in heat capacity criterion. In fact the behavior observed has much in common with a molecular liquid which has been heated to far above its T_g (where its excess heat capacity has fallen to a fraction of its value at T_g). This itself is understandable since the between-chain interactions must be highly anharmonic in both crystal and liquid states at temperatures in excess of twice the glass transition temperature of the monomer liquid. An energy hypersurface consequence should be the existence of many low energy modes of escape from any given minimum and a less well-defined landscape limit (hence dynamical crossover temperature) than in the case of fragile molecular liquids. This may be the reason that polymers tend to be more fragile than molecular liquids (43), and tend to obey the WLF or VFT equations over wider ranges of relaxation times and with more consistent pre-exponents (hence C_1 values) than do fragile molecular liquids.

The extension of the crystalline polypropylene data to the melting point shows the difference in liquid and crystalline heat capacity almost disappearing, giving an appearance akin to that of the glass-forming liquid metals (44), the crystalline states of which melt to liquids of very low viscosity.

All considered, it is not obvious how to obtain an appropriate ΔC_p to use in a landscape height calculation or indeed if the concept of a landscape limit is a useful one for polymers (although the coincidence of T_c and $\alpha-\beta$ bifurcation temperatures seen in Fig. 2 would suggest that it should be). It is possible that one could use T_c/T_g ($= T_x/T_g$) value to back-calculate an effective contributing ΔC_p value. This is an area for further work, in which an investigation of the effect of diluent concentration on ΔC_p for the binary polymer diluent solutions, could play a useful role.

Concluding remarks.

The "polymer problem" brought out in the above discussion is provocative and deserves further exploration. Understanding could be helped considerably by a systematic investigation of the entropies and Kauzmann temperatures of a series of easily crystallized n-mers of increasing n value, and a concurrent determination of their fragilities by the $F_{1/2}$ criterion *(45)*.

Acknowledgements. This work was supported by the NSF under Solid State Chemistry grant no. DMR9614531. The assistance of Vesselin Velikov in identifying the effect of chain length on the jump in heat capacity at T_g from the Russian literature is gratefully acknowledged. We also thank P.G. Santangelo and C. M. Roland for sharing information on their current studies of heat capacity and fragility in polymers in advance of publication.

Literature Cited

1. Anderson, P. W. In *Ill-Condensed Matter*; Balian, R.; Maynard, R.; Toulouse, G., Ed.; Les Houches: North Holland, 1979; pp 171.
2. Götze, W. In *Liquids, Freezing, and the Glass Transition,* Hansen, J.-P.; Levesque, D., Ed.; NATO-ASI; Les Houches: North Holland (Amsterdam), 1989: Götze, W., and Sjögren, L., *Rep. Progr. Phys.*, **1992**,*55,* 241,
3. Goldstein, M. *J. Chem. Phys.* **1969**, *51*, 3728. The configuration space is a space of 3N + 1 dimensions, N the number of particles.
4. Gibbs, J. H. In *Modern Aspects of the Vitreous State*; McKenzie, J. D., Ed.; Butterworths: London, 1960; ch. 7.
5. Stillinger, F. H.; Weber, T. A. *Science* **1984**, *225*, 983 (1984); **1995**, *267*, 1935.
6. Ball, K. D.; Berry, R. S.; Kuntz, R. E.; Li, F.-Y.; Proykova, A.; Wales, D. J. *Science* **1996**, *271*, 963.
7. Speedy, R. J.; Debenedetti, P. G. *Mol. Phys.* **1996**, *88*, 1293.
8. Kauzmann, W. *Chem. Rev.* **1948**, *43*, 219.
9. Angell, C. A.; Sichina, W. *Ann. N.Y. Acad. Sci.* **1976**, *279*, 53.
10. Stickel, F.; Fischer, E. W.; Richert, R. *J. Chem. Phys.* **1996**, *104*, 2043.
11. Angell, C. A. APS Symposium Proceedings, *J. Res. NIST* (in press).
12. Rössler, E.; Sokolov, A. P. *Chem. Geol.* **1996**, *128*, 143.
13. Novikov, V. N.; Rössler, E.; Malinovsky, V. K.; Surovtev, N. V. *Europhys. Lett.* **1996**, *35*, 289.
14. Angell, C. A. *J. Phys. Chem. Sol.* **1988**, *49*, 863.
15. Angell, C. A. *J. Non-Cryst. Sol.* , **1991**, *131-133*, 13.
16. Angell, C. A.; Tucker, J. C. (to be published).
17. Angell, C. A., in "Complex Behavior of Glassy Systems" (Proc. 14th Sitges Conference on Theoretical Physics, 1996) Ed. M. Rubi, Springer, 1997, p. 1.
18. Clarke, J. H. R. *Trans. Faraday Soc. 2* **1976**, *76*, 1667.
19. G. Li, W. Du, A. Sakai, and H. Z. Cummins, *Phys. Rev A* **1992**, *46*, 3343

20. W. Petry, E. Bartsch, F. Fujara, M. Kiebel, H. Sillescu, and B. Ferrago, *Z. Phys. B*, **1991**, *83*, 175
21. M. Goldstein, *J. Chem. Phys.*, **1976**, *64*, 4767.
22. E. Rössler, *Phys. Rev. Lett.* **1990**, *65*, 1595
23. P. W. Anderson. Ann. N. Y. Acad. Sci, **1986**, *484*, 241
24. C. A. Angell, E. Williams, K. J. Rao and J. C. Tucker, *J. Phys. Chem.*, **1977**, *81*, 238
25. Angell, C. A. *Polymer*, **1997**, *38*, 6261.
26. Frick, B., and Richter, D, Science, **1995**, *267*, 1939, (The D value $(D = B/T_0)$ extracted from the Vogel-Fulcher parameters provided in this ref. is 10.9, comparable to polypropylene oxide, however the T_0 value is somewhat further below T_g than the common finding $T_0 = T_g - 50$). Other assessments (see ref. 28) assign PBD a very high fragility.
27. Angell, C. A., Monnerie, L., and Torell, L. M., Symp. Mat. Res. Soc. Ed. J. M. O'Reilly, **1991**, *215*, 327.
28. Plazek, D. J. and Ngai, K.-L., Macromolecules, **1991**, *24*, 1222
29. Bershtein, V. A.; Egorov, V. M. *Differential Scanning Calorimetry of Polymers: Physics, Chemistry, Analysis, Technology;* Ellis Horwood: New York.
30. Jacobsen, P., Borjesson, L., and Torell, L. M., J. Non-Cryst. Sol.
31. Sahoune A., and Piche, L., in "Structure and Dynamics of Glasses and Glassformers, Eds., Angell, C. A., Ngai, K.-L., Kieffer, J., Egami, T., and Nienhaus, G. U., MRS. Symp. Proc. **1997**, *455*, 183
32. Santangelo, P. G., and Roland, C. M., (preprint, 1997)
33. Angell, C. A., J. Non-Cryst. Sol. **1985**, *73*, 1: Lee and Tatsumisago, M., *J. Non-Cryst. Sol.*, **1996**,
34. Tatsumisago, M., B. L Halfpap, J. L. Green, S. M. Lindsay, and C. A. Angell, *Phys. Rev. Lett.*, **1990**, *64*, 1549
35. Gaur and Wunderlich, B., J. Phys. Chem. Ref. Data, 10, 1001, (1981)
36. Wunderlich, B., Polymer, 29, 1485, (1988)
37. S.Tanaka, O. Yamamuro, and H. Suga, J. Non-Cryst. Sol. 171,259, (1994)
38. Comparison with 43% of the heat capacity of 2,4, dimethyl heptane would arguably be a better choice but the C_p data are not available.)
39. Angell, C. A.*J. Am. Ceram. Soc.*, **1968**, *51*, 117: C. A. Angell, P. H. Poole, and J. Shao, *Nuovo Cimento*, **1994**, *16D*, 993 : J. Shao and C.A. Angell, *Proc. XVIIth Internat. Congress on Glass*, (Beijing) **1995**, *Vol. 1*, P. 311
40. Ngai, K. L., and Roland, C. M., in "Structure and Dynamics of Glasses and Glassformers, Eds., Angell, C. A., Ngai, K.-L., Kieffer, J., Egami, T., and Nienhaus, G. U., MRS. Symp. Proc. **1997**, *455*, 81
41. Sokolov, A. P., "Structure and Dynamics of Glasses and Glassformers, Eds., Angell, C. A., Ngai, K.-L., Kieffer, J., Egami, T., and Nienhaus, G. U., MRS. Symp. Proc. **1997**, *455*, 69
42. Buchenau, U., *Phil. Mag.*, **1992** *65*, 303-315.
43. Bohmer, R., Ngai, K-.L., Angell, C. A., Plazek, D., J.,*J. Chem. Phys.*, **1993**, *99*, 4201
44. H. S. Chen and D. Turnbull, *Appl. Phys. Lett.* **1967**, *10*, 284,
45. R. Richert and C. A. Angell, *J. Chem. Phys.* (in press).

Chapter 4

Dynamic Properties of Polymer Melts above the Glass Transition: Monte Carlo Simulation Results

Jörg Baschnagel

Institut für Physik, Johannes-Gutenberg Universität, 55099 Mainz, Germany

Monte Carlo simulation results for the non-equilibrium and equilibrium dynamics of a glassy polymer melt are presented. When the melt is rapidly quenched into the supercooled state, it freezes on the time scale of the simulation in a non-equilibrium structure that ages physically in a fashion similar to experiments during subsequent relaxation. At moderately low temperatures these non-equilibrium effects can be removed completely. The structural relaxation of the resulting equilibrated supercooled melt is strongly stretched on all (polymeric) length scales and provides evidence for the time-temperature superposition property.

If a liquid is cooled, it usually crystallizes at the melting point unless crystallization is prevented by rapid quenching or by the complex structure of the liquid (1–4). The latter criterion is met by many polymers. They prefer to remain in the amorphous state, even when cooled very slowly. Typical cooling rates range between $10^{-4} - 10^{-1} Ks^{-1}$ (5), which already comes close to a quasi-static cooling process. This property makes polymers (almost) ideal systems to study the glass transition and also explains the technical relevance of such an investigation. Many modern polymeric materials are glasses, and an understanding or even a predictability of their properties is therefore of high technological importance.

During the cooling process the structural relaxation time, τ, of the polymer melt and of other fragile glass formers (6) first rises gradually up to $\tau \approx 10^{-9}$ s and then very steeply over about eleven orders of magnitude ($\tau \approx 10^2$ s) before the liquid vitrifies (1,2,6–8). Such a temperature dependence of the relaxation time is characteristic of glass forming materials. It is not observed during the crystallization process. Above the glass transition temperature T_g (defined as the temperature corresponding to $\tau \approx 10^2$ s (6)) the liquid is in (metastable) equilibrium, whereas it freezes in a (non-equilibrium) amorphous structure at T_g. The properties below the glass transition depend on the cooling rate and "physically age" during further equilibration. As for the strong increase of the structural relaxation time above T_g, physical aging is a fairly general property of structural glass formers $(4,9)$.

The present paper discusses this non-equilibrium and the equilibrium dynamic behavior for a simple, computer generated model of a supercooled glassy polymer melt. The next section introduces the bond-fluctuation (lattice) model, describes how the glassy behavior is built in, and gives some details on the simulation methodology. Then we present some representative results for the cooling rate dependent freezing and the physical aging of the model, whereas the final section deals with the equilibrium dynamic properties in the supercooled state.

Coarse-Grained Lattice Simulations for Glassy Polymer Melts: Bond-Fluctuation Model and Monte Carlo Approach

The bond-fluctuation model $(10\text{--}12)$ is an intermediate description between a highly flexible continuum treatment and traditional lattice models of self-avoiding walks (13). It shares with the latter models the simple – and from a computational point of view highly efficient (14) – lattice structure, but distinguishes itself from them by exhibiting a multitude of bond vectors. Due to the lattice structure a monomer of the model does not directly correspond to a chemical monomer, but rather to a group of chemical monomers (typically, such a group comprises $3 - 5$ monomers for simple polymers, such as polyethylene; see Figure 1 (15)). Since a lattice bond should thus be interpreted as the vector joining the mass centers of these groups, its length and direction will fluctuate. The bond-fluctuation model maps this idea onto a simple cubic lattice by associating a monomer with a unit cell of the lattice. In order to impose local self-avoidance of the monomers (excluded volume interaction) and to guarantee uncrossability of the bond vectors during the simulation (no phantom chains) the allowed bond vectors are $\{(2,0,0),(2,1,0),(2,1,1),(2,2,1),(3,0,0),(3,1,0)\}$. This yields 108 different bond vectors.

Energy Function and Geometric Frustration. In addition to the excluded volume interaction an energy function $\mathcal{H}(\boldsymbol{b})$ is associated with each bond vector $\boldsymbol{b}$, which favors bonds of length $b = 3$ and directions along the lattice axes ($i.e.$, $\mathcal{H}(\boldsymbol{b}) = 0$) in comparison to all other available bond vectors ($i.e.$, $\mathcal{H}(\boldsymbol{b}) = \epsilon$) $(16\text{--}18)$. Figure 2 illustrates the effect of this energy function. When the temperature decreases each bond tries to reach the ground state ($i.e.$, a bond with $\mathcal{H}(\boldsymbol{b}) = 0$) and thereby blocks four lattice sites for other monomers. This loss of available volume generates a competition between the energetically driven expansion of a bond and the packing constraints of the melt. Due to this competition some bonds are forced to remain in the excited state (see Figure 2). They are *geometrically frustrated* $(16\text{--}18)$. The development of the geometric frustration during the cooling process causes the glassy behavior of the model and corresponds to the

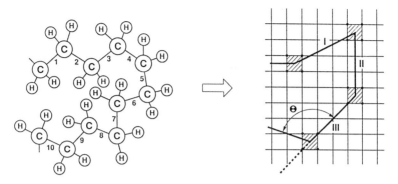

Figure 1. Schematic illustration of the construction of a coarse-grained model for a macromolecule such as polyethylene. In the example shown here, the subchain formed by the three C–C bonds labeled 1,2,3 is represented by the effective bond labeled as **I**, the subchain formed by the three bonds 4,5,6 is represented by the effective bond labeled as **II**, etc. In the bond-fluctuation model the length b of the effective bond is allowed to fluctuate in a certain range $b_{min} \leq b \leq b_{max}$, and excluded-volume interactions are modeled by assuming that each bond occupies a plaquette (or cube) of 4 (8) neighboring lattice sites which then are all blocked for further occupation. From (17).

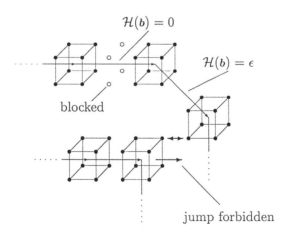

Figure 2. Sketch of a possible configuration of monomers belonging to different chains in the melt in order to illustrate the effect of the model's energy function and the concept of geometric frustration. All bond vectors shown in this picture have the energy ϵ except the vector (3,0,0) which is in the ground state. This vector blocks four lattice sites (marked by o) which are no longer available to other monomers, since two monomers may not overlap. Due to the excluded volume interaction the jump in direction of the large arrow is also forbidden. This leads to geometric frustration. From (17).

driving force which the Gibbs-Di Marzio theory makes responsible for the glass transition of polymer melts (19).

Monte Carlo and Cooling Procedure. As a result of the competition between energy and packing constraints the structural relaxation time strongly increases during the cooling process if the usual bond-fluctuation dynamics is used. This kind of dynamics consists of the following steps:

- choose a monomer at random

- choose a lattice direction at random

- check whether the new lattice sites are empty (excluded volume)

- check whether the new bonds are in the allowed set (chain connectivity)

- calculate the energy change ΔE caused by the move and accept it with probability $\min(1, \exp[-\beta \Delta E])$ (Metropolis criterion (20)) ,

where $\beta = 1/k_{\mathrm{B}}T$ is the reciprocal temperature (the temperature is measured in units of ϵ/k_{B}). These *local* moves are supposed to mimic a random force exerted on a monomer by its environment. They lead to Rouse-like dynamics which is typical of a polymer in a dense melt (21,22).

In order to cool the melt from high ($\beta = 0$) to low temperatures (lowest temperature $\beta_{\mathrm{max}} = 20$), β is increased according to the cooling schedule (16–18)

$$\beta = \beta_{\mathrm{max}} \Gamma_Q t . \tag{1}$$

Here Γ_Q is the cooling rate, and t the Monte Carlo time (measured in Monte Carlo steps (MCS); see the definition below). In the simulation the cooling rate was varied over two orders of magnitude from $\Gamma_Q = 4 \times 10^{-5}$ to $\Gamma_Q = 4 \times 10^{-7}$ (measured in MCS^{-1}).

Computer versus Experimental Time Scale. In order to obtain a feeling how these cooling rates compare to experimental cooling times one has to take into account that a random hopping event in the bond-fluctuation model should be interpreted as a conformational change of a subgroup of chemical monomers, which typically takes 10^{-11} s (15,23) at high temperatures. Requiring therefore that the average relaxation time at high temperatures of the present model should be 10^{-11} s, and using the high temperature value for the monomer move acceptance rate of the bond-fluctuation model (*i.e.*, 10^{-1} (14)) to account for the influence of the melt's density, the unknown prefactor in the Monte Carlo transition rate, the Monte Carlo time step (MCS), can be translated to 1 MCS $\approx 10^{-12}$ s. Although this identification of the abstract Monte Carlo time unit with the physical time unit is based on a plausibility argument only, it yields the right order of magnitude, as attempts to simulate bisphenol-A-polycarbonate (24,25) and polyethylene (26) by the bond-fluctuation model showed. Using therefore this estimate one can draw two conclusions: First, a cooling process with the above mentioned rates is finished after about $10^{-8} - 10^{-6}$ s. The corresponding cooling rates are much higher than those generally used in experimental studies, but are slow in the field of computer simulations (27,28). Second, the fast cooling forces the simulated melt to freeze on a time scale which is many orders of magnitude smaller than that of the experimental glass transition.

Real versus Artificial Dynamics. Due to the inevitable high cooling rates in simulations non-equilibrium effects already emerge at moderately supercooled temperatures for the studied model. In some case these effects can completely mask the underlying equilibrium properties. An example will be given in the next but one section (see Figure 8). In order to circumvent this problem and to achieve equilibration one can exploit an advantage of the Monte Carlo technique, namely that the elementary move may be adapted at will. Since the final equilibrium state is independent of the way by which it was reached, the *realistic* (glassy) dynamics may be replaced by an *artificial* one which uses *non-local* moves. A non-local move involves a collective motion of all monomers of a chain. Such a collective motion may be realized by the so-called *slithering snake dynamics* (13,29), for instance. In the slithering snake dynamics one attempts to attach a randomly chosen bond vector (from the set of allowed bond vectors) to one of the ends of a polymer (both also randomly chosen). If the attempt does not violate the excluded volume restriction, the move is accepted with probability $\exp[-\Delta E/k_\mathrm{B}T]$, where ΔE now represents the energy difference between the newly added bond and the last bond of the other end of the chain, and the last bond is removed. Whereas the local dynamics propagates the chains for one lattice constant if a move is accepted, the slithering-snake dynamics shifts the whole chain for one lattice constant. Therefore one expects the slithering-snake algorithm to be faster by a factor of the order of the chain length already at high temperatures. This expectation was confirmed in the simulation, and furthermore it was found that the algorithm is also very efficient in equilibrating low temperatures, in which we are particularly interested (30).

Simulation Parameters. For the present study a chain length of $N = 10$ was used. If one takes into account that a lattice monomer roughly corresponds to a group of three to five chemical monomers (see above) our simulation deals with fairly short, oligomeric chains. The cubic simulation box (of linear dimension $L = 30$) contained $K = 180$ chains so that the volume fraction of occupied lattice sites is $\phi = 8NK/L^3 = 0.5\bar{3}$. This value is a compromise between two requirements: it is high enough for the model to exhibit the typical behavior of dense melts at high temperatures (14) as well as pronounced frustration effects at low temperatures (16–18) and low enough to allow for a sufficient acceptance rate of monomer (or chain) moves to make the equilibration of the melt in the interesting temperature range possible (*i.e.*, sufficiently high insertion probability for a new bond to make the slithering-snake dynamics superior to the local dynamics) (30). In order to improve the statistics 16 independent simulation boxes were treated in parallel. Thus the total statistical effort involves 28800 monomers, which ensures a high accuracy of the results.

Cooling Rate Effects

The quench rates quoted in the previous section encompass cooling times which range from 2.5×10^4 MCS's to 2.5×10^6 MCS's. Whereas these times are much larger than the structural relaxation time of a polymer, the Rouse time τ_R, at infinite temperature ($\tau_R(\infty) \approx 5380$ MCS's), the cooling process must reach a temperature, at which $\tau_R(T) \approx \Gamma_Q^{-1}$. Then the melt freezes on the time scale of the simulation. The present section discusses the properties of this cooling rate dependent freezing by some representative examples.

An Example: The Mean-Squared Bond Length. Figure 3 (*17*) depicts the temperature dependence of the mean-squared bond length, b^2, for five representative cooling rates (see equation 1). In the high temperature region ($T \in [0.6, 2.0]$) the curves for the different rates nicely collapse. This shows that the melt is in a thermally equilibrated liquid state. In this state the melt is mobile enough to respond easily to the speed, by which the temperature is changed. The effect of the finite cooling rate starts to be felt below $T \approx 0.5$ and is accompanied by a strong expansion of the mean bond length. The increase of b^2 continues until $T \approx 0.2$, where the curves level off. In this temperature range the intrinsic relaxation times of the melt become comparable to the observation time. Then the melt falls out of equilibrium, and b^2 gets locked at a value depending on the cooling rate. The smaller the rate, the more time the melt has to relax. Therefore b^2 increases with decreasing cooling rate in the low temperature region ($T \leq 0.2$). However, b^2 is always smaller than 9, the value expected if all bonds were in the ground state. This result shows that the model indeed introduces strong topological constraints leading to geometric frustration.

Cooling Rate Dependence of the Glass Transition Temperature. The behavior depicted in Figure 3 is not restricted to the mean-squared bond length, but also typical of other quantities (mean bond angle, radius of gyration, etc. (*16–18*)). From these curves one can extract the cooling rate dependence of the glass transition temperature T_g defined by the intersection point of a linear extrapolation from the liquid and from the glassy side. The results of such an analysis are shown in Figure 4 (*17*). Whereas older experimental studies found a linear relationship between T_g and $\ln \Gamma_Q$, we obtain a non-linear dependence, which can be well fitted by the following variant of the Vogel-Fulcher equation

$$T_g(\Gamma_Q) = T_K + \frac{A}{\ln(B/\Gamma_Q)} \, . \tag{2}$$

More recent experiments (*5*) and extensive molecular dynamics simulations on both Lennard-Jones (*31*) and silica glasses (*32*) yielded the same non-linear relationship. Therefore the original linear dependence seems to be questionable and must presumably be attributed to a too narrow span of studied cooling rates.

In equation 2 T_K denotes the freezing temperature at an infinitely slow cooling rate. For this limit the abbrevation T_K was chosen to remind of the Kauzmann paradox (*6,33*), which originates from a similar extrapolation procedure. T_K will

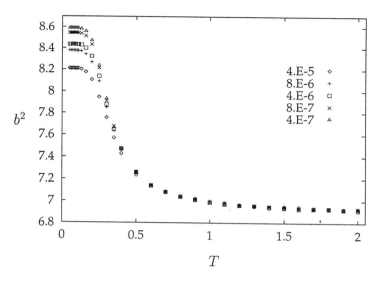

Figure 3. Plot of the mean-square bond length vs T for five different cooling rates: $\Gamma_Q = 4 \times 10^{-5}$ ($\diamond$), $\Gamma_Q = 8 \times 10^{-6}$ (+), $\Gamma_Q = 4 \times 10^{-6}$ ($\square$), $\Gamma_Q = 8 \times 10^{-7}$ ($\times$) and $\Gamma_Q = 4 \times 10^{-7}$ ($\triangle$). From (17).

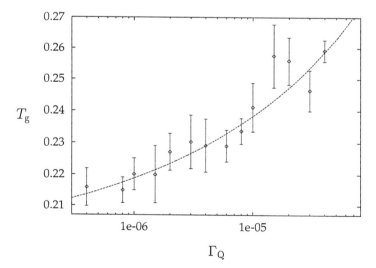

Figure 4. Glass transition temperature T_g vs Γ_Q. The dashed line represents a fit by equation 2. From (17).

therefore also be referred to as Kauzmann temperature in the following. The fit of the simulation data to equation 2 yields: $T_K = 0.17 \pm 0.02$, $A = 0.467 \pm 0.200$ and $B = 5.85 \times 10^{-3} \pm 2.5 \times 10^{-3}$. The numerical value of T_K coincides within the error bars with the Vogel-Fulcher temperature T_0 that one obtains by fitting the Vogel-Fulcher equation, *i.e.*,

$$D(T) = D_\infty \exp\left[-\frac{C}{T - T_0}\right] , \tag{3}$$

to the chain's diffusion coefficient, *provided* one limits the fit interval to that temperature range, where the diffusion coefficient decrease by about two orders of magnitude from its high temperature limit, *i.e.*, to the same time interval as covered by the cooling rates.

The same coincidence, $T_K \approx T_0$, is also found, for an impressive number of different glass formers, in experiments when comparing the results of extrapolations of the structural relaxation time and of transport coefficients with that of the cooling rate dependence of the entropy (*34*). The agreement of T_K and T_0 is usually interpreted as a clear evidence for an underlying (thermodynamic) glass transition which is hidden in the experimentally inaccessible temperature range, but the signature of which becomes visible at (the cooling rate dependent) T_g. However, the present simulation results challenge this view because an extension of the Vogel-Fulcher fit to lower temperatures makes T_0 drop significantly (see next section). Similar results are also obtained in the above mentioned simulation on silica glass (*32*). This finding raises the question whether $T_K \approx T_0$ is merely a consequence of the fact that different quantities are monitored over a comparable time interval. Of course, experiments use considerably larger time intervals for the extrapolation, which should make the outcome much more reliable. But still the extrapolation has to cover many orders of magnitude in time (below T_g), and quite generally the predictions of extrapolations are the less accurate, the larger the range is they have to cover. Perhaps the established Vogel-Fulcher (or Kauzmann) temperature is therefore not a characteristic material constant of a (fragile) glass former, but only a consequence of an empirical, though very expedient, high temperature description of the structural relaxation. A recent experimental reanalysis of the temperature dependence of the structural relaxation time for a variety of glass formers (*35,36*) admits the same interpretation. This study shows that for some glass formers a high temperature Vogel-Fulcher fit is possible, which overestimates the traditional Vogel-Fulcher temperature considerably.

Physical Aging Behavior

The discussion of the previous section suggests that the model should exhibit physical aging in the temperature range where cooling rate effects are pronounced. Dating back to the pioneering work by Struik (*9*), "physical aging" means the change of the properties of the glassy state with increasing annealing time in a certain temperature region below T_g. This temperature interval is limited from below by a temperature T_{min}, below which the glass can be considered as a (disordered equilibrium) solid for all practical purposes (no further structural relaxation), and from above by a temperature T_{max} ($T_{max} \approx T_g$), above which all non-equilibrium effects vanish. Whereas this temperature interval is essentially limited to temperatures below T_g in experiments, one expects it to extend into the supercooled region far above T_g in simulations due to rapid cooling.

An Example: The Radius of Gyration. Figure 5 (*37*) exemplifies this expectation for the present model. The figure compares the temperature dependence of the radius of gyration R_g^2 for the fastest ($\Gamma_Q = 4 \times 10^{-5}$) and the slowest cooling rate ($\Gamma_Q = 4 \times 10^{-7}$) studied in the previous section with that obtained after annealing the melt for 4×10^5 MCS's at each temperature. This annealing time is about two orders of magnitude larger than the cooling time needed to cross the temperature interval $0.1 < T \leq 0.6$ for $\Gamma_Q = 4 \times 10^{-5}$, but comparable to that needed to cross the same temperature interval for $\Gamma_Q = 4 \times 10^{-7}$. Therefore one expects pronounced effects for the fastest cooling rate. Figure 5 shows that the influence of annealing becomes visible at $T_{max} \approx 0.5 > T_g$. Indeed, the fixed annealing time is sufficient for the configurations prepared by the fastest cooling rate to adapt to the temperature of the ambient heat bath for $0.3 \leq T \leq 0.5$ so that all aged data collapse onto a common (equilibrium) curve. For smaller temperatures, however, the effects of the finite cooling time emerge again. Depending on the originally used cooling rate, the aged curves first pass through a maximum (at $T \approx 0.25$ for $\Gamma_Q = 4 \times 10^{-5}$ and at $T \approx 0.2$ for $\Gamma_Q = 4 \times 10^{-7}$) and then return to the initial data. Below $T_{min} \approx 0.1$ the applied annealing has no discernible effect. These low temperature configurations represent the frozen solid phase in practice. Qualitatively, these results are therefore comparable to those obtained in experiments. In addition, they show that very slow inital cooling is superior to fast initial quenching followed by long annealing runs to generate configurations close to equilibrium.

Andrade Behavior. In order to investigate the influence of different aging times the annealing was extended over one decade to longer times in the temperature interval $0.17 \leq T \leq 0.23$. Figure 6 (*37*) exemplifies the results for the radius of gyration (starting configuration: data of $\Gamma_Q = 4 \times 10^{-7}$). The non-equilibrium state of the melt in this temperature region becomes obvious from the figure. Whereas R_g^2 should increase with decreasing temperature (stiffening of the chains), the opposite temperature dependence is observed at all annealing times. Therefore the studied range of annealing times is still much shorter than the actual relaxation time of R_g^2 in this temperature range. Motivated by experimental findings (*9*), we tried to describe the increase of R_g^2 with the annealing time t by

$$R_g^2(T, t) = R_g^2(T, 0) \left[1 + \left(\frac{t}{\tau_g} \right)^{1/3} \right] , \qquad (4)$$

which is supposed to be valid for $t/\tau_g \leq 1$. Equation (4) implies a time-temperature superposition principle (*1,7,8*) for the aging properties of a glass former at short annealing times. The solid lines in Figure 6 show that equation (4) actually provides a good description, yielding an Arrhenius-like temperature dependence of

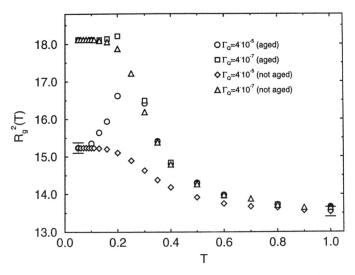

Figure 5. Temperature dependence of the radius of gyration, R_g^2, for different cooling rates, Γ_Q, and aging times ($N = 10$, $d = 3$). The data labeled "not aged" are generated by cooling the melt according to equation 1 with Γ_Q-values as specified in the figure. The data labeled "aged" are the results obtained after aging the melt for 4×10^5 MCS's at each temperature. From (37).

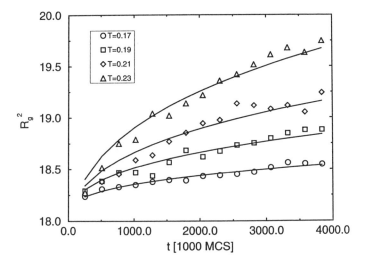

Figure 6. Dependence of the radius of gyration R_g^2 on the aging time for different temperatures. The solid curves are fits to equation 4. From (37).

the relaxation time τ_g (see Figure 7 (37)). This quantitative description does not only work for the radius of gyration, but also for quantities sensitive to shorter length scales, such as the bond length or the bond angle (data not shown here) (37). It therefore seems to represent the general short-time aging behavior of the present model in the temperature region close to T_g.

Physical Aging and Correlation Functions. Another consequence of physical aging is shown in Figure 8. The figure depicts the dependence of a correlation function, of the incoherent intermediate scattering function $\Phi_q^s(t)$, on the extent to which the initial configuration (*i.e.*, the configuration at $t = 0$) was aged. The curves were obtained by first cooling the melt from infinite temperature to $T = 0.16$ and then equilibrating further for a certain time period to generate the initial configuration, from which $\Phi_q^s(t)$ was calculated at the maximum of the static structure factor (*i.e.*, at $q = 2.92$ measured in units of the lattice constant) according to

$$\Phi_q^s(t) = \frac{1}{M} \sum_{m=1}^{M} \left[\left\langle \cos\left(q\left[r_m(t) - r_m(0)\right]\right) \right\rangle \right]_q . \tag{5}$$

The sum in equation (5) runs over all monomers in the melt, which are at time t at position $r_m(t)$, and the symbols $\langle \bullet \rangle$ and $[\bullet]_q$ stand for the thermal average over the monomer positions (38–40) and the lattice analogue of a spherical average in the continuous reciprocal space (16). The aging time of the initial configuration is then defined as the sum of the cooling time to reach $T = 0.16$ and the subsequent equilibration time. For the lowest, middle and upper curves it was 6×10^3 MCS's, 4×10^6 MCS's and 1.4×10^7 MCS's, respectively.

From the figure two conclusions may be drawn: First, the better the initial equilibration, the longer the (final) structural relaxation time. Similar results are also observed in simulations of spin glasses (41) and spin systems without permanent frustration (42), for which analytical theories have been proposed, which can (partly) rationalize these findings (43–45), as well as in molecular dynamics simulations of a glassy binary Lennard-Jones mixture (46). Second, not only quantitative, but also qualitative properties of the relaxation behavior may be blurred by non-equilibrium effects. Whereas the upper two curves clearly show a two-step decay, this feature is completely missing in the lower, least aged curve. Again, this observation finds – to some extent – its counterpart in the above mentioned molecular dynamics study. Since the two-step relaxation of $\Phi_q^s(t)$ is a major prediction of the mode-coupling approach to the structural glass transition (7,8,38), our simulation data stress how important careful equilibration is if one wants to test this modern (and any other equilibrium) theory.

Equilibrium Dynamic Properties

By virtue of the slithering-snake algorithm it is (hitherto) possible to remove all non-equilibrium effects up to $T \approx 0.16$ so that one can study the equilibrium dynamic properties of the model in the supercooled state. During supercooling the structural relaxation time increases over several orders of magnitude, and dynamic

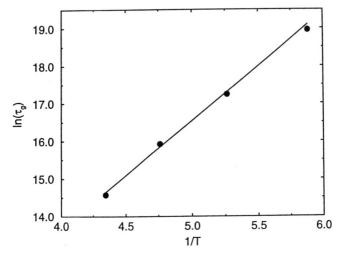

Figure 7. Arrhenius plot of the characteristic aging time τ_g of equation (4) for the radius of gyration. From (37).

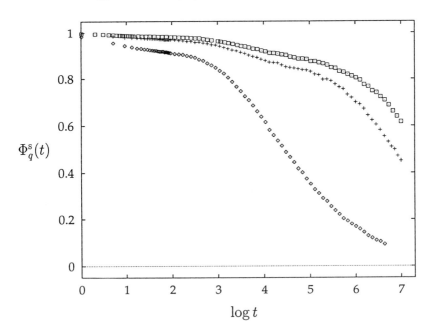

Figure 8. Comparison of the decay of $\Phi_q^s(t)$ calculated at $T = 0.16$ and $q = 2.92$ after the melt has been equilibrated for 6×10^3 MCS's ($\diamond$), 4×10^6 MCS's ($+$) and 1.4×10^7 MCS's ($\square$). Whereas a two-step relaxation process is completely hidden by non-equilibrium effects for the curve with the smallest equilibration time, such a two-step process is visible after 4×10^6 MCS's and remains present, if the melt is further equilibrated. From (40).

correlation functions become strongly stretched. These features are not specific to polymer melts, but typical of all (fragile) glassforming systems (47).

The theoretical concepts to rationalize these generic properties are still rather controversial. An often advocated view is that the dynamic behavior in the supercooled state reflects the growth of some kind of spatial correlations which are supposed to become long-range close to the glass transition temperature T_g. It is speculated that the molecules of the glass former group into clusters which percolate at the (thermodynamic) glass transition (*i.e.*, at $T_0 < T_g$) (48–50). However, up to now the evidence to support this speculation by experiments (51,52) or simulations (28) is still fairly limited.

On the other hand, the mode-coupling approach to the structural glass transition (7,8,38) challenges the idea of the formation of growing clusters. It rather explains the glassy behavior as a purely dynamic phenomenon, which is caused by an underlying critical temperature $T_c > T_g$, without that a length scale simultaneously increases. Although many of the theoretical predictions are found in experiments and simulations (28,38,47), it is still a matter of debate to what extent the present state of the theory can describe the dynamic behavior in the supercooled state (for instance, the temperature dependence of the structural relaxation time (35,36)).

In this section we want to present various results, by which we tried to test some of the above sketched theoretical ideas with our model.

An Example: The End-to-End Vector Correlation Function. In order to exemplify the strong increase of the structural relaxation time during supercooling Figure 9 shows a scaling plot of the end-to-end vector correlation function (53). This correlation function is defined by

$$\Phi_{ete}(t) = \frac{\langle \boldsymbol{R}(t)\boldsymbol{R}(0)\rangle - \langle \boldsymbol{R}(t)\rangle\langle \boldsymbol{R}(0)\rangle}{\langle \boldsymbol{R}(0)^2\rangle - \langle \boldsymbol{R}(0)\rangle^2} , \qquad (6)$$

where $\boldsymbol{R}(t)$ is the vector joining the ends of a chain at time t. From the Rouse model (see (21,22) and below) one expects $\Phi_{ete}(t)$ to exhibit a time-temperature superposition property for all times. This means that the shape of $\Phi_{ete}(t)$ remains unaffected by a variation of temperature. Only the relaxation time, τ_{ete}, changes so that $\Phi_{ete}(t)$ can be written as $\Phi_{ete}(t) = \Phi_{ete}(t/\tau_{ete})$. Figure 9 shows that this expectation is borne out by the simulation data in the temperature interval from $T = 0.23$ (moderately supercooled) to $T = 1.0$ (normal liquid-like), and that the master curve can be well described by a stretched exponential function,

$$\Phi_{ete}(t) = \exp\left[-\left(\frac{t}{\tau_{ete}}\right)^{\beta_R}\right] , \qquad (7)$$

with $\beta_R = 0.67$. The scaling was obtained by shifting the correlation functions, calculated at different temperatures, onto an arbitrarily chosen reference curve ($T = 0.4$ in this case). Such a time-temperature superposition property is also found for other correlation functions, such as for the Rouse modes (see below) and for the intermediate scattering function at late times (53). Whereas the

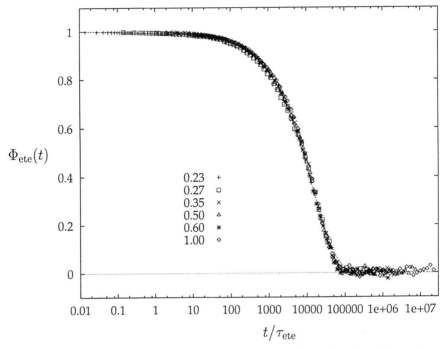

Figure 9. Scaling plot of the end-to-end vector correlation function. The scaling time, τ_{ete}, is defined as the factor which is needed to shift the curves for $T = 0.23, 0.27, 0.35, 0.50, 0.60, 1.00$ onto the curve for $T = 0.40$ (arbitrarily chosen reference curve). The figure also includes a stretched exponential fit shown as a dotted line (stretching exponent $\beta_R = 0.67$). From (53).

latter finding agrees with one of the predictions of the idealized mode-coupling theory (7,8,38) and may thus be interpreted as a result of the glassy dynamics, the superposition of $\Phi_{ete}(t)$ or of the Rouse-mode correlation function is rather a consequence of the chain connectivity. Contrary to simulations for simple liquids (28,55,56), it is therefore not easy to distinguish for the present model between a glassy or a polymeric origin of this property.

Temperature Dependence of the Relaxation Time. The temperature dependence of the resulting relaxation time is depicted in Figure 10 and compared with that of the chain's diffusion coefficient D. In the studied temperature interval both $1/\tau_{ete}$ and D decrease by about $2 - 3$ orders of magnitude, and their temperature dependence can be fairly well fitted by a Vogel-Fulcher equation (see equation 3). The fit results yield different amplitude factors, but (almost) the same Vogel-Fulcher activation energy ($C \approx 0.76$ for D, $C \approx 0.77$ for $1/\tau_{ete}$) and Vogel-Fulcher temperature ($T_0 \approx 0.129$ for D, $T_0 \approx 0.132$ for $1/\tau_{ete}$). That the values of C and T_0 agree for both quantities is expected (and gratifying) because the Vogel-Fulcher activation energy and temperature should be characteristic of the glassy properties of the melt and independent of the quantity from which they are determined.

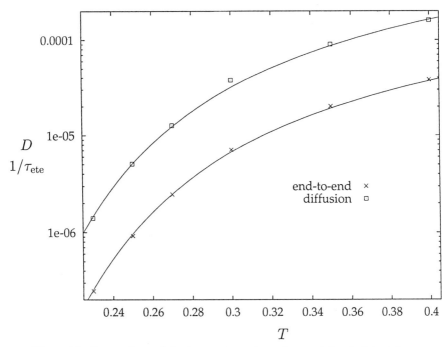

Figure 10. Comparison of the temperature dependence of the chain's diffusion coefficient, D ($\times$), and the end-to-end vector scaling time, τ_{ete} ($\diamond$). Both quantities can be fitted by a Vogel-Fulcher equation (solid lines). Within the error bars the Vogel-Fulcher activation energy and temperature agree with each other for D and τ_{ete}. From (53).

One of the theoretical concepts to explain the strong increase of the structural relaxation time is to assume a static glass correlation length ξ, which grows beyond any bound as (an ideal) T_{g} (*i.e.*, T_0) is reached, and which scales with the relaxation time as ξ^z (z is a (constant) dynamic critical exponent). Such a relationship is found for spin glasses (57). Motivated by that finding one could try to reveal the presence of ξ for structural glasses by measuring, for instance, the diffusion coefficient for different lattice sizes L. Since $D \propto \tau^{-1} \sim \xi^{-z}$, one expects D to level off as soon as $\xi \simeq L$. Such a test is shown in Figure 11 for $L = 5 - 30$ (Binder, K.; Baschnagel, J.; Böhmer, S.; Paul, W. *Phil. Mag. B*, in press.). Obviously, there is no pronounced effect on D, at least in the studied temperature window. Therefore we must conclude that if there is a growing glass correlation length in our model, it does not couple to the diffusion coefficient. However, there might be other collective time scales feeling the growth of ξ, just as in a critical binary mixture, where the collective diffusion and not the self-diffusion, measured here, is affected. A possible test of this conjecture would be to calculate the coherent intermediate scattering function, for instance, and to determine the dependence of its final structural relaxation time on L.

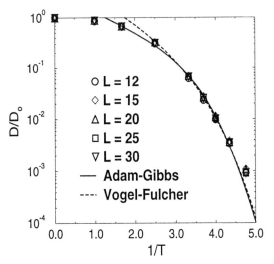

Figure 11. Self-diffusion coefficient plotted vs. temperature, for several lattice sizes. The solid and the dotted line are fits to the Vogel-Fulcher and Adam-Gibbs equation, respectively. From (Binder, K.; Baschnagel, J.; Böhmer, S.; Paul, W. *Phil. Mag. B*, in press.).

Figure 11 also reproduces the Vogel-Fulcher fit of Figure 10 and additionally shows a fit to the Adam-Gibbs equation (48),

$$D \propto \exp\left[-\frac{C}{TS(T)}\right] , \qquad (8)$$

where $S(T)$ was calculated by an analytical approximation which is accurate in the studied temperature interval (54). In this interval the Adam-Gibbs formula fits the diffusion coefficient (at least) as well as the Vogel-Fulcher equation. Since a comparison of the simulated entropy with the Gibbs-Di Marzio theory (see (19,58)), suggested $S(T) > 0$ for all finite temperatures and thus $T_0 = 0$ on the basis of the Adam-Gibbs theory, the comparable fit quality of the Vogel-Fulcher and the Adam-Gibbs formula shows that accurate diffusion data at much lower temperatures would be required to distinguish between them or to rule them out.

Rouse Model. In 1953 P. E. Rouse proposed a simple model to describe the dynamics of a polymer chain in dilute solution (21,59). The model considers the chain as a sequence of Brownian particles which are connected by harmonic springs. Being immersed in a (structureless) solvent the chain experiences a random force by the incessant collisions with the (infinitesimally small) solvent particles. The random force is assumed to act on each monomer separately and to create a monomeric friction coefficient. The model therefore contains chain connectivity, a local friction and a local random force. All non-local interactions between monomers distant along the backbone of the chain, such as excluded-volume or hydrodynamic interactions, are neglected. This neglect is not justified in dilute solutions (21,22). However, as the polymer concentration increases, the chains begin to screen both interactions mutually. Therefore it was conjectured (60) that the Rouse model could apply to the motion of a polymer in a dense melt. If the polymers are small (smaller than the entanglement length), there are many evidences in favor of this conjecture from experiments (61–63) and simulations (64,65). Therefore it is generally assumed that the Rouse (or a Rouse-like) model accurately describes the essential (long-time) dynamic properties of short chains in dense melts at high temperatures. It is an interesting question to what extent this remains true if the melt is progressively supercooled towards its glass transition.

Analysis of the Rouse Modes. In order to test the applicability of the Rouse model we calculated the basic quantities, the Rouse modes, and compared the simulation results with the theoretical predictions. The Rouse modes are defined as the cosine transforms of the position vectors, r_n, to the monomers. For the discrete polymer model under consideration they can be written as (66)

$$X_p(t) = \frac{1}{N} \sum_{n=1}^{N} r_n(t) \cos\left[\frac{(n-1/2)p\pi}{N}\right] , \quad p = 0, \ldots, N-1 . \tag{9}$$

The time-correlation function of the Rouse modes is given by

$$\Phi_{pq}(t) = \langle X_p(t) X_q(0) \rangle = \langle X_p(0) X_q(0) \rangle \exp\left[-\frac{t}{\tau_p(T)}\right] , \quad p = 1, \ldots, N-1 \tag{10}$$

with $(p, q \neq 0)$

$$\langle X_p(0) X_q(0) \rangle = \frac{b^2}{8N[\sin(p\pi/2N)]^2}\delta_{pq} \quad \text{and} \quad \tau_p(T) = \frac{\zeta(T)b^2}{12k_\mathrm{B}T[\sin(p\pi/2N)]^2} , \tag{11}$$

where b and ζ are the effective bond length (only weakly temperature dependent) and the monomeric friction coefficient, respectively. According to equations 10 and 11 the Rouse modes should have the following properties: (1.) They are orthogonal at all times. (2.) Their correlation function decays exponentially. (3.) The normalized correlation functions for different mode indices, p, and temperatures can be scaled onto a common master curve when the time axis is divided by $\tau_p(T)$. Let us see to what extent these properties are realized by the studied model.

Figure 12 depicts, as a representative example, the initial correlation function, $\Phi_{1p}(0)$, $p = 1, \ldots, 9$, to test the orthogonality of the Rouse modes. It shows that the modes remain statically uncorrelated down to the lowest studied temperatures. In the normal high temperature state of the melt this result is either assumed due to the good agreement of theoretical predictions, derived from the Rouse modes, with experiments and simulations or verified directly by simulations (65). However, for supercooled melts the orthogonality of the Rouse modes has, to the best of our knowledge, not been observed before. Of course, the lowest temperature of our

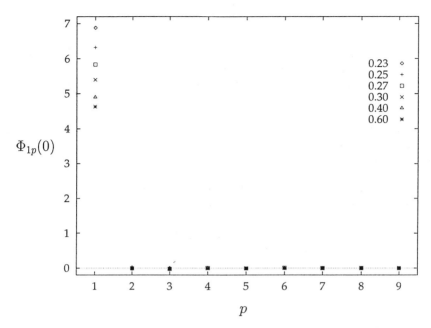

Figure 12. Test of the orthogonality of the Rouse modes at $t = 0$ for various temperatures ranging from the normal liquid-like ($T = 0.6$) to the moderately supercooled regime ($T = 0.23$). The figure shows a representative example for the correlation of the first Rouse mode with all modes (except $p = 0$), *i.e.*, $\Phi_{1p}(0)$ for $p = 1, \ldots, 9$. From (*53*).

simulation, $T = 0.23$, still belongs to the moderately supercooled regime which is yet outside the temperature interval, where mode-coupling effects become visible. This interval starts at $T \approx 0.21$ (see below and (*39,40*)). Therefore the present study cannot exclude that static correlations among the Rouse modes might develop when the temperature is further decreased into this interval. Whether this is true or not remains to be answered by future work. But we note that from high temperatures to $T = 0.23$ the relaxation times have already increased by about three decades.

Figure 13 tests another prediction of the Rouse model, the time-temperature superposition property. Again, a representative example is shown, *i.e.*, the correlation function of the third Rouse mode. As the theory anticipates, it is indeed possible to superimpose the simulation data, obtained at different temperatures, onto a common master curve by rescaling the time axis. The required scaling time, τ_3, is defined by the condition $\Phi_{pp}(\tau_3) = 0.4$. The choice of this condition is arbitrary. Since the Rouse model predicts that the correlation function satisfies equation (10) for all times, any other value of $\Phi_{pp}(t)$ could have been used to define τ_3. This scaling behavior is in accordance with the theory. However, contrary to the theory, the correlation functions do not decay as a simple exponential, but as

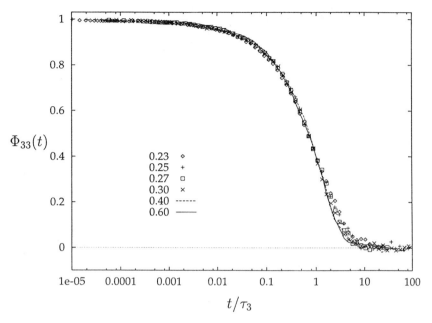

Figure 13. Time-temperature superposition property for the Rouse mode correlation function, exemplified by the third mode, $\Phi_{33}(t)$, for the temperatures $T = 0.23, 0.25, 0.27, 0.30, 0.40$ and 0.60. The decay of the master curve is stretched (stretching exponent $\beta_3 \approx 0.8$). From (53).

a stretched exponential function with a stretching exponent, β_p, that depends on the mode index p. Figure 14 shows that the exponent increases with increasing p ($\beta_1 \approx 0.88$ for $p = 1$ and $\beta_6 \approx 0.74$ for $p = 6$). Since the pth mode probes the dynamic processes of subchains of size N/p, the latter result implies that the motion of a subchain is the more cooperative, the smaller the subchain if one interpretes a decrease of β_p as an increase of cooperativity. A possible interpretation of this finding could be that the presence of the surrounding chains strongly influences the motion of a chain locally, but that this influence diminishes if more and more monomers are involved. This interpretation also means that the interactions between the chains in the present model do not merely rescale the monomeric friction coefficient, as it is inferred from other tests of the Rouse model by simulations (65), but that they also lead to qualitative deviations from the Rouse description. Whether this is true is still an open question which deserves more investigations.

Finally, Figure 15 shows the temperature dependence of the inverse relaxation time for the first three Rouse modes. As for the diffusion coefficient and the relaxation time of the end-to-end vector, $1/\tau_p$ decreases by about $2 - 3$ orders of magnitude in the studied temperature interval and may be fitted by the Vogel-Fulcher equation with a common (but, compared to D and $1/\tau_{\mathrm{R}}$, slightly higher)

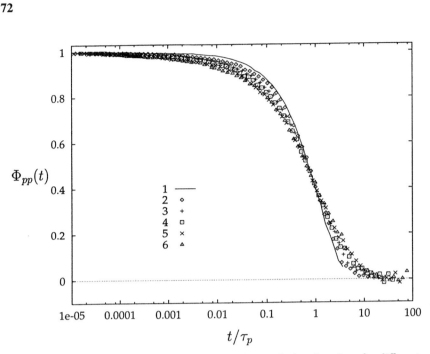

Figure 14. Attempt to scale the Rouse-mode correlation functions for different mode indices, $p = 1, \ldots, 6$, at a given temperature (here $T = 0.23$). The decay of the curves is always stretched, and the degree of stretching increases with increasing p. From (53).

activation energy ($C \approx 0.86$) and Vogel-Fulcher temperature ($T_0 \approx 0.127$). These results, taken together with those for D and $1/\tau_R$, suggest that the Vogel-Fulcher temperature of the model lies around $T_0 \approx 0.12 - 0.13$.

Mode-Coupling Analysis

During the last twelve years a new theoretical approach to the structural glass transition, the mode-coupling theory (MCT) (38), has been developed. The physical picture of this theory may be summarized as follows: Particles in a liquid sit in cages formed by their neighbors. For temperatures close to the triple point the enclosed particles can easily leave the cages. This mobility entails a short structural relaxation time. If the liquid is supercooled, the cages gradually tighten. Thus the structural relaxation time increases and would diverge at a critical temperature T_c ($T_c > T_g$ in general), if the relaxation was only determined by the *cage effect*. However, most glass formers do not freeze completely at T_c. Additional transport processes become therefore effective close to and below T_c. MCT provides such an additional relaxation channel in terms of *hopping processes*. The version of the theory which deals with the interplay of cage effect and hopping processes is called *extended MCT* to distinguish it from the *idealized MCT* which ignores the hopping

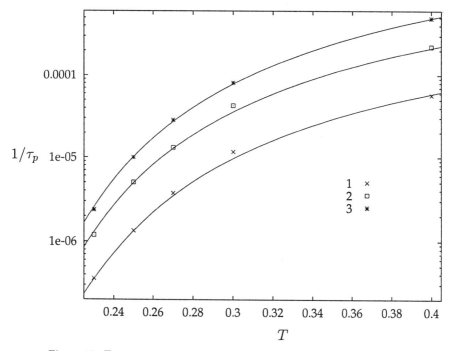

Figure 15. Temperature dependence of the Rouse-mode relaxation times for the first three Rouse modes, $p = 1, 2, 3$, and comparison with the Vogel-Fulcher equation (solid lines). As for D and τ_{ete} (see Figure 9), the Vogel-Fulcher activation energy and temperature agree for the three relaxation times reasonably well. From (53).

contributions. The appeal of the theory stems from the fact that it derives novel general predictions about the structural relaxation of supercooled liquids in the vicinity of T_c, which may be tested quantitatively in experiments and simulations (for reviews see (38,28) and Kob, W. In *Experimental and Theoretical Approaches to Supercooled Liquids: Advances and Novels*; Fourkas, J., Kivelson, D., Mohanty, U., Nelson, K., Eds.; ACS Books: Washington, DC, 1997; in press.).

Extended MCT-Analysis of the Incoherent Scattering Function. Such a test was performed for the present model. The main results are: (1.) At low temperatures there is an intermediate time window (β-relaxation regime), in which the incoherent intermediate scattering function $\Phi_q^s(t)$ exhibits a two-step relaxation (see Figure 16). (2.) If analyzed by the idealized MCT, the studied temperature interval is split into high- and low-temperature parts. At high temperatures the idealized theory accurately describes the decay of $\Phi_q^s(t)$ in the β-relaxation regime, whereas at low temperatures it overestimates the freezing tendency of the melt at long times (38,39). (3.) This discrepancy can be removed quantitatively by the extended MCT. The result is presented in Figure 16 (38,40). Although the

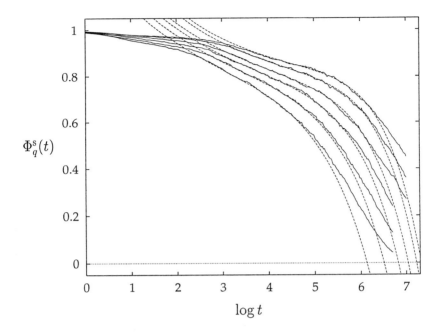

Figure 16. Incoherent intermediate scattering function, $\phi_q^s(t)$, versus the logarithm of time (solid lines) for the temperatures $T = 0.16, 0.17, 0.18, 0.19, 0.20, 0.21$ (temperature increases from right to left in the figure). The dashed lines are fits by the extended mode-coupling theory. From (40).

extended analysis improves the fit considerably, it does not change the idealized fit results of the essential theoretical parameters. The fit shown in Figure 16 contains five open parameters, four of which are temperature independent. The temperature dependence enters only via the microscopic time scale, determined from the initial decay of $\Phi_q^s(t)$ before the β-relaxation window by a fit with the Rouse model, and via the reduced distance to the critical point, $T - T_c$. The fit result for the critical mode-coupling temperature is $T_c \approx 0.15$, which is larger than T_0, as in experiments. (4.) In addition, the relaxation of $\Phi_q^s(t)$ at times before and after the β-relaxation window can be fitted by the Rouse-theory and by a Kohlrausch function with a temperature independent exponent (time-temperature superposition principle), respectively (39,40).

Conclusions

The bond-fluctuation model, combined with a simple two-level energy function which favors long bonds, reproduces many features of experimental glass formers. As the temperature decreases, the structural relaxation time increases by several orders of magnitude so that the melt freezes on the time scale of the simulation in an amorphous structure as soon as the relaxation time becomes comparable to the inverse cooling rate. The corresponding freezing temperatures depends non-linearly on the logarithm of the cooling rate. If the melt is further equilibrated in this low temperature regime, it ages physically. The physical aging of the model is an extremely stretched process, the initial stage of which is compatible with an Andrade behavior, implying an aging-time-temperature superposition property. Circumventing this slow relaxation by the slithering-snake algorithm, it is possible to study the equilibrium dynamic behavior of the model in the supercooled state. The supercooled dynamics is characterized by stretched relaxation functions, by the time-temperature superposition property (at least at late times) and by a non-Arrhenius-like increase of the structural relaxation time.

These phenomenological features were compared with several theoretical ideas. The non-Arrhenius-like increase of the structural relaxation time is compatible with both a Vogel-Fulcher and an Adam-Gibbs equation, but does not seem to be connected to a growing glass correlation length. The absence of such a correlation length is a result of the mode-coupling approach to the structural glass transition, which provides a fairly accurate description of the incoherent scattering function above the critical temperature. Drawbacks of this MCT analysis are still that the analysis can at present not be extended to temperatures below T_c, that an understanding of the fit parameters in terms of the model's properties is missing, and that it is unclear whether and – if yes – how the mode-coupling effects, which cause the two-step relaxation of $\Phi_q^s(t)$, also emerge in other (more polymeric) relaxation functions.

Acknowledgments

This work was performed in fruitful collaboration with K. Okun, M. Wolfgardt, E. Andrejew, S. Böhmer, W. Paul, M. Fuchs and K. Binder. It is a pleasure to thank them for their support. A generous grant of computer time on the CRAY-YMP from Höchstleistungsrechenzentrum (HLRZ) at Jülich and the Regionales Hochschulrechenzentrum Kaiserslautern (RHRK) is gratefully acknowledged. This work was financially supported by the Deutsche Forschungsgemeinschaft (DFG) under the grant numbers SFB 262, Ba 1554/1 and Bi314/12.

Literature Cited

(1) Jäckle, J. *Rep. Progr. Phys.* **1986**, *49*, 171.

(2) Zallen, R. *The Physics of Amorphous Solids*; Wiley: New York, NY, 1983.

(3) Rössler, E.; Sillescu, H. In *Material Science and Technology*; Zarzycki, J., Ed.; VCH: Weinheim, 1991; Vol. 9; pp 574–618.

(4) McKenna, G. B. In *Comprehensive Polymer Science*; Booth, C.; Price, C., Eds.; Pergamon Press: New York, 1989, Vol. 2; pp 311–362.

(5) Brüning, R.; Samwer, K. *Phys. Rev. B* **1992**, *46*, 11318.

(6) Angell, C. A. *J. Non-Cryst. Solids* **1991**, *131–133*, 13.

(7) Götze, W. In *Liquids, Freezing and the Glass Transition*; Hansen, J. P.; Levesque, D.; Zinn-Justin, J., Eds.; North-Holland: Amsterdam, 1990, Part 1; pp 287–503.

76

(8) Götze, W.; Sjögren, L. *Rep. Prog. Phys.* **1992**, *55*, 241.

(9) Struik, L. C. E. *Physical Aging in Amorphous Polymers and Other Materials*; Elsevier: Amsterdam, 1978.

(10) Carmesin, I.; Kremer, K. *Macromolecules* **1988**, *21*, 2819.

(11) Deutsch, H.-P.; Binder, K. *J. Chem. Phys.* **1991**, *94*, 2294.

(12) Wittmann, H.-P.; Kremer, K. *Comp. Phys. Comm.* **1990**, *61*, 309; Wittmann, H.-P.; Kremer, K. *Comp. Phys. Comm.* **1992**, *71*, 343.

(13) Sokal, A. D. In *Monte Carlo and Molecular Dynamics Simulations in Polymer Science*; Binder, K., Ed.; Oxford University Press: New York, NY, 1995; pp 47–124.

(14) Paul, W.; Binder, K.; Heermann, D. W.; Kremer, K. *J. Phys. II* **1991**, *1*, 37.

(15) Binder, K. In *Monte Carlo and Molecular Dynamics Simulations in Polymer Science*; Binder, K., Ed.; Oxford University Press: New York, NY, 1995; pp 3–46.

(16) Paul, W.; Baschnagel, J. In *Monte Carlo and Molecular Dynamics Simulations in Polymer Science*; Binder, K., Ed.; Oxford University Press: New York, NY, 1995; pp 307–355.

(17) Baschnagel, J.; Binder, K.; Wittmann, H.-P. *J. Phys.: Condens. Matter* **1993**, *5*, 1597.

(18) Baschnagel, J.; Binder, K. *Physica A* **1994**, *204*, 47.

(19) Gibbs, J. H.; Di Marzio, E. A. *J. Chem. Phys.* **1958**, *28*, 373.

(20) Binder, K.; Heermann, D. W. *Monte Carlo Simulation in Statistical Physics: An Introduction*; Springer: Heidelberg, 1992.

(21) Doi, M.; Edwards, S. F. *Theory of Polymer Dynamics*; Clarendon Press: Oxford, 1986.

(22) Dünweg, B.; Stevens, M.; Kremer, K. In *Monte Carlo and Molecular Dynamics Simulations in Polymer Science*; Binder, K., Ed.; Oxford University Press: New York, NY, 1995; pp 125–193.

(23) Rosenke, K.; Sillescu, H.; Spiess, H. W. *Polymer* **1980**, *21*, 757.

(24) Paul, W. *AIP Conf. Proc.* **1992**, *256*, 145.

(25) Paul, W. *J. Non-Crys. Solids* **1994**, *172-174*, 682.

(26) Tries, V.; Paul, W.; Baschnagel, J.; Binder, K. *J. Chem. Phys.* **1997**, *106*, 738.

(27) Clarke, J. H. R. In *Monte Carlo and Molecular Dynamics Simulations in Polymer Science*; Binder, K., Ed.; Oxford University Press: New York, NY, 1995; pp 272–306.

(28) Kob, W. In *Annual Reviews of Computational Physics*, Stauffer, D., Ed.; World Scientific: Singapore, 1995, Vol. 3; pp 1–43.

(29) Kremer, K.; Binder, K. *Comp. Phys. Rep.* **1988**, *7*, 259.

(30) Wolfgardt, M.; Baschnagel, J.; Binder, K. *J. Phys. II (France)* **1995**, *5*, 1035.

(31) Vollmayr, K.; Kob, W.; Binder, K. *J. Chem. Phys.* **1996**, *105*, 4714.

(32) Vollmayr, K.; Kob, W.; Binder, K. *Phys. Rev. B* **1996**, *54*, 15808.

(33) Kauzmann, W. *Chem. Rev.* **1948**, *43*, 219.

(34) Angell, C. A. *J. Res. Natl. Inst. Stand. Technol.* **1997**, *102*, 171.

(35) Stickel, F.; Fischer, E. W.; Richert, R. *J. Chem. Phys.* **1995**, *102*, 6251.

(36) Stickel, F.; Fischer, E. W.; Richert, R. *J. Chem. Phys.* **1995**, *104*, 2043.

(37) Andrejew, E.; Baschnagel, J. *Physica A* **1996**, *233*, 117.

(38) *Transport Theory and Statistical Physics*; Yip, S.; Nelson, P., Eds.; Marcel Dekker, Inc.: New York, NY, 1995; Vol. 24, No. 6–8.

(39) Baschnagel, J. *Phys. Rev. B* **1994**, *49*, 135.

(40) Baschnagel, J.; Fuchs, M. *J. Phys.: Condens. Matter* **1995**, *7*, 6761.

(41) Cugliandolo, L. F.; Kurchan, J.; Ritort, F. *Phys. Rev. B* **1994**, *49*, 6331.

(42) Rieger, H.; Kisker, J.; Schreckenberg, M. *Physica A* **1994**, *210*, 326.

(43) Cugliandolo, L. F.; Kurchan, J.; Parisi, G.; Ritort, F. *Phys. Rev. Lett.* **1995**, *74*, 1012.

(44) Franz, S.; Mézard, M. *Physica A* **1994**, *210*, 48.

(45) Marinari, E.; Parisi, G. *J. Phys. A* **1993**, *26*, L1149.

(46) Kob, W.; Barrat, J.-L. *Phys. Rev. Lett.* **1997**, *78*, 4581.

(47) Ngai, K. L., Ed.; *J. Non-Cryst. Solids* **1994**, *172-174*.

(48) Adam, G.; Gibbs, J. H. *J. Chem. Phys.* **1965**, *43*, 139.

(49) Glotzer, S. C.; Coniglio, A. *Comp. Mat. Sci.* **1995**, *4*, 325.

(50) Grest, G. S.; Cohen, M. H. In *Advances in Chemical Physics*; Prigogyne, I.; Rice, S. A., Eds.; Wiley: New York, NY, 1981, Vol. 48; p 455.

(51) Donth, E. *Relaxation and Thermodynamics of Polymers: Glass Transition*; Akademie-Verlag: Berlin, 1992.

(52) Fischer, E. W. *Physica A* **1993**, *201*, 183.

(53) Okun, K.; Wolfgardt, M.; Baschnagel, J.; Binder, K. *Macromolecules*, **1997**, *30*, 3075.

(54) Wolfgardt, W.; Baschnagel, J.; Paul, W.; Binder, K. *Phys. Rev. E.* **1996**, *54*, 1535.

(55) Kob, W., Andersen, H. C. *Phys. Rev. Lett.* **1994**, *73*, 1376.

(56) Kob, W.; Andersen, H. C. *Phys. Rev. E* **1995**, *53*, 4134; *ibid.*, *51*, 4626.

(57) Binder, K; Young, A. P. *Rev. Mod. Phys.* **1986**, *58*, 801.

(58) Di Marzio, E. A.; Yang, A. J. M. *J. Res. Natl. Inst. Stand. Technol.*, **1997**, *102*, 135.

(59) Rouse, P. E. *J. Chem. Phys.* **1953**, *21*, 1272.

(60) Ferry, J. D. *Viscoelastic Properties of Polymers*; Wiley: New York, NY, 1980.

(61) Graessley, W. W. *Adv. Polym. Sci.* **1974**, *16*, 1.

(62) Pearson, D. S.; Fetters, L. J.; Graessley, W. W.; ver Strate, G.; von Meerwall, E. *Macromolecules* **1994**, *27*, 711.

(63) Richter, D.; Willner, L.; Zirkel, A.; Farago, B.; Fetters, L. J.; Huang, J. S. *Phys. Rev. Lett.* **1993**, *71*, 4158.

(64) Carmesin, I.; Kremer, K. *J. Phys. France* **1990**, *51*, 915.

(65) Kremer, K.; Grest, G. S. *J. Chem. Phys.* **1990**, *92*, 5057.

(66) Verdier, P. H. *J. Chem. Phys.* **1966**, *45*, 2118.

Chapter 5

The Thermal Glass Transition Beyond the Time Trap

K.-P. Bohn, and J. K. Krüger

FB 10.2 Experimentalphysik, Universität des Saarlandes, 66041 Saabrücken, Germany

High performance Brillouin spectroscopy was used to investigate the temperature dependent behavior of the longitudinal acoustic mode, the refractive index, and the hypersonic longitudinal mode Grüneisen parameter around the quasi-static glass transition temperature T_g of Polyvinylacetate. In the glass transition zone the longitudinal mode Grüneisen Parameter γ_L shows an apparent anomaly with a step-like anomaly. Whereas the temperature position of this anomaly clearly depends on the thermal history of the sample, the amplitude $\Delta\gamma_L = \gamma_L(T > T_g) - \gamma_L(T < T_g)$ does not. Additional studies of the coupling between the α-process and the thermal glass transition of epoxies will be presented. As a sensitive measure for α-relaxations we introduce the opto-acoustic dispersion function. We show that close but above T_g the related relaxation spectrum still contains components within the microwave frequency region and finally is truncated by the thermal glass transition.

In literature, the nature of the thermal glass transition (TGT) from the liquid to the glassy state as well as the nature of the glassy state itself still are a matter of debate (*1 - 15*). The main difficulties in understanding the TGT are extremely slow kinetic and relaxation processes in the vicinity of T_g, which generally interfere with typical experimental time scales and which therefore are not at all easy to measure. As a consequence the main questions in the understanding of the nature of the TGT are: i) does the TGT simply reflect an unavoidable cross-over of intrinsic relaxation times with typical time constraints of the experimental technique and the scientist's patience ("time trap") or ii) does there exist an underlying phase transition beyond the "time trap" describing a glass transition into an ideal glassy state (*1*) ?

As an experimental approach to investigate the temperature behavior of extremely slow structural glass relaxation processes (α-relaxation) we resently presented measurements with time domain Brillouin spectroscopy (TDBS) performed in the glass transition region of Polyvinylacetate (PVAC) (*14, 15*). In the gigahertz regime this special application of Brillouin spectrosopy is able to measure the temporal evolution of the longitudinal frequency f_L after a temperature perturbance $\Delta T = T_j - T_{j+1}$ (j = 0,1, ...) of the sample. The temperature behavoir of the relaxation process was deduced from the temporal evolution of the longitudinal hypersonic frequency f_L after a series of temperature perturbances ΔT of the sample starting at a temperature T_0 well above the glass transition. The temporal evolution of f_L after a temperature jump from T_j to T_{j+1} could be described by a Kohlrausch-Williams-Watts (KWW) function (*16*) $f_L(t) = f_L^{\infty} - (f_L^{\infty} - f_L^i) \cdot \exp\{-(t/\tau)^{\beta}\}$, where t is the time, τ denotes the relaxation time, β is a parameter describing the relaxation time distribution, f_L^i is the instantaneous frequency response, and f_L^{∞} is the fully relaxed frequency. Figure 1a shows a typical relaxation curve $f_L(t)$ after a temperature jump from 303.1 K to 300.4 K. A new temperature jump from T_j to T_{j+1} was started after the hypersonic frequency f_L of the sample has reached a constant value $f_{L,j}^{\infty}$ at T_j. The resulting f_L^{∞} data are shown in Figure 2a as a function of temperature. $f_L^{\infty}(T)$ shows a kink at the operative glass transition temperature $T_g = 297.3$ K. The only data point below T_g showing a relaxation behavior (T = 294.6 K) does not relax to the extrapolated frequency value $f_L^{fl}(294.6$ K) of the liquid.

From the fit parameters τ and β of the KWW function the related mean relaxation time $\langle\tau\rangle$ can be calculated. The relaxation frequencies $f_r = \langle\tau\rangle^{-1}$ are shown in Figure 1b. In contrast to the Vogel-Fulcher-Tamman (VFT) behavior (*17-19*) of many glass formers the data of the relaxation frequencies show an Arrhenius-like behavior ending at T = 294.6 K (● in Figure 1b). This behavior clearly indicates a cut-off of the related relaxation process at the TGT and leads to the conclusion that an underlying transition from the equilibrium liquid phase into a non-ergodic glass phase really exists (*14, 15*). Following the idea of a cross-over effect causing the glass transition we forced the KWW fit at T = 294.6 K to the extrapolated frequency value $f_L^{fl}(294.6$ K). Then however the resulting relaxation frequency (□ in Figure 1b) deviates by two orders of magnitude from the Arrhenius behavior.

In the liquid state the instantaneous hypersonic frequency response per Kelvin, $g_j^i = (f_{L,j+1}^i - f_{L,j}^{\infty})/\Delta T$, was found to equal $\Delta f_L/\Delta T$ of the glassy state (*14, 15*). The frequency response resulting from the temperature step from T = 297.5 K ($> T_g$) to T = 294.6 K ($< T_g$) however revealed a g^i (294.6 K) which is too small compared to the respective values of the liquid and the glassy state (*14, 15*).

From this discrepancy of g^i close but below T_g, the kink anomaly of $f_L(T)$, and the cut-off of the Arrhenius-like behavior of the relaxations, a model of an intrinsic glass transition was developed (*14, 15*): due to an instantaneous coagulation of clusters of minimal free volume at the intrinsic glass transition temperature T_{gs} a network of coagulated cluster chains is built. The last relaxation of f_L below T_g results from the ultimate freezing of the residual liquid in the voids of the cluster network. Following the suggested coagulation model, the discontinuity in g^i disappears if T_{gs} becomes the target of a temperature jump from above – only in this case $T_g = T_{gs}$ holds.

As an independent experiment to verify the findings of the TDBS we also performed quasi adiabatic measurements of the specific heat capacity c_p (*14*) using modulated differential scanning calorimetry (MDSC) (*20-22*). Usually the glass transition is characterized by a step like behavior of $c_p(T)$. The thermal glass transition

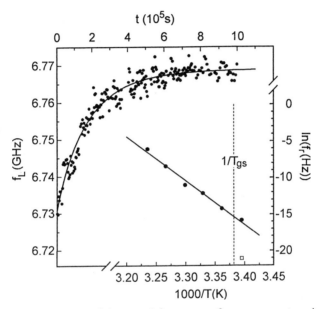

Figure 1. a) Relaxing part of the sound frequency after a temperature jump from T = 303.1 K to T = 300.4 K as a function of time. b) Arrhenius plot of the main structural glass relaxation frequencies f_r.

temperature T_g^{cal} measured by calorimetry often is defined by the inflection point of the $c_p(T)$ curve. From our MDSC data (Figure 2c) we found i) that the c_p transition interval is by a factor of about 5 larger than that of the TDBS measurements and ii) that the intrinsic glass transition temperature T_{gs} is reflected by the temperature where the $c_p(T)$ curve merges with its low temperature asymptote G^a (see the residuals $c_p(T) - G^a(T)$ in (14)). Even at quasi adiabatic measurements T_g^{cal} was found to be about 16 K above T_{gs}. In the case of PVAC the caloric glass transition temperature therefore is not related to the intrinsic glass transition directly. The behavior of $c_p(T)$ in the T_g region however is a clear sign of the stability limit of the glassy state at T_{gs}. The increase of $c_p(T)$ above T_{gs} simply reflects the increasing amount of Brownian motion at temperatures $T \geq T_{gs}$. Hence the MDSC results support the idea of an intrinsic glass transition in PVAC.

From the interpretation of the intrinsic glass transition as a hidden phase transition there arises the question about the order parameter (OP). As the glass transition mainly affects the mechanical properties (see Figures 1a and 2a) one could argue that these properties couple to the susceptibility of the hypothetical OP. In the case of spin glasses the linear order parameter susceptibilities just show a cusp like behavior at T_g. The nonlinear OP susceptibilities however have been found to diverge in the vicinity of the glass transition temperature (23). Therefore we present studies of nonlinear elastic properties of PVAC at the glass transition. As nonlinear elastic properties usually are not easy to measure we investigate the longitudinal mode Grüneisen parameter γ_L (LMGP) (24). This quantity reflects nonlinear elastic properties on the one hand (25) and can be determined exclusively from Brillouin spectroscopic data (7, 9).

In addition to the investigations of the influence of the glass transition on the LMGP of PVAC we study the coupling between the α-process and the thermal glass transition in a glass-forming mixture of epoxies (EPON) by means of the opto-acoustic dispersion function (OADF). Elastic properties of EPON are published in (31). We will show that the certain temperature behavior of the OADF in the glass transition region of EPON supports the suggested transition hypothesis. We shall demonstrate that the α-relaxation spectrum measured at Brillouin frequencies extends until T_g but is truncated by the TGT.

Experimental

Sample Preparation. The Polyvinylacetate under study is the same atactic polymer recently investigated with time domain Brillouin spectroscopy (14, 15). PVAC belongs to the class of ideal glass formers as it shows a glass transition but does not crystallize on any time scale. Above its glass transition PVAC therefore can be held in thermal equilibrium even on the time scale of the involved glass relaxation process. The molecular weight of PVAC is $M_w = 91$ kg/mole. A plate like sample of 1 mm thickness and 12 mm diameter was well annealed in an oven under vacuum. In order to get a stress free sample with optical quality of the relevant surfaces we pressed the sample between two plates of silicon rubber having surfaces of optical quality. The plate like sample was put into a special cuvette having inner dimensions slightly larger (10 μm) than the sample disk. From earlier investigations of the glass transition of Polymethylmethacrylate and Polystyrene (7, 8) it is evident that every mechanical constraint imposed from outside the sample seriously influences mechanical and

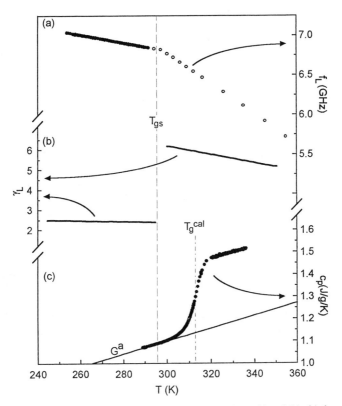

Figure 2. a) f_L measured on cooling using TDBS (see (*14, 15*)), b) longitudinal mode Grüneisen parameter (see (*33, 34*)), c) specific heat capacity measured with quasi-adiabatic MDSC. G^a: linear behavior of $c_p(T)$ in the glassy state; T_{gs}: intrinsic glass transition temperature: T_g^{cal}: caloric glass transition temperature.

thermal properties at the glass transition. In order to avoid such constraints (e.g. sticking of the sample to the sample holder) we embedded the sample in a thin film of partially polymerized silicon rubber. The degree of polymerization was chosen as such high that the molecules of the resulting viscous liquid could not penetrate into the PVAC sample but that the liquid was fluid enough to remove every friction between the cuvette and the sample. Prior to the Brillouin measurements the sample was aged within its sample holder at T_g during several months and then slowly cooled down to about 240 K. The inspection of the sample with an optical microscope and with a laser beam gave no hints for swelling due to penetration of the partially polymerized silicon rubber.

The epoxy under study is a mixture consisting of Diglycidyl ether of bisphenol A,

$$CH_2 CHCH_2 \left[O - \langle\rangle - \underset{CH_3}{\overset{CH_3}{C}} - \langle\rangle - OCH_2 \underset{OH}{CHCH_2} \right]_n O - \langle\rangle - \underset{CH_3}{\overset{CH_3}{C}} - \langle\rangle - OCH_2 CHCH_2$$

(EPON) with a known distribution of molecular weights: 86% n = 0, 13% n = 1, 1% n = 2. First acoustic results have been reported previously (*31*). For the measurements of the OADF the fluid sample was put into a special cuvette of about 1 mm thickness and a diameter of 12 mm. The inner surfaces of the cuvette had a special coating in order to avoid any sticking of the glass forming sample. Thus the sample surfaces had a high optical quality in the entire temperature range of the investigations.

Brillouin Measurements. Our Brillouin measurements were made either on stepwise heating or stepwise cooling. The measurements on PVAC have been performed with a five-pass Fabry-Pérot spectrometer. In the case of EPON we used a six-pass tandem Fabry-Pérot spectrometer. Both spectrometers are described elsewhere (*9*). In order to obtain a high resolution together with the required long time stability the temperature of the spectrometer including the spectrometer control was stabilized to better than 0.1 K. Moreover, the spectrometer control including the temperature control of the sample, the data collection procedure and the evaluation of the data has been completely automated yielding immediately the final Brillouin data after each measurement. The spectrometer is able to run completely stable without the need of any human interaction over many months (*7, 9*). The sample was positioned within a home-made top-loading thermostat. The temperature control was maintained with a PID-controller (ITC-4, Oxford Instruments) using a Rhodium-Iron resistance. The temperature of the sample was measured with a thin chromel-alumel thermo-couple positioned within the sample slightly above the scattering volume. In order to control the influence of local heating the sample by the laser beam we made Brillouin measurements as a function of the laser power. The final Brillouin measurements were performed with an incident laser power of about 10 mW.

Results and Discussion

Longitudinal Mode Grüneisen Parameter of PVAC. In an isotropic solid the longitudinal mode Grüneisen parameter (LMGP) γ_L relates the frequency f_L of a longitudinal sound mode of a given wave vector **q** to the mass density ρ of the sample (*24-26*):

$$\gamma_L(q) = (\partial \ln f_L(q))/(\partial \ln \rho). \tag{1}$$

In literature (25-30) it is anticipated that i) the concept of the LMGP can be extended to acoustic waves propagating in liquids, ii) the LMGP varies only slightly in a given phase, and iii) the LMGP of a liquid still reflects the anharmonic properties of the material.

The determination of γ_L needs the knowledge of the sound frequency $f_L(q)$ at a given wave vector q as a function of the mass density ρ. The 90A-scattering geometry fulfills exactly the measurement condition for $f_L(q)$ because $q^{90A} = 17.271 \ 10^{-3} \ nm^{-1}$ is strictly constant, even if the temperature varies. The change of $f_L(q)$ and ρ may be induced by pressure or temperature changes. We have chosen the second method. The temperature dependence of ρ may be determined independently from the hypersonic frequency measurements or, alternatively, can be deduced exclusively from Brillouin measurements. The advantage of the latter method is that all information is obtained simultaneously from the same small information volume of only $10^{-7} \ cm^3$, i.e. all data suffer exactly the same thermal history and ambiguities about different preparation conditions are avoided. The calculation of the mass density from the longitudinal frequencies f_L^{90A} and f_L^{90R}, measured in 90A and 90R scattering geometry respectively, with the help of the opto acoustic dispersion function (9) and the Lorenz-Lorentz equation (32) is described in (33, 34).

The calculated LMGP γ_L of PVAC is shown in Figure 2b as a function of temperature. $\gamma_L(T)$ indicates a kind of discontinuous behavior in the glass transition region. $\gamma_L(T)$ is higher in the liquid phase than in the glassy phase, indicating a stronger anharmonicity within the glassy state. The step like anomaly in $\gamma_L(T)$ indicates a rather abrupt change of the average molecular interaction force at the ideal glass transition. Whereas $\gamma_L(T)$ nearly is constant in the glassy phase, $\gamma_L(T)$ significantly increases in the fluid phase on approachimg the glass transition. From a comparison to $\gamma_L(T)$ of other liquids outside the glass transition region (33) we followed that the negative slope in $\gamma_L(T)$ above T_g of PVAC might be a fingerprint of the ideal glass transition. In (34) we investigated the influence of different thermal histories on the LMGP. Whereas the temperature position of the $\gamma_L(T)$ anomaly might be shifted on the temperature axis due to different thermal treatments of the sample, the difference $\Delta\gamma_L = \gamma_L(T > T_g) - \gamma_L(T > T_g)$ was found to be an invariant of the sample under study. In (33) we found a similar behavior of the thermal Grüneisen parameter $\gamma_{th} = v\alpha B_s/c_p$.

Opto-acoustic Dispersion Function of EPON. Using Brillouin spectroscopy as an experimental technique to study the static as well as the dynamic glass transition it turns out, that only the Brillouin frequency f is a sensitive probe for the quasi-static glass transition whereas the Brillouin linewidth Γ is not (cf. (9)). Although the frequency f of a Brillouin line can be measured much more precisely than ist linewidth Γ, f as a function of temperature T is rather insensitive in order to detect the temperature dependent evolution of acoustic relaxations as e.g. that of the α-relaxation process. A suitable experimental probe for changes of static as well as of dynamic acoustic properties provides the OADF which can be calculated from the hypersonic frequencies f^{90A} and f^{90R} measured for the two wave vectors q^{90A} and q^{90R}, respectively. Using the scattering geometries 90A and 90R simultaneously, we measured the hypersonic frequencies f^{90A} and f^{90R} (Figure 3) of EPON. As in the case

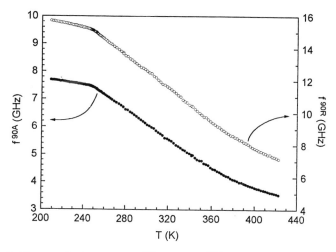

Figure 3. Hypersonic frequencies f^{90A} (●) and f^{90R} (O) of EPON as a function of temperature.

of other polymers the TGT of EPON is indicated by a defined kink in both hypersonic frequency curves at $T_g = 247.8$ K.

For the OADF $D(q^{90A}, q^{90R}, T)$ the equation

$$D\left(q^{90A}, q^{90R}, T\right) = \sqrt{\frac{c'\left(q^{90R}, T\right)}{c'\left(q^{90A}, T\right)}\left(n^2(T) - \frac{1}{2}\right) + \frac{1}{2}} \qquad (2)$$

holds (c.f. (9)), where $c'(q^{90A})$ and $c'(q^{90R})$ are the real parts of the complex elastic constant c^* for the corresponding wave vectors q^{90A} and q^{90R}, and n is the refractive index. $D(q^{90A}, q^{90R}, T)$ can be calculated from the frequencies f^{90A} and f^{90R} by

$$D\left(q^{90A}, q^{90R}, T\right) = \sqrt{\frac{1}{2}\left\{\left(\frac{f^{90R}(T)}{f^{90A}(T)}\right)^2 + 1\right\}} \qquad (3)$$

(e.g. (9)). In isotropic materials q^{90A} and q^{90R} are symmetry equivalent real quantities defined by the scattering angle, the vacuum laser wavelength λ_0 and the phase velocity of the light within the sample.

As q^{90A} is strictly independent of the temperature, equation (2) gives in the fast motion ($2\pi f\tau \ll 1$) as well as in the slow motion case ($2\pi f\tau \gg 1$) the refractive index n(T) at the laser wavelength λ_0 (c.f. (9)). In the acoustic relaxation regime $D(q^{90A}, q^{90R}, T)$ exeeds n(T). In (35) we present different theoretical curves of $D(q^{90A}, q^{90R}, T)$ calculated under the assumption, that the real part of the complex elastic constant $c^*(q, T)$ can be written in the form $c'(q, T) = c^\infty(T) - \Delta c/\{1 + 4\pi^2 f^2(q, T)\tau^2(T)\}$. For the exponent $\beta < 1$ this formular describes a Cole davidson function. The relaxation time τ was assumed to follow a VFT law. Under these conditions the OADF deviates from n(T) only well above the TGT and $D(q^{90A}, q^{90R}, T)$ shows a defined kink at T_g. As a typical example for the OADF of polymers we present $D(q^{90A}, q^{90R}, T)$ of PVAC in Figure 4.

The OADF of EPON (Figure 5) however behaves quite different from that of PVAC (Figure 4): Although the sound frequency-curves (Figure 3) shows only one kink, indicative for one TGT at $T_g = 247.8$ K, the measured $D(q^{90A}, q^{90R}, T)$-function clearly shows a two-maximum structure and a low temperature tail which ends definitely at T_g. In (35) we have shown that this behavior of the measured $D(q^{90A}, q^{90R}, T)$-data is incompatible with a pure relaxation ansatz. The simultaneous appearance of two $D(q^{90A}, q^{90R}, T)$-peaks indicates the existence of two local α-relaxators which independently couple to the longitudinal sound mode. Therefore it seems to us, that the α-relaxations initially are independent of the TGT but that the onset of the TGT renormalizes the α-processes and definitely truncate them at T_g. That means not the slowing down of the α-process drives the TGT but the opposite case seems to be true. In this sense the ideal glass transition is not due to the cross-over of intrinsic and experimental time scales although this slowing down influences the experimental results (14, 15, 33, 34).

Summary

We summarize our results as follows: i) Using TDBS we found a phase transition like phenomenon at a well defined intrinsic glass transition temperature T_{gs}. The fully relaxed longitudinal frequency shows in the limit of infinite small temperature jumps

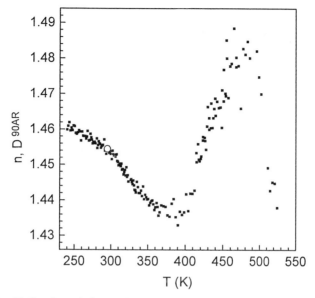

Figure 4. Refractive index n(T) and opto-acoustic dispersion function D($\mathbf{q}^{90A}$, $\mathbf{q}^{90R}$, T) of PVAC (○, ■). ◆ n(T = 295 K) measured with an Abbé refractometer at λ = 514.5 nm.

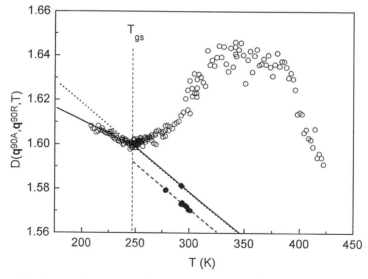

Figure 5. Refractive index n (•, ◆) and opto-acoustic dispersion function D90AR () of EPON as a function of temperature T. Solid line: refractive index at λ = 514.5 nm (•, ◆: measured with an Abbé refractometer); dashed line: refractive index for white light (•: measured with an Abbé refractometer).

a sharp thermoreversible kink at T_{gs}. ii) The discontinuity in g^i at T_g disappars if T_{gs} becomes a target of a temperature jump from above. iii) Close to T_{gs} the average relaxation time of the α-process behaves Arrhenius-like reaching values of about $7 \cdot 10^6$ s and showing a cut-off. iv) The existence of the intrinsic glass transition is confirmed by the deviation of the $c_p(T)$ curve from the linear asymptote at T_{gs} measured under quasi adiabatic conditions by MDSC. v) The transition from the liquid to the glass is characterized by an anomaly in the LMGP describing a significant change of the average molecular interaction forces. The position of the anomaly depends on the thermal history whereas the amplitude of the anomaly is an invariant of the system. vi) In EPON the onset of the TGT even renomalizes the α-process and truncates this relaxation process at T_g.

Literature Cited

1. Gibbs, J. H.; DiMarzio, E. A. *J. Chem. Phys.* **1958**, *28*, 373
2. Gee, G. *Contemp. Phys.* **1970**, *11*, 313
3. Wong, J.; Angel, C. A. *Glass Structure by Spectroscopy*; Marcel Dekker, New York, **1976**
4. Kovacs, A. J.; Aklonis, J. J.; Hutchinson, J. M.; Ramos, A. R. *J. Polym. Sci.: Polym. Phys. Ed.* **1979**, *17*, 1097
5. Zarzycki, J. *Les Verres et l'Etat Vitreux*; Masson, Paris, **1982**
6. Jäckle, J. *Rep. Progr. Phys.* **1986**, *49*, 171
   Jäckle, J. *J. Phys. Cond. Matter* **1989**, *1*, 267
7. Krüger, J. K.; Roberts, R.; Unruh, H.-G.; Frühauf, K.-P.; Helwig, J.; Müser, H. E. *Progress in Colloid & Polymer Science* **1985**, *71*, 77
8. Frühauf, K.-P.; Helwig, J.; Müser, H. E.; Krüger, J. K.; Roberts, R. *Colloid & Polymer Science* **1988**, *266*, 814
9. Krüger, J. K. in "*Optical Techniques to Characterize Polymer Systems*", Ed. by H. Bässler, Elsevier, Amsterdam, Oxford, New York, **1989**
10. Elliott, S. R. *Physics Of Amorphous Materials*; Longman, **1990**
11. Götze, W.; Sjögren, L. *Rep. Prog. Phys.* **1992**, *55*, 241
12. Donth, E. *Relaxation and Thermodynamics in Polymers, Glass transition*; Akademie-Verlag, Berlin, **1992**
13. Ngai, K. L. *Universal Patterns of Relaxations in Complex Correlated System* in *Disorder Effects on relaxation Processes*, Richert/Blumen, Springer-Verlag, Berlin, **1994**
14. Krüger, J. K.; Bohn, K.-P.; Jiménez, R. *Condensed Matter News* **1996**, *5*, 10
15. Krüger, J. K.; Bohn, K.-P.; Jiménez, R.; Schreiber, J. *Colloid & Polymer Science* **1996**, *274*, 490
16. Williams, W.; Watts, D.C. *Trans. Faraday Soc.* **1979**, vol 66, 80
17. Vogel, H. *Phys. Z* **1921**, *22*, 645
18. Fulcher, G. S. *J. Am. Chem. Soc.* **1925**, *8*, 339
19. Tammann, G. *Der Glaszustand*, Leopold Voss, Leipzig, **1933**
20. Boller, A.; Jin, Y.; Wunderlich, B. *Journal of Thermal Analysis* **1994**, *42*, 307
21. Sauerbrunn, S. R.; Crowe, B. S.; Reading, M. *21st Proc. NATAS Conf. in Atlanta*, GA, pp. 137 - 144, **1992**
22. Reading, M.; Elliot, D.; Hill, V. *21st Proc. NATAS Conf. in Atlanta*, GA, pp. 145 - 150, **1992**
23. Prejean, J. J. in *Dynamics of Disordered Materials, Springer Proc. Phys.* **37**, p. 242, ed. Richter, D.; Dianoux, A. J.; Petry, W.; Teixera, J.; Springer, Berlin, **1989**
24. Grüneisen, E. *Ann. Phys. Leipzig* **1908**, *26*, 211
25. Grimvall, G. *Thermophysical Properties of Materials*, North-Holland, Amsterdam, **1986**

26. Barron, T. H. K.; Collins, J. G.; White, G. K. *Adv. Phys.* **1980**, *29*, 609
27. Brody, E. M.; Lubell, C. J.; Beatty, C. L. *J. Polym. Sci. Polym. Phys. Ed.* **1975**, *13*, 295
28. Ludwig, W. *Festkörperphysik*, Akademische Verlagsgesellschaft, Leipzig, **1978**
29. Leibfried, G.; Ludwig, W. *Solid State Phys.* **1961**, *12*, 275
30. Frenkel, J. *Kinetic Theory of Liquids*, Dover, New York, **1975**
31. Matsukawa, M; Ohtori, N; Bohn, K.-P.; Krüger, J. K. *Conference Proceedings of the 17th symposium on Ultrasonic Electronics* p. 219, Yonezawa, Japan, **1989**
32. Lorentz, H. A. *Wied. Ann. Phys.* **1880**, *vol 9*, 641
    Lorenz, L. V. *Wied. Ann. Phys.* **1880**, *vol 9*, 70
33. Krüger, J. K.; Bohn, K.-P.; Pietralla, M.; Schreiber, J. *J. Phys.: Condens. Matter* **1996**, *8*, 10863
34. Krüger, J. K.; Bohn, K.-P.; Schreiber, J. *Phys. Rev. B* **1996**, *54*, 15767
35. Krüger, J. K.; Bohn, K.-P. *Phase Transitions* **1997** (in press)

Chapter 6

Dynamic Monte Carlo Simulation and Dielectric Measurements of the Effect of Pore Size on Glass Transition Temperature and the State of the Glass

N. Haralampus[1], J. P. Northrop[1], C. Scordalakes[1], E. Ashmore[1], W. Martin[1], D. Kranbuehl[1], and P. H. Verdier[2]

[1]Department of Chemistry, College of William and Mary, Williamsburg, VA 23187–8795
[2]Polymers Division, National Institute of Standards and Technology, Gaithersburg, MD 20899

Results of experimental and theoretical studies directed at gaining a better understanding of the effect of confinement on T_g are reported. In experimental studies, the dielectric rotational relaxation spectrum has been measured for several glass forming small molecules both in the bulk and in 4 nm porous Vycor glass. Theoretical studies have been conducted using Monte Carlo simulation techniques to study the effect of confinement in cavities on the formation of the glassy state as the system is cooled from the melt using different cooling rates.

With the increasing interest in nanostructures has come an interest in the effect of confinement on the dynamic behavior and properties of these materials (*1-8*). Much of this interest has been generated by the recent studies of Jackson and McKenna (*1-2*) which showed that glasses formed in small pores have a lower T_g than in the bulk. They observed further that this reduction in T_g increased as the pore size decreased. Since that study, other workers have become interested in the effect of confinement on the dynamics and structure of a glass forming liquid as it approaches the glass transition temperature. These additional studies have led to the view that confinement does affect the structure and dynamics as a result of at least two effects, the large increase in surface area and the small dimensions of the cavity. It is well known that the length scale of the cooperatively of the α-relaxation process during the onset of glass formation is large, involving many molecules. Furthermore it increases dramatically as the temperature approaches T_g. It is not surprising therefore that there is a significant effect of confinement on the bulk dynamic properties which are separate from the effects of the surface. This has been verified through dielectric relaxation experiments on a simple glass forming liquid (*6*).

This report describes recent preliminary results of both experimental dielectric relaxation studies and theoretical Monte Carlo model simulation studies. Both studies focus on the onset of a glass formation in an environment where the molecules are spacially confined. The studies were undertaken to better understand the effect of confinement on T_g and on the dynamics as the liquid approaches the glass transition. The motivation begins in part because the McKenna result of a decrease in T_g due to confinement on first analysis might seem anomalous since confinement should hinder motion and thereby increase T_g.

The dielectric rotational relaxation spectrum of two glass forming small molecules both in the bulk and in 4 nm porous Vycor glass have been measured. Theoretical studies have been conducted using Monte Carlo simulation techniques on short polymer chains to study the effect of confinement on glass formation as a polymer system is cooled from the melt using different rates of coding.

Experimental

Dielectric relaxation measurements were made using a 3-terminal parallel plate capacitance cell. The plates were compressed onto a sheet of Corning 7930 Vycor glass, which is characterized by Corning as having 40Å pores. The glass disks were 82 mm in diameter and 1.59 mm thick. Before running, the glass was cleaned by boiling in nitric acid at 105° to 110°C for 10 hours. The glass was then rinsed with distilled water, dried in a desiccator and impregnated with the polar organic liquid while in the capacitance cell. Bulk polar organic liquid measurements were made in the same cell with the identical spacing between the plates as with the porous glass, 1.59 mm.

A Solartron 1260 Impedance Analyzer was used to make measurements of C, capacitance, and G, conductance, over a range of frequencies form 5 Hz to 100 kHz both during cooling and heating. From the known geometry of the cell, values of the complex permittivity $\varepsilon^* = \varepsilon' - i\varepsilon''$ were calculated where $\varepsilon' = C/C_o$ and $\varepsilon'' = G/\omega\varepsilon_o C_o$ where $\omega = 2\pi$ frequency, C_o is the measured air-filled capacitance of the cell with a 1.59mm gap, and $\varepsilon_o = 8.854 \times 10^{-14}\,C^2J^{-1}cm^{-1}$.

Measurements of the clean, empty Vycor glass were made. The dielectric loss ε'' was negligible compared to the values measured when the glass was impregnated with the polar liquid.

Two liquids were studied. Glycerol (cat. no. 24068-0 of 99+% purity) and dimethyl phthalate (Aldrich cat. No. 24068-0 of 99% purity) were purchased from Aldrich. Both were used as received. In each case the dielectric cell was heated with the plates apart in an oven at approximately 50°C. The liquid was added to the cell and the Vycor glass was inserted with the plates still apart. The cell was left in the oven overnight to allow sufficient time for penetration. The following day the plates were pressed up against the glass and the cell was cooled with N_2 gas from a dewar.

Monte Carlo Simulation Model

In our off-lattice random-coil Monte Carlo model, a single polymer chain N-1 units long is modeled by a string of N impenetrable beads (9-12). The vectors or "sticks" connecting bead centers along the chain are each one unit in length. Each bead-stick pair is taken to represent one "statistical segment," i.e. a moderately large number of chemical monomers in a real polymer chain. The beads are of unit diameter, touching spheres. They are not constrained to lie on a lattice, and no restriction is placed on the angle between successive connection vectors along the chain (See Figure 1). Brownian motion of a real chain is replaced in our model by sequences of elementary moves, which we shall call move cycles. In the simulations reported here, each move cycle consists of selecting a single bead of the chain at random and attempting a local move of that bead and its connecting vectors. The attempted move is to a position obtained by rotating the selected bead about an axis passing through the centers of its immediate neighbors along the chain, through an angle chosen at random in the range $(-\pi, \pi)$. See Figure 2. For an end bead, with only one neighbor, the rules are modified by first generating the center coordinate of a "phantom" bead at unit distance and randomly chosen direction form the selected end bead. The potential energy function is a hard-sphere repulsion of the bead radius (excluded volume), a finite attractive potential V/kT between $r + 1.0$ and zero for larger distances between beads. The decision to move or not move the chosen bead is determined by calculating the difference ΔE in "normalized energy" between the selected bead and all the other beads before and after the proposed move; then moving it with a probability p where $\Delta E = V/kT(\text{final}) - V/kT(\text{initial})$ and

$$p = 1 \text{ if } \Delta E < 0$$
$$p = \exp(-\Delta E) \text{ if } \Delta E > 0$$

In the present study, freely jointed statistical segment polymer chains with 10 hard sphere beads were studied in 3 environments. The three environments are shown in Figure 3. The first environment is a confining tube of radius 3, length 10 with rounded ends. The second is a similar tube of radius 3 and length 16. The third is a sphere of radius 8. The densities in all 3 environments were approximately equal with bead volume to total volume ratios of 0.185, 0.186, and 0.176, respectively. Tube one contained eight chains, tube two contained fourteen chains, and the sphere contained 72 chains.

Nine equilibrium configurations were obtained for each environment with interaction-temperature-potential values of $V/kT = 1.0$. That is, the potential well was repulsive at the equilibrium starting temperature. This was an attempt to extend the "temperature" range of the quench. Each of the nine equilibrium configurations for each environment was generated by running the simulation once in equilibrium for an additional time, estimated to be at least 10 times the limiting relaxation time (9-12). The quench was then simulated by increasing the value of V/kT at four different rates designated 0.1x, x, 10x, and 100x. Thus, as the value of V/kT is increased or the temperature is lowered, the well becomes more attractive. This attraction draws the beads closer together, decreasing the volume and the mobility of the chain beads. Eventually this decrease in volume or increase in density, is so great that the beads or molecules become locked up to form a material in which the volume change with temperature is much less (i.e. a glass).

Experimental Results and Discussion

Dielectric measurements of the complex permittivity $\varepsilon^* = \varepsilon' - i\varepsilon''$ were made as a function of the temperature after a relatively rapid quench to approximately -60°C at 3°C/min during a more gradual heating rate of less than 1°C/min. Figures 4 and 5 display the loss factor ε'' versus frequency during slow heating of the quenched polar liquids at three different temperatures and for runs on the neat bulk system and the molecules impregnated in the glass. These figures clearly show that the width of the relaxation

Polymer Chain

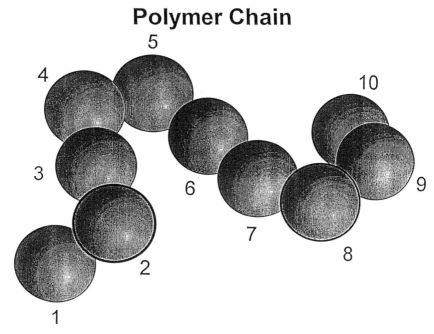

Figure 1. Representation of the chain as it is free to move to any conformation in 3-dimensional space.

Move Cycle

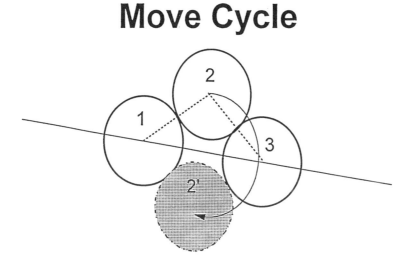

Figure 2. Rotational movement of bead 2 about the axis from bead 1 to 3 to its new position 2' chosen at random from the angles -π to π.

3 Systems

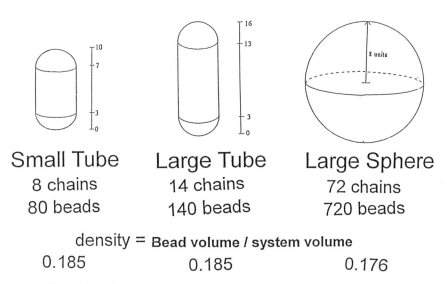

Small Tube	Large Tube	Large Sphere
8 chains	14 chains	72 chains
80 beads	140 beads	720 beads

density = **Bead volume / system volume**

| 0.185 | 0.185 | 0.176 |

Figure 3.   Diagram of shape of the three confining environments.

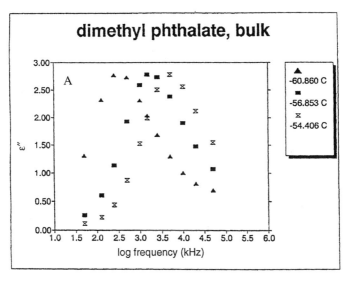

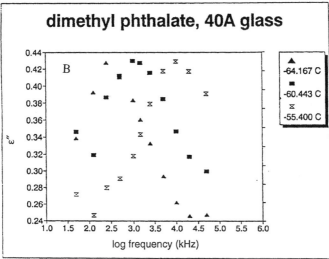

Figure 4.   Plot of ε'' log (frequency) for dimethyphthalate (a) in the bulk and (b) in the pore glass.

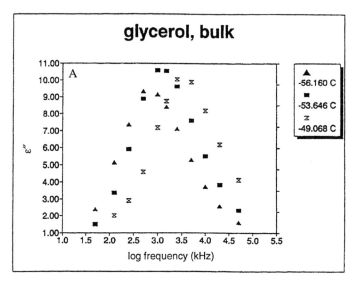

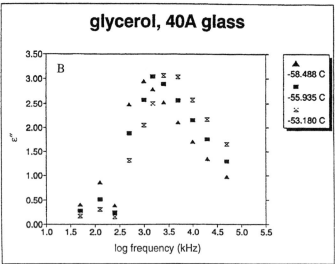

Figure 5.   Plot of ε″ versus log (frequency) for glycerol (a) in the bulk and (b) in the pore glass.

spectrum is broader in the confining pores of the Vycor glass than in the bulk for both systems. It is generally assumed that this characteristic broad distribution in the frequency dependence of the loss factor ε'' in the highly viscous pre-glass state is well characterized by a stretched exponential relaxation function, $\exp[-(t/\tau_o)^\beta]$ (4,6-8,13). Assuming this functional dependence, the distribution parameter, β, for dimethyl phthalate decreases from $\beta = 0.55$ ($\beta = 1$ for simple exponential decay and is often observed in non-associating low viscosity small molecule liquids) to $\beta \approx 0.25$ in the porous glass. Similarly, β for glycerol decreases from $\beta \approx 0.65$ in the bulk to $\beta \approx 0.55$ in the confining pores.

Particularly important is the dependence of $\varepsilon''(max)$ on the temperature. A characteristic relaxation time, $\tau = 2 \pi f_{max}$, can be defined from the frequency at which ε'' achieves its maximum value. Figures 6 and 7 plot the temperature at which ε'' is a maximum for each frequency as f_{max} versus $1/T$ for both of the systems in the bulk and in the porous glass. In both cases the molecules in the pores appear to experience slightly different temperature dependence than in the bulk. At low temperatures, low frequencies, the molecules are rotating more rapidly in the confining pores. This higher mobility, shorter relaxation time, suggests the temperature necessary to quench this motion to that of the glassy state will be lower in the pores than in the bulk. Thus the effective glass transition temperature as characterized by this long range, highly cooperative α dielectric relaxation time as determined from the maximum in ε'' at each frequency is lower in the pores than in the bulk. This result is in agreement with the earlier differential scanning calorimetry T_g data of McKenna.

Monte Carlo Simulation Results and Discussion

Figure 8 shows for the smaller tube the variations in density for each of the slowest rates of change in $\Delta E = V/kT$, i.e. quench rates. As the "temperature" is lowered by increasing V/kT a distinct transition in the rate of change in density with V/kT occurs. This is indicative of the liquid-to-glass transition. In order to characterize this rate of change in density, five lines were fit to the data. First using the slowest rate of quench, $0.1x$, which shows a change in density ρ close to that of the next fastest speed, x, the change in density ρ versus V/kT for the liquid region was fit to a straight line in the range of $\Delta E = +0.5$ to -0.5. Then the rate of change in ρ with V/kT for the final glassy state was determined at the lowest temperatures over the range $V/kT = -1.5$ to -4.

Similar simulations were conducted on the larger tube cavity of identical radius but greater length. The results of these simulations are shown in Figure 9. Liquid and glass coefficients of the rate of density change with V/kT were determined as in the small tube.

For the large sphere, the radius is 8 compared to 3 for both the small and large tube cavity. This geometry was chosen to approach the bulk state. The results of these simulations are shown in Figure 10. Because the system was so large it was run only to $V/kT = -2.0$. The rate of change in density with V/kT was determined in the glass range of -1.5 to -2.0 and the liquid line was fit in the 0 to 0.5 region for the slowest $0.1x$ quench rate.

For all three systems, an effective glass transition value of V/kT was determined from the intersection of the liquid and glass rate of change in density with change in V/kT

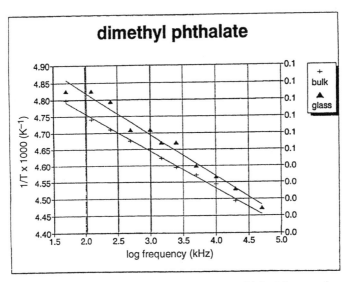

Figure 6. Plot of 1/temperature versus frequency at which ε'' is a maximum. A characteristic relaxation time $\tau = 2\pi f$ is determined at this temperature for dimethyl phthalate.

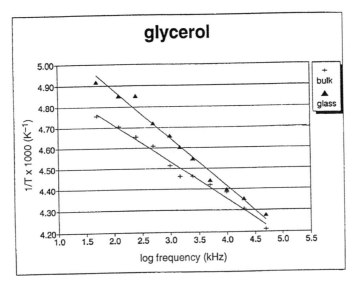

Figure 7. Plot of 1/temperature versus frequency where $\tau = 2\pi f$ at this temperature for glycerol.

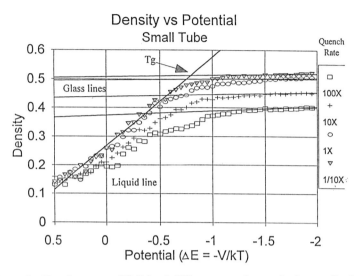

Figure 8. Density versus V/kT for 4 different quench rates in the small tube.

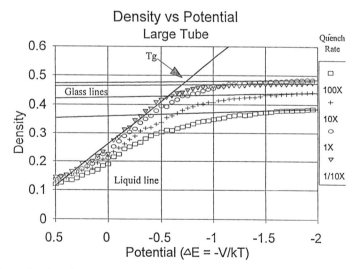

Figure 9. Density versus V/kT for 4 different quench rates in the large tube.

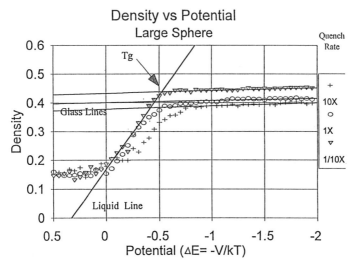

Figure 10. Density versus V/kT for 3 different quench rates in the large sphere.

TABLE 1	values = V/kT @ Tg		
	Small Tube	Large Tube	Large Sphere
1/10th X	0.79	0.70	0.52
1X	0.75	0.73	0.46
10X	0.54	0.56	0.42
100X	0.36	0.35	——

TABLE 2	Density (bead volume/ unit volume)		
	Small Tube	Large Tube	Large Sphere
1/10th X	0.52	0.47	0.45
1X	0.50	0.49	0.42
10X	0.45	0.44	0.40
100X	0.40	0.39	—

lines. The values of V/kT at each of these effective glass transitions are reported in Table I. The results indicate a clear trend for V/kT to decrease with quench rate. The V/kT values are related to the reciprocal of the temperature at the glass transition. Thus the results in Table I first show T_g is decreasing as the quench rate becomes slower for all three systems. Second the results in Table I show that T_g decreases as the size of the cavity decreases. Hence confinement is decreasing the glass transition temperature as observed by the dielectric measurements and by the calorimetry measurements of McKenna.

Finally, Table II reports values of the final density in the glass for each quench rate in the three environments at $V/kT = 2.0$. As expected, the results show an increase in density as the quench rate decreases and T_2, as measured by V/kT, decreases.

Conclusions

Both the experimental and the theoretical results provide some insight into the seemingly anomalous results of Jackson and McKenna that T_g decreases in confined space and the intuitive concept that small pores hinder motion and thereby increase T_g. The dielectric relaxation experimental results show that confinement changes the activation energy of the relaxation time such that at a low temperature, long times, the molecule relaxes more quickly in the pore, but at higher frequencies and higher temperatures it relaxes more quickly in the bulk. The Monte Carlo polymer chain theoretical results suggest that the smaller the confining space, the lower the T_g and the glass being formed is more dense. The smaller space affects the mobility and motion in these simulations such that $d\rho/d(1/T)$ is smaller and a more dense glass forms at a lower temperature.

Both results suggest the measurement of T_g is a non-equilibrium phenomenon in which the time scale of the experiment and the dynamical behavior of the material are coupled, making it very difficult to draw direct conclusions from T_g measurements alone.

Acknowledgment

DEK appreciates the support from the NSF Science and Technology Center at Virginia Polytechnic Institute and State University contract DMR 912004 and from conversations with Greg McKenna.

Literature Cited

1. Jackson, C. L.; McKenna, G. B., *J. Chem. Phys.* 1990, *93*, pp. 9002.
2. Jackson, C. L.;McKenna, G. B. *J. Noncrystl Solids* **1991**, pp. 221.
3. Pissis, P.; Daoukako-Piammanti, D.; Apekis, L.; Christodoulides, C. *J. Phys. Cond. Matter* **1994**, *6*, pp. 325.
4. Stickel, F.; Kremer, F.; Fisher, E. W. *Physica A*, **1993**, *201*, pp. 318.
5. Zhang, J.; Liu, G.; Jonas, J. *J. Phys. Chem.*, **1992**, *92*, pp. 3478.
6. Schuller, J.; Richert, R.; Fischer, E. W. *Phys. Rev B*, **1995**, *52* pp. 15232.
7. Streck, C.; Mel'nichenko, Y.; Richert, R. *Phys. Rev. B*, **1996**, *53*, pp. 5341.
8. Mel'nichenko, Y.; Schuler, J.; Richert, R.; Ewen, B. *J. Chem. Phys.*, **1995**, *103*, pp. 2016.

9.   Kranbuehl, D. E.; Eichinger, D.; Verdier, P. H. *Macromolecules*, **1991**, *24*, pp. 2419.

10. Verdier, P.; Kranbuehl, D. E. *Poly Mat. Sci. Eng.*, **1994**, *71*, pp. 643.

11. Kranbuehl, D. E.; Verdier, P. H. *J. Chem Phys*, **1997**, *106(11)*, pp. 4788.

12. Kranbuehl, D. E.; Verdier, P. H. *J. Chem. Phys.* **1997**, *106(11)*, pp. 4788-4796.

13. Moynihan, C. T.; Boesch, L. P.; Laberge, N. L. *Phys. Chem. Glasses* **1973**, *14(6)*, pp. 122.

Chapter 7

Temperature-Modulated Calorimetry of the Frequency Dependence of the Glass Transition of Poly(ethylene terephthalate) and Polystyrene

Bernhard Wunderlich[1] and Iwao Okazaki[2,3]

[1]Department of Chemistry, The University of Tennessee, Knoxville, TN 37996–1600
[2]Chemical and Analytical Sciences Division, Oak Ridge National Laboratory, Oak Ridge, TN 37831–6197

Glass transitions involve mainly the onset or freezing of cooperative, large-amplitude motion and can be studied using thermal analysis. Temperature-modulated calorimetry, TMC, is a new technique that permits to measure the apparent, frequency-dependent heat capacity. The method is described and a quasi-isothermal measurement method is used to derive kinetic parameters of the glass transitions of poly(ethylene terephthalate) and polystyrene. A first-order kinetics expression can describe the approach to equilibrium and points to the limits caused by asymmetry and cooperativity of the kinetics. Activation energies vary from 75 to 350 kJ/mol, dependent on thermal pretreatment. The preexponential factor is, however, correlated with the activation energy.

In this paper, the freezing of the cooperative, large-amplitude motion at the glass transition temperature is analyzed with the newly developed temperature-modulated differential scanning calorimetry (TMDSC) (*1*). This technique adds determination of the time-dependence to the thermal analysis capabilities. Using irreversible thermodynamics and kinetic models, kinetic parameters and their dependence on the nature of the glass can be determined. Data on atactic polystyrene and amorphous and semicrystalline poly(ethylene terephthalate)s are used as first examples (*2-4*). Activation energies and preexponential factors were gained by analyzing first-order kinetics of quasi-isothermal measurements (*5*) in the glass transition region. These parameters are then applied to interpret the reversing, apparent heat capacity from TMDSC with an underlying heating rate $<q>$ (*6*) using model calculations (*7*).

---

[3]Current address: Toray Industries, Inc., Otsu, Shiga 520, Japan.

The Method of TMDSC

The instrument for the research is a heat-flux-based colorimeter of TA Instruments, Inc. (MDSC™) (*1*). It is modulated at the block temperature, T_b, with a sinusoidally changing amplitude, governed by the temperature measured at the sample position, T_s:

$$T_S = T_0 + \langle q \rangle t + A \sin(\omega t + \varepsilon) \tag{1}$$

where T_0 is the initial temperature; $<q>$, the underlying heating rate set for the experiment; A, the maximum modulation amplitude of the sample temperature, also fixed for each experiment; ω, the given modulation frequency represented by $2\pi/p$, with p representing the modulation period in s; and ε, the phase difference of the sample temperature relative to the set reference frequency.

The main measured quantity discussed in this paper is the *reversing heat flow*. It is proportional to the temperature difference between reference calorimeter (empty) and sample calorimeter ($\Delta T = T_r - T_s$). Representing the instantaneous heat flow as a Fourier series with ν representing an integer running from 1 to ∞ and p:

$$HF(t) = b_0 + \sum_{\nu=1}^{\nu=\infty} \left[a_\nu \sin \frac{2\pi\nu}{p} t + b_\nu \cos \frac{2\pi\nu}{p} t \right] \tag{2}$$

where b_0 is the *total heat flow* ($= <HF(t)>$), averaged over full modulation periods, so that any modulation effect of the chosen frequency ω and its higher harmonics vanishes. The maximum reversing heat-flow amplitude is the first harmonic contribution, given as:

$$\langle A_{HF}(t) \rangle = \sqrt{a_1^2 + b_1^2} \tag{3}$$

and extracted from the experimental data by the supplied software. The reversing heat capacity is then (with $A_\Delta \propto A_{HF}$, and K the appropriate heat-flow constant) (*6*):

$$|(C_s - C_r)| = \frac{A_\Delta}{A} \sqrt{\left(\frac{K}{\omega}\right)^2 + C_r^2} \tag{4}$$

where C_s and C_r are the sample and reference heat capacities, respectively.

Typical run parameters are maximum modulation amplitudes A of 0.5 to 1.5 K and modulation frequencies of 0.06 to 0.2 radians s^{-1} (p = 100-30 s). For *quasi-isothermal runs*, $<q>$ is zero (*5*), so that the modulation is about a fixed T_0. Separate experiments are done at different values of T_0 to cover the glass-transition range. At each T_0 sufficient time is spent to reach steady state and collect statistically significant data for an additional 10 min. Data for fully amorphous PET and PS are reported in (*13*), and a wide range of partially crystallized and drawn films of PET in (*4*). *Standard TMDSC runs* have heating rates $<q>$ of 0.1 to 2.5 K min^{-1} and must be analyzed considering the two time scales, that of the continuous heating and the modulation frequency.

Quasi-Isothermal TMDSC in the Glass-Transition Region

Heat capacities of liquids can be divided into a fast-responding part due to vibrations, C_{p_o}, and a slow, cooperative part due to large-amplitude motion that leads to an equilibrium number, $N*$, of configurations of energy ε_h. The value of $N*$ can be taken as an internal variable of the system (8):

$$C_p(liquid) = C_{p_o} + \varepsilon_h\left(\frac{dN*}{dT}\right) = C_{p_o} + \varepsilon_h \alpha \tag{5}$$

In the glass transition region the approach to equilibrium may be approximated by first order kinetics as long as the distance from equilibrium is small. The instantaneous number of high-energy configurations is represented by N and the relaxation time by τ:

$$\left(\frac{dN}{dT}\right) = \frac{1}{\tau}(N*-N) \tag{6}$$

Under quasi-isothermal conditions and at steady state, the solution of Equation (6) can be written with constants $A_N = A\alpha/N_o$ and $A_\tau = A\varepsilon_j/(RT_o^2)$, where ε_j is the activation energy for the formation of the high-energy configurations, assumed to be describable by an Arrhenius expression [$\tau = B \exp \varepsilon_j/(RT)$] (9):

$$\frac{N - N_o^*}{N_o^*} = \frac{A_N A_\tau}{2} + (A_N + A_\tau)\cos\omega\sin(\omega t - \gamma) - \frac{A_N A_\tau}{2}\cos 2\beta \cos 2(\omega t - \beta) \tag{7}$$

From Equations (2) and (3) it is obvious that the reversing heat capacity, Equation (4), makes use only of the middle term on the right-hand side of Equation (7). The first term is constant with time and contributes only to b_o, the last is a second harmonic and contributes only to a_2 and b_2 of Equation (2). The phase shift γ is linked to the relaxation time τ at T_o via $\tan\gamma = \omega\tau$, and the apparent heat capacity, which is measured as the reversing C_p, is equal to:

$$C_p^\#(liquid) = C_{p_>} + N_o\varepsilon_h[(A_N + A_\tau)/ A]\cos\gamma \tag{8}$$

From this equation one sees that the large-amplitude motion contributes fully in the liquid state ($\gamma = 0$) and not at all in the glassy state ($\gamma = \pi/2$). The glass transition temperature, defined at the temperature of half-vitrification or devitrification, occurs at $\gamma = \pi/3$. An example of the data treatment as indicated in Equation 8 is shown in Figure 1.

106

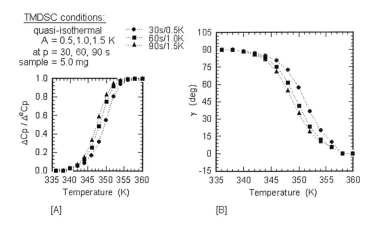

Figure 1 Quasi-isothermal analysis of the glass transition of poly(ethylene terephthalate).

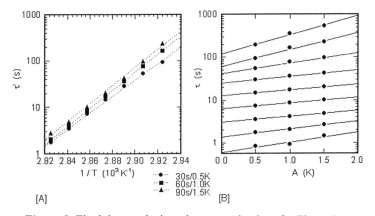

Figure 2 Final data analysis and computed values for Figure 1.

The curves of Figure 1A are plotted with $C_p^{\#}(\text{liquid}) - C_{po} = \Delta C_p$. The experimental data of ΔC_p were, in addition, first normalized to the equilibrium difference of the liquid and vibrational heat capacities. Equation (8) allows then the transformation to γ shown in Figure 1B. Using the values of γ of Figure 1B, a plot of $\ln \tau'$ vs. $1/T_o$ can be drawn, as shown in Figure 2A {$\cos \gamma = (1/\tau)/[(1/\tau)^2 + \omega^2]^{\frac{1}{2}}$ }. Clearly, different modulation amplitudes give different average values of τ and, on extrapolation, different activation energies. This observation points to an important difference between TMDSC and dynamic-mechanical analysis, DMA. In DMA the stress or strain is modulated, keeping the temperature, and with it the relaxation time, constant (as long as the strain is sufficiently small to keep the sample in the range of linear viscoelasticity). In TMDSC even changes in temperature as small as 1 K move the experiment out of the range of linear response. It is thus necessary to extrapolate the data to zero modulation amplitude A, as shown in Figure 2B. Next Figure 3A shows the extrapolation of the relaxation times at zero amplitude to give ε_j and the pre-exponential factor B. With B and ε_j, the apparent heat capacity can be calculated for any frequency, as shown in Figure 3B. Carrying out this data analysis for a number of different amorphous and semicrystalline glasses gives the parameters B and ε_j listed in Table I (3,4).

Table I. Glass Transition Parameters for PS and PET

Sample (type and treatment, w_c = crystallinity)	ε_j (kJ/mol)	B (s)
PS, amorphous, no special pretreatment	345.5	1.88×10^{-46}
PET, amorphous, melt-quenched	328.19	5.59×10^{-49}
PET 8% w_c by cold cryst. 1 h at 370 K	350.57	2.76×10^{-52}
PET 17% w_c by cold cryst. 1.5 h at 370 K	329.74	3.98×10^{-49}
PET 26% w_c by cold cryst. 2 h at 370 K	173.31	2.55×10^{-25}
PET 44% w_c by cooling from the melt, 5 K/min	152.85	2.45×10^{-22}
PET film, biaxially drawn, 42% w_c	78.44	1.78×10^{-10}

The table illustrates the large change of ε_j with pretreatment, correlated to the common observation that crystallization and drawing broadens the glass transition region. Another observation is that the activation energies and the preexponential factors are strongly coupled. They can be written for PET as:

$$\tau = \tau_1 \exp\left[\varepsilon_j\left(\frac{1}{RT} - \frac{1}{RT_1}\right)\right] \tag{9}$$

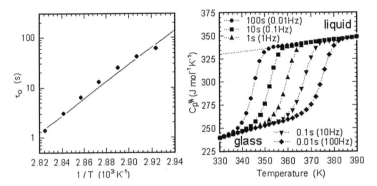

Figure 3 Data analysis for amorphous PET.

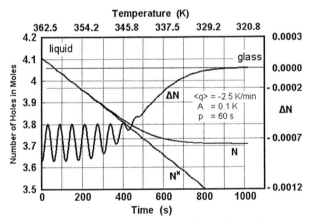

Figure 4 Computed change of high-energy configurations using Data of Table 1.

where $\tau_1 = 132.5$ s and $T_1 = 341.1$ K. The temperature parameter T_1 is only a little lower than the lowest measured glass transition (346.5 K). From the simple kinetics, one expects close correspondence only in the vicinity of equilibrium. The empirical corrections developed over the years to account for the asymmetry of approach to equilibrium and for the cooperativity of the large-amplitude motion (TNM, VF, WLF, KWW equations) (10) need, at present, too many parameters to be fitted quantitatively to the experimental data. Qualitative agreement has been achieved with assumed parameters (11).

TMDSC Compared to Model Calculations

In standard TMDSC, an additional underlying heating rate complicates the analysis of Equation (6). It takes now the form (12):

$$\left(\frac{dN}{dT}\right) = \frac{N_o^*\left(1 + A_N \sin \omega t + q_N t\right) - N_o}{\tau_o\left(1 - q_\tau t - A_\tau \sin \omega t\right)} \tag{10}$$

with the two new parameters describing the changes due to the underlying heating rate $[q_N = <q>\alpha/N_o^*$ and $q_\tau = <q>\varepsilon_j/(RT_o^2)]$. Although possible, the solution of Equation (10) is rather cumbersome and numerical solutions are more convenient. Figure 4 shows the numerical integrations of the changes of N with time and temperature for amorphous PET for the given TMDSC parameters as expressed by Equation(10). The curve ΔN is the change in N per second, the step of the numerical integration, and N^* is calculated from the equilibrium, given in Equation (5) (24).

Figure 5 shows the first harmonic of the solution of Equation (10) which represents the reversing heat capacity as computed by the TA Instruments, Inc. software [see Equations (3) and (4)]. The heavy line is the total C_p, as derived from b_o of Equation (2). Outside of the glass transition this heat capacity is equal to the heat capacity measured with standard DSC of the same cooling rate $<q>$. In the glass transition region it only approximates the apparent heat capacity of standard DSC because of contributions of the type seen in Equation (7). The reversing C_p decreases at higher temperature than the total C_p because of its faster time scale of measurement. The bell-shaped curve is the difference, between the total and the reversing heat capacity, called the nonreversing C_p. This nonreversing C_p is not the dissipative part of the complex heat capacity, although it has a similar shape.

Figure 6 illustrates that the second harmonics is a minor, but not negligible correction. Subtracting the second harmonic from the total C_p yields a small change from the total heat capacity, as shown in the figure.

Of additional interest are the remaining small ripples of the various C_p-plots. The deconvolution using Equation (2) should have removed all periodic contributions of frequency ω and higher harmonics. Inspection of Equation (10) shows, however, that the underlying heating rate causes a small frequency shift of the type of a Doppler effect, as found in the analysis of sound from moving sources, quantitatively assessable through the model calculations (9). On heating, the oscillation frequency of the heat flow into the

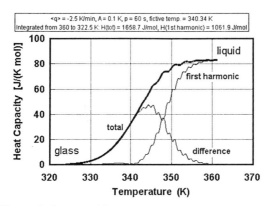

Figure 5 Computed heat capacity for amorphous PET.

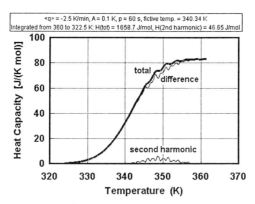

Figure 6 Second harmonic of the heat capacity contribution.

sample is higher than ω, because of the increasingly faster response caused by the underlying heating rate <q>. The reverse is true on cooling (12). Trying, then, to represent the heat flow with the Fourier series [Equation (2)] with the slightly different frequency ω gives rise to the observed ripples. Experimental data may not show the ripple because of the additional smoothing by the commercial software, omitted in the presentation of the model calculations in Figures 5 and 6.

Figure 7 shows four sets of experimental data on heating and cooling, compared with the quasi-isothermal measurements derived as for Figure 3. At low <q> the reversing C_p approaches the quasi-isothermal data. At increasing <q> the results from cooling and heating experiments separate increasingly. Close to the liquid state, model calculations, as in Figures 4-6, correspond closely to the experiment. The cross-over at larger distance from the liquid (equilibrium) state is, however, not modeled (12). It is caused by an "autocatalytic" effect on heating and a "self-retarding" effect on cooling as has been found also by DMA and volumetric experiments about the kinetics of the glass transition (10). This is clear evidence of the cooperative nature of the glass transition that needs to be corrected by a more detailed kinetics expression than given in the present description.

A similar effect that illustrates the need to introduce a cooperative kinetics which uses a relaxation time in Equation (10) that depends not only on temperature, but also on the number of frozen high-energy conformations is shown in Figure 8. The curves represent, as in Figure 7, TMDSC traces on heating with a fixed underlying heating rate. The different glasses of polystyrene were produced by annealing at several temperatures in the glass transition for various times (2). This process changes the frozen high-energy configurations N at the beginning of the heating experiment. One can clearly see that the better annealed samples of smaller N show a higher T_g. The differences disappear as equilibrium in the form of the liquid state is approached, and are relatively small.

The apparent total heat capacity, as represented by b_o, contains the enthalpy relaxation in form of an endotherm. This well-known hysteresis effect (8) can be separated to the degree of precision of the representation of Equation (10) by the TMDSC software, as is shown in Figure 9 (2). The apparent, nonreversing heat capacity is the total heat capacity minus the apparent, reversing heat capacity, shown in Figure 8. The major contribution of the endotherms arises from non-modulated relaxation of N on heating, as long as ΔN of Figure 4 is much larger than the modulation-caused changes and the glass-transition temperature is approached.

Complex Thermodynamic Quantities

The similarity of TMDSC and dynamic-mechanical analysis (DMA) raised the question whether a complex heat capacity would be of use in analogy to the stress/strain ratio (14). One writes for the reversing heat capacity of Equation (4) the complex expression for C_p as:

$$C^* = C' - iC'' = C_o \exp(-i\theta) \qquad (11)$$

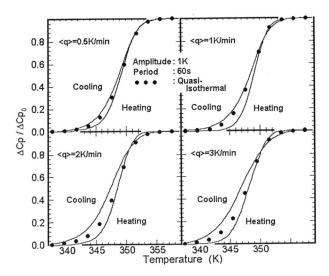

Figure 7 Relative heat capacity of PET by standard TMDSC.

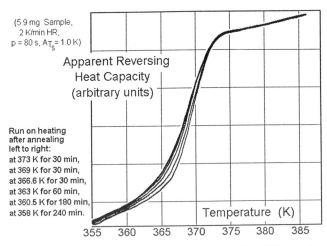

Figure 8 Apparent reversing heat capacity of PS after setting different thermal histories.

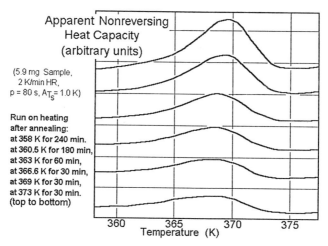

Figure 9 Apparent non-reversing heat capacity of PS for samples as in Figure 8.

where C' is the real part, or the instantaneous response, to a change in temperature (analogous to the storage modulus of DMA), and C'' is the imaginary, dissipative, delayed part (analogous to the loss modulus of DMA). It follows that the absolute amount of the complex heat capacity is:

$$|C_o| = \sqrt{(C')^2 + (C'')^2} \qquad (12)$$

As for the modulus in DMA, which is the ratio of peak stress to peak strain, C_o is proportional to the ratio of peak heat flow to peak temperature-amplitude (multiplied with the square-root expression) and equals the reversing C_p [see Equation (4)].

In a sample with an apparent C_p, one determines C_o via Equation (4), which can be separated into real and dissipative components with a phase angle θ. [The phase angle θ is different from γ of Equation (8), since it considers only the small temperature range of ±A, and is derived from the lags of ΔT and T_s, δ and ε, respectively, after correction for instrument effects]. The real part of the heat capacity accounts for a storage part. Its heat is transferred in-phase with the temperature change. The slow kinetics given by Equation (6) limits the heat transfer. Not all of the high-energy configurations N^* are created or collapsed in the given time interval. As temperature and time change with modulation, the additional configurations needed to reach internal equilibrium (N^*) contribute to the heat flow at a higher or lower temperature, accounting for the out-of-phase heat flow.

The real parts of the heat capacity is linked to the enthalpy gain of the sample that behaves as if the heat flow were instantaneous. The integral of this part of the complex heat capacity over one modulation period is obviously zero. The question remains what is the imaginary part? In the case of the DMA experiment, the complex modulus is linked to work done on and recovered from the sample. By integration of the complex stress σ over the sinusoidal strain ε, one finds the energy dissipated over one cycle, $\Delta W''$, is $G''\varepsilon_o^2\pi$, while $\Delta W' = 0$. The integral for heat dissipated over one cycle, in turn would be: $\_C_o \sin(\omega t - \theta)\, d(A \sin \omega t)$ with θ representing the phase lag between heat capacity and temperature. This results in an enthalpy of $A\pi C'' = A\pi C_o \sin \theta$ from the out-of phase part of C^*, i.e. the imaginary part of the heat capacity violates the first law of thermodynamics and, thus, is not a valid assessment of the enthalpy change. To check on the enthalpy balance, the *heat flow* must be integrated over one period, not the complex heat capacity C^* over T_s. One may, however, interpret the enthalpy exchange at $T_s(t)$ as an entropy production, since C'' is exchanged largely at a different than the equilibrium temperature.

A more detailed discussion of the importance and history of the complex heat capacity has been given recently at the Fourth Lähnwitz Seminar and will be published in *Thermochimica Acta*. Further clarification of this question based on a detailed derivation of the irreversible thermodynamics of the glass transition is in preparation and expected to be published in 1998 (*15*).

Conclusions

With the development of TMDSC it has become possible to study the kinetics of the freezing and unfreezing in the glass transition region. As one would expect, limitations exist in the generation of quantitative information. Quasi-isothermal data extrapolated to zero temperature-modulation amplitude need to be generated to characterize a sample, as illustrated in Figures 1-3. This extrapolation to make the kinetic expression of Equation (6) linear, corresponds to the limits of the description of DMA to linear viscoelasticity.

Using the kinetic parameters of Table 1, one can attempt to calculate the thermal behavior in the glass transition region by numerical integration of Eq (10), as shown in Figures 4-6. These analyses clarify the limitations of present TMDSC practices. Close to the liquid state, Equation (10) with temperature dependent τ and N^* is valid. Further from equilibrium, deviations occur, which are based on the cooperativity and asymmetry not included in Equation (6), but are seen in the cross-over of the TMDSC curves of Figure 7. At least one more parameter is needed to account for the change of the relaxation time with N, as also seen in Figure 8.

Interpretation of Equations (7) and (10) permits further, in connection with Figures 4-7, to establish the limits of using the first harmonic of the Fourier series Equation (2) as reversing heat flow and as a tool to establish the enthalpy relaxation, as shown in Figure 9. The constant and second harmonic terms of Equation (7) are not included in the reversing C_p. The first is included in the total C_p, the latter in the nonreversing C_p. Furthermore, a small change of the oscillation frequency when using a non-zero $<q>$ causes a periodic mismatch of the separation of the first harmonic (ripples in Figures 5 and 6).

Overall, the analysis of the TMDSC and quasi-isothermal TMC of polystyrene and PET have shown that as a first approximation the reversing heat capacity is the value of the complex heat capacity at the given frequency [Equation (12)]. Although it is possible with recent software modifications to directly determine the phase angle for a separation of the dissipative and real parts of the heat capacity, the resulting data are approximations because of the loss of linearity in the glass transition region. Important progress is expected in the understanding of the glass transition by studying these small deviations from linearity and comparing the data with mechanical and dielectric modulated techniques on the same, well-characterized materials.

Acknowledgments

This work was supported by the Division of Materials Research, National Science Foundation, Polymers Program, Grant # DMR-9703692 and the Division of Materials Sciences, Office of Basic Energy Sciences, U.S. Department of Energy at Oak Ridge National Laboratory, managed by Lockheed Martin Energy Research Corp. for the U.S. Department of Energy, under contract number DE-AC05-96OR22464.

Literature Cited

1. Reading, M. *Trends in Polymer Sci.*, **1993**, *8*, 248.
2. Boller, A.; Schick, C.;Wunderlich, B. *Thermochim. Acta*, **1995**, *266*, 97.
3. Boller, A.; Okazaki. I.; Wunderlich, B. *Thermochim. Acta*, **1996**, *284*, 1.
4. Okazaki, I.; Wunderlich, B. *J. Polymer Sci.: Part B: Polymer Phys.*, **1996**, *34*, 2941.
5. A. Boller, A.; Jin, Y.; Wunderlich, B. *J. Thermal Analysis*, **1994**, *42*, 307.
6. Wunderlich, B.; Jin, Y.; Boller, A. *Thermochim. Acta*, **1994**, *238* , 277.
7. Wunderlich, B. *J. Thermal Anal.*, **1997**, 48, 207.
8. Wunderlich, B.; Bodily, D. M.; Kaplan, M. H. *J. Appl. Phys.*, **1964**, *34*, 95.
9. Wunderlich, B.; Boller, A.; Okazaki, I.; Kreitmeier, S. *J. Thermal Analysis*, **1996**, *47*, 1013.
10. Matsuoka, S. *Relaxation Phenomena in Polymers*. Hanser Publishers, Munich, Germany, 1994.
11. Hutchinson, J. M.; Montserrat, S. *Thermochim. Acta*, **1997**, *47*, 263; and *J. Thermal Analysis*, **1996**, *47*, 103.
12. Thomas, L. C.; Boller, A.; Okazaki, I.; Wunderlich, B. *Thermochim. Acta*, **1997**, *291*, 85.
13. Wunderlich, B.; Okazaki, I. *J. Thermal Analysis*, **1997**, *49*, 57.
14. see, for example, Schawe, J. E. K. *Thermochim. Acta*, **1995**, *261*, 183.
15. Baur, H.; Wunderlich, B. *J. Thermal Analysis*, submitted.

MOLECULAR MOBILITY AND RELAXATIONS IN THE GLASSY STATE

Chapter 8

Polymer Glasses: Thermodynamic and Relaxational Aspects

Robert Simha

Department of Macromolecular Science, Case Western Reserve University, Cleveland, OH 44106–7202

Starting in the melt, proceeding to the transition region and continuing into the glassy state, sets of equilibrium, and non-equilibrium processes are considered. We examine the consequences of a unified view derived from a lattice-hole model, involving a hole fraction h to account for the structural disorder. The role of h as a free volume quantity is explored in various processes. A comparison of h with a free volume quantity directly measured by positron spectroscopy shows a gratifying concordance in the melt. Possible resolutions of differences in the glass are examined. Correlations between equation of state and other physical properties in melt and glass with h as the connecting bridge are extended to time dependent processes.
Finally we discuss relaxation or physical aging processes with relaxing free volume functions and their distributions as the determinants. Based on stochastic formulations, simulations of the cooling process from melt to glass yield T_g as a function of cooling rate and most importantly, of the characteristic interaction parameters of the polymer.

The purpose of this paper is to discuss certain property sets from a unified point of view, which results in correlations between such sets. This point of view is based on a particular statistical model of a molecularly disordered state. We start with the melt, proceed to the glass transition region, and thence to the glassy state. Here the concerns are both the steady state and the relaxational regimes.

Equilibrium Melt: Equation of State

The fluid is modeled as an assembly of lattice sites occupied by chain segments in contact with a fraction h of empty sites or holes. The h-function is to account for the temperature, pressure (and stress) dependent structural disorder in the system. What is required is the configurational free energy F *(1)*

$$F = F[V, T; h(V, T)] \qquad (1)$$

The equation of state (eos) is derived from the two coupled equations

$$P = -(\partial F / \partial V)_T ; (\partial F / \partial h)_{V,T} = O \qquad (2)$$

The second of these equations eliminates h and expresses the required minimum condition on F at equilibrium. We omit explicit expressions. *(1)*

The Characteristic System Parameters. The intersegmental interaction potential is defined by a minimum attraction ε^* and a repulsion volume υ^*. A quantity 3c is to indicate the number of volume dependent degrees of freedom of the molecule. This parameter will be of the order of the chainlength s in a flexible chain. In terms of these the variables of state can be scaled through the combinations

$$P^* = qz\varepsilon^* / s\upsilon^* ; T^* = qz\varepsilon^* / kc \qquad (3)$$

with qz the number of intermolecular contacts of the s-mer in a lattice of coordination number z. Scaling and in effect the theory's quantitative success without additional assumptions rest on the constancy of these parameters. This requirement is obvious for ε^* and υ^*. As has been demonstrated by now for many polymer melts, *(2)* the theory is quantitatively accurate and thus the scaling parameters are known, all this with the implication of constant ε^*, υ^* and moreover c. Departures are seen in the third derivatives of the partitions function, such as the T-dependence of the thermal expansivity. *(3)*

The h-Solution. At atmospheric pressure, or scaled pressure $\tilde{P} \to O$, equation 2 yields *(3)*

$$h = [V - K(T / T^*)V^*]V \qquad (4)$$

with K a very slowly varying function of temperature, $K \approx 0.95$. Equation 4 clearly displays the character of a free volume fraction with an excluded volume KV^*. In what follows we shall have occasion to adopt this view of the h-function. We note moreover the close relationship between V^* and Bondi's hard core van der Waals volumes derived from structural geometry and atomic volumes, *(4)* and consistent with the definition of υ^*.

The Transition Region

An important point is clearly the behavior of the h-function at the boundary between melt and glass as seen in the V - T plane under different pressures. All properties in the non-equilibrium glass however are not uniquely defined but depend on the formation history. Granting "rapid" experimentation and thus negligible relaxation, and a fixed cooling rate, different histories may be distinguished by differences in the pressure history. In particular consider on the one hand the application of a series of pressures in the melt followed by cooling into the glassy domain. This generates a series of different glassy structures and a pressure coefficient dT_g/dP. As an alternative cool the melt under a single pressure and then subject the glass so formed to a series of pressures. Thus a glass of a single structure, characteristic of the formation pressure results, with a pressure coefficient dT_g^+/dP and $T_g^+ \neq T_g$ except in special cases. It has been shown *(5)* that the first Ehrenfest relation

$$dT_g^+ / dP = \Delta\kappa / \Delta\alpha \qquad (5)$$

must hold, with Δ indicating changes along the transition line and κ and α isothermal compressibility and isobaric thermal expansivity respectively. The relation between the two pressure coefficients is given by (6)

$$dT_g / dP = dT_g^+ / dP - \kappa' / \Delta\alpha \qquad (6)$$

To define κ' we recall the classical densification experiment where a glass is formed under a pressure P' and, once having arrived sufficiently far in the glassy domain, is depressurized. Due to the memory effect the final volume is smaller at the identical T and P than the virgin glass. Thus there exists a compressibility κ' defined as (7)

$$\kappa' = -(1 / V)(\partial V / \partial P')_{T,P} \qquad (7)$$

Equations 5-7 provide a test of the theory, based on the behavior of the h-function. From the eos we obtain (6,7)

$$dT_g^+ / dP = (\partial T / \partial P)_h F_P / F_T \qquad (5')$$

where $F_T = 1 - (\partial h / \partial T)_{P,g} / (\partial h / \partial T)_{P,\ell}$ and an analogous definition of F_P. F_P and F_T represent the degree of freeze-in of the h-function at T_g, (3) with g and ℓ indicating glass and liquid respectively. The h derivatives in the melt are obtained by evaluating equation 2. For the glass we continue to operate with the lattice hole model, and as a minimum requirement, discard the equilibrium condition. Thus equation 2 now becomes

$$-P = (\partial F / \partial V)_{T,h,c} + (\partial F / \partial h)_{V,T,c}(\partial h / \partial V)_T$$

$$+(\partial F / \partial c)_{V,T,h}(\partial c / \partial V)_T \qquad (2')$$

In equation 2' we have allowed for the possibility that h is not frozen at T_g and also that c does not remain constant in the glassy state (see below). Making the assumption that c = const., we invoke experiment to derive h (3,8). Figure 1 illustrates the result for 2 PVAc glasses generated under pressures of 1 and 800 bar respectively. In the melt there is full coincidence between theory and experiment. That is, the h-values extracted from experiment by means of the pressure equation 2 agree with the solid line defined by the minimization of the free energy. In the glass, the black circles indicate results of equation 2' as applied to the eos of the two glasses. The open circles represent a simplification which neglects the second term on the r · h· s. The important conclusion to be drawn from this and other examples is that in terms of the h-function, T_g is not an iso-free volume temperature. From the intermediate location of the "glassy" points between extrapolated liquid and "frozen" i.e. horizontal lines F_T and by analogous constructions F_P can be determined. Since at T_g^+ T, P, h are continuous, it can be shown (7) that F_P should equal F_T. For the two instances where experimental information is available, i.e. PVAc (8) and selenium, (9) this is found to be valid in good approximation. Hence, according to equation 5', h should be constant along the T_g^+ - P transition line. Indeed this is found to be the case, *(10)* notwithstanding the fact that h is <u>not</u> frozen in the glass. The utility of equation 5' in combination with equations 5 and 6 is seen by computing the densification rate κ' without requiring recourse to $\Delta\kappa$ and thus the eos of the glass. McKinney has presented a broad investigation of several combinations of experimental data with theory for 23 polymers of widely varying T_g. *(7)* Values of κ' are also employed to estimate densification rates of refractive indices.

Freeze-in at T_g. Equation 5' introduced the freezing fraction F_T. It has been examined for a series of polymers ranging from poly (dimethyl siloxane) to poly (α-methyl styrene) *(3)*. Figure 2 shows a linear correlation relation with Tg, at atmospheric pressure. Most recently, Utracki *(11)* has investigated the effect of pressure on F_T, based on V-T data between 100 and 4000 bar for a single polymer, polystyrene. *(12)* It is noteworthy that the line in Figure 2 also represents an adequate linear fit to the T_g- dependence generated by pressure variations. The general conclusion is the requirement of a larger mobile hole or free volume fraction to pass into the liquid state at a larger T_g.

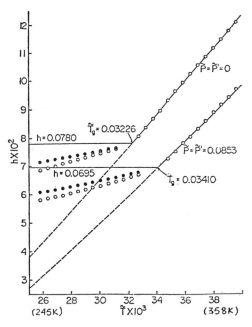

Figure 1.   Hole fraction h as function of (scaled) temperature for 2 poly(vinylacetate) melts and their glasses formed under pressures of 1 and 800 bar. Black circles, equation 2', open circles, equation 2 for melt and approximation for glass (see text). Horizontal lines mark the magnitude of h at T_g, as derived from melt theory. Reproduced from ref. 8. Copyright 1977 American Chemical Society.

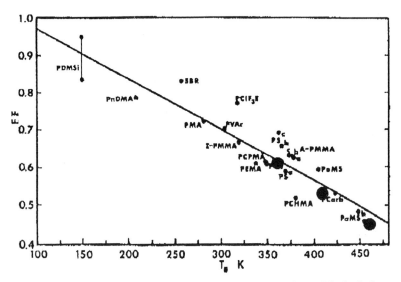

Figure 2.   Correlation of frozen fraction FF = F_T and T_g. Black circles polystyrene under different pressures. Adapted from refs. 3 and 11.

Heat Capacity. A direct test of the theory is the difference $C_p - C_v = TV\alpha^2/\kappa$. For PVAc at $T_g = 304K$ we find experimental values of 0.261 and 0.069 J/gK respectively, compared with calculations of 0.250 and 0.055 for liquid and glass respectively. Considering the eos derivatives and the numerical solutions of equation 2' required with the input of experimental data, these findings are to be considered satisfactory. The next quantities of interest are the heat capacities C_p and C_v. These are anticipated to be smaller than is observed, since we are counting only the contributions of volume dependent degrees of freedom. Indeed, fractions of less than 20% and even smaller for C_v are derived. These ratios are further reduced in the glassy state. Assuming that only the 3c degrees of freedom referred to earlier contribute to the change ΔC_p at T_g, we find that only 47% of ΔCp are accounted for in PVAc. This Cp-deficit implies a deficit in entropy fluctuations and in ΔS. It suggests a consideration of the entropy relation corresponding to equation 2', i.e.

$$-S = (\partial F / \partial T)_{V,h,c} + (\partial F / \partial h)_{V,T,c}(\partial h / \partial T)_V$$

$$+(\partial F / \partial c)_{V,T,h}(\partial c / \partial V)_T \qquad (2\text{''})$$

with the temperature and volume or pressure coefficients of c now changing at T_g. Accordingly then, to the input of eos must be added entropy contributions. These can be derived by means of the standard relations

$$(\partial S / \partial P)_T = -(\partial V / \partial T)_P ; C_p = T(\partial S / \partial T)_P \qquad (8)$$

Integration of these relations will yield the configurational entropy S (P,T) based on the eos and the temperature dependence of the configurational portion of C_p. Equations 2' and 2'' then provide the functional dependence of both h and c. The problem is the C_p contribution. We postpone a further discussion and resume the subject below in connection with results of positron spectroscopy.

Steady State Glass: EOS and Related Properties

We have discussed the application of the lattice hole model which extracts the structure function h from the experimental eos, see equation 2'. The consistency of the approch is now to be explored by the prediction of several other properties. If successful this then establishes sets of property correlations. Unless otherwise indicated the assumption of a constant c is maintained.

Density Fluctuations. A formal statistical mechanics in non-equilibrium states has been developed by Fischer and Wendorff *(13)* in terms of the affinity A as a conjugate quantity to an ordering parameter. Their result is:

$$< \delta\rho^2 > / \rho^2 = (kT / N^2) / (\partial\mu / \partial N)_{T,V,A} \qquad (9)$$

with μ the chemical potential and N the number of molecules in a volume V. At equilibrium A = 0 and equation 9 reduces to the familiar result

$$< \delta\rho^2 > / \rho^2 = (kT / V)\kappa(T) \qquad (9')$$

with κ the isothermal compressibility of the melt. The authors suggest that not too far into the glassy domain, linearity in T would prevail and $\kappa(T)$ be replaced by $\kappa(T = T_g)$.
We pursue this matter by choosing in equation 9 h as the ordering function and thus

$$A = -(\partial F / \partial h)_{V,T}$$

With the aid of the previous analysis of equation 2', A and equation 9 can be evaluated. *(14)* Experimental data have been presented by Wiegand and Ruland. *(15)* In Figure 3 results for polystyrene are depicted. We note first the agreement between theory and experiment in the melt. Below T_g there are four items. First, the experiment. Second and third, the theory in two versions, one of which simplifies the solution of equation 2' by neglecting the second term on the r $\cdot$ h $\cdot$ s $\cdot$, and finally the solid line from equation 9' with T = T_g. The latter tends to overestimate the fluctuations, but the theory performs satisfactorily. We turn next to two properties which emphasize the character of the h-function as a free volume quantity.

Thermo-Elasticity. This concerns the elastic moduli as functions of T and P, and of course, formation history. A single elastic constant needs to be computed, once the bulk modulus is known. The theory *(16)* considers two consequences of an anisotropic stress. First the distortion of spherical into ellipsoidal cell symmetry under an uniaxial strain. A second contribution to the free energy arises from changes in the free volume function h. At temperatures below 40-50K where h is practically frozen, *(17)* only the first effect is significant. Figure 4 displays the results for one of the glasses seen in Figure 1, i.e. shear (G), Young (Y) and Poisson (μ) constants. The theory predicts an increase in G and Y in the densified glass at the identical T and P. *(16)* A successful comparison between experiment *(18)* and prediction based on eos data for PMMA *(3,19)* has been presented.

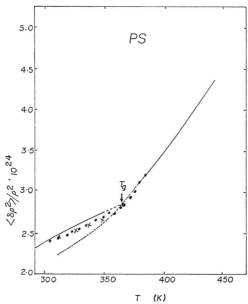

Figure 3. Density fluctuations in polystyrene melt and glass. Black circles, experiment, ref. 15. Crosses, theory, equations 2' and 9. Dotted line, approximation in equation 2' (see text). Solid line, equation 9', $T = T_g$. Reproduced from ref. 14. Copyright 1982 American Chemical Society.

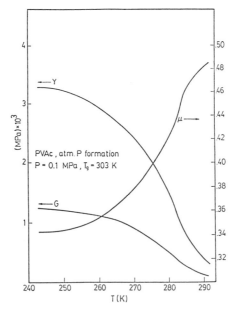

Figure 4. Shear (G) and Young (Y) moduli, and Poisson constant μ of a poly (vinyl acetate) glass as functions of temperature at atmospheric pressure. Reproduced from ref. 16. Copyright 1988 American Chemical Society.

Positron Annihilation Life Time Spectroscopy (PALS) and Free Volume.
In the present context the essential feature is the use of the o-positronium
species as a structural probe. The pertinent relation is:

$$h_p = C I_3 < v_f(\tau_3) > \qquad (10)$$

Here h_p is the volume fraction of cavities, (v_f) the mean cavity volume, a
function of the probe's lifetime τ_3, I_3 the corresponding intensity and CI_3 the
number density of cavities. When h_p is measured as a function of temperature,
it becomes relevant to inquire into possible connections between h_p and h. An
assumed identity at one temperature determines the proportionality factor C
and allows for comparisons over the range of temperatures in melt and glass.
Results for four polystyrene fractions *(20)* are seen in Figure 5. In the melt
there is once more full concordance between experimental and theoretical free
volumes. In the glass we observe two features. The probe assesses a
transition, but T_g is lowered by 2-7 degrees. Most importantly, h_p is lower
than h as obtained from the eos, and equation 2'. Various conjectures
regarding the origins of this discrepancy have been suggested. *(20)* Here
however we wish to focus on the theory and the assumption of a constant c in
the glassy state. This issue has already come up in connection with the ΔC_p
question.

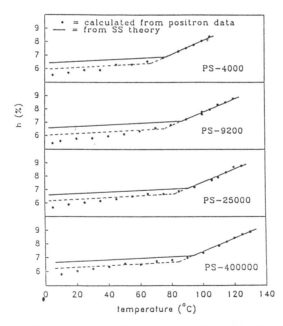

Figure 5.   Comparison of o-Ps h_p (equation 10) and eos h as functions of
temperature for 4 polystyrene fractions. Points, experiment. Solid lines,
theory, equations 4 and 2'. For explanation see text. Reproduced with
permission from ref. 20. Copyright 1994 John Wiley & Sons, Inc.

Analysis. This was undertaken by Carri *(21)* whose focus was the h_p-h discrepancy in the glass. He accepted accordingly the previous assumptions in the melt. The results there will then provide boundary conditions on any modified values of h and c then obtained.

Formally, coincidence between the two free volume functions can be established simply by rescaling the temperature for the positron experiment. This, according to equation 3 can be interpreted as a temperature dependent c with c actually increasing with T. However, in addition to the empiricism of the approach, it does not deal with the eos.

Thermodynamics. *(21)* This is to determine the volume and entropy as functions of T and P by using equation 8. It involves the eos and the temperature dependent configurational heat capacity C_p at atmospheric pressure. The polymer of choice was Bisphenol-A-Polycarbonate. The eos and positron data are available. *(22,23)* The total heat capacity C_p of the glass is linear in T. *(24)* The assumption then is a corresponding expression for C_p (config.).
Thus one has

$$c_p(\text{config.}) = \alpha T + c_p(0)$$

This, combined with the eos, determines the entropy surface. Equations 2' and 2" are then solved numerically. *(21)* The continuity conditions of h and c at T_g determine the parameter $c_p(0)$ and an integration constant in the entropy, with

α remaining a disposable parameter. An upper limit of 1.0463 is dictated by the value for the total heat capacity. The solutions for the h-function indeed are successful in generating an increased temperature coefficient and this result

is practically independent of α. Moreover, allowing for a shift of the T_g obtained from DSC to that observed in the positron experiment, improves the agreement further, as is seen in Figure 6. Solutions for the normalized c-

function at a series of α-values are displayed in Figure 7. The overall increase

of c with increasing α is expected.

In this manner it becomes possible to reconcile not only in the melt, but also in the glass thermal and positron results, by generating concordance between the free volume function h and the cavity fraction h_p. We note the maximum of c in Figure 7. It could be indicative of two opposing tendencies, namely the loosening of the structure with increasing temperature, followed by a decrease of intermolecular interactions at elevated (average) free volume levels. Now, since the original introduction by Prigogine and colleagues, the physical meaning of c has been as the number of volume dependent degrees of freedom or "soft" modes. It is difficult to understand why this number should be larger in the glass than in the melt. Mathematically, this arises from the coupling of h and c in equations 2' and 2". A reduction of h below the level appropriate for constant (liquid-like) c, is compensated by an increased level of c for a specified eos. It may be argued that a complete theory should allow for a temperature and pressure dependent c in both melt and glass, with different derivatives. The extensive comparisons with experimental eos do not require this, unless higher derivatives are to be accurately fitted. Moreover, the current form of the partition function does not yield a minimum condition on c. Additional assumptions regarding the coupling between c and h would have to be introduced. Such a coupling can be derived from a consideration of a Flory flex factor with an h-dependent rotational energy bias E, which competes with

Hole Fraction vs. Temperature

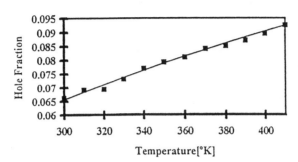

Figure 6.   Positron cavity fraction h_p for polycarbonate (Lexan) as function of temperature. Points, experiment. Line, solution of equations 2' and 2" (see text). Adapted from ref. 21.

External Degrees of Freedom vs. Temperature

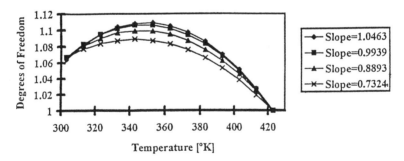

Figure 7.   Normalized c quantity as function of temperature for polycarbonate (Lexan). Lines, solutions of equations 2' and 2" for different values of α-parameter (see text). Adapted from ref. 21.

the intermolecular energy ε^*. (25) The consequences of this proposition for the glassy state, in particular the ΔC_p problem, remain to be explored. As for the differences in the PALS area, the foregoing discussion has ascribed these differences below T_g entirely to the thermodynamic theory. Is it possible, on the other hand, that radiation and rate effects come into play at low temperatures, depending on the manner of experimentation?

Dynamics: Free Volume Relations in the Melt.

We proceed to a consideration of relaxation processes. These may encompass rheological and transport or extensive thermodynamic functions. A question here is the relation between a free volume quantity f and a characteristic relaxation quantity. This question has been repeatedly addressed since Batschinski's early effort (26) in terms of viscosity and diffusion and a van der Waals free volume. In more recent times Doolittle's (27) equation for viscosity has received frequent attention. Generalizing to a mobility measure M, e.g. a diffusion coefficient, a fluidity or a relaxation frequency, the relation is

$$\ell n M / M_r = B(1/f - 1/f_r) \qquad (11)$$

B is a constant and r signifies a reference state. Two related questions arise. First, the interpretation of f. We consider the h-function and thus a connection to the configurational thermodynamics. Second there is the validity of equation 11. Utracki's analysis (28) of Newtonian viscosities for n-paraffins over wide ranges of P and T and of the eos yields master curves in 1/h but no linearity according to equation 11. Linearity can be restored in terms of a modified hole fraction $h + \Delta$. Turning next to the viscoelastic temperature shift factor, equation 11 with $f = h$ is satisfied for PMMA and PVAc melts. (29,30) An extensive series of steady state viscosities, viscoelastic and dynamic dielectric measurements for several polymers was analyzed by Vleeshouwers. (31) Determining parameters from one set of experiments yields predictions of another set. Encompassing both T and P-dependences, the result is

$$\ell n a_{TP} = B /(h + CT - A) + const \qquad (11')$$

The constant is determined by the reference state. At atmospheric pressure C = O, but there is no consistent extension into elevated pressures since the numerical values of A and B are not invariant. From this analysis it is not clear how the opposing T and P coefficients of h enter into the dynamics. We conclude that over a sufficient range of variables, a correlation between dynamics and thermodynamics through the h-function persists, but is not of the simple Doolittle form. It would be interesting in this connection to explore possible correlations between the degree of freeze-in, see equation 5' and Figure 2, and the temperature coefficient of viscosity in the approach to T_g.

Dynamics: Glassy State

With results for the steady state glass at hand, it is natural to consider again the volume and the h-function. Assuming the aging glass to pass through a series of steady states, we assume the continued validity of equation 2'. With the B parameter of equation 11 obtained for PMMA and PVAc, it becomes possible to derive a good prediction of the aging viscoelastic a_T shift factor, based on measured volume relaxation data. (29,30) It would be of considerable interest to apply this procedure to a prediction of relaxing elastic moduli, based on the theory of Ref. 16.

The last correlations rest on equation 11, applied to the thermal free volume function h under isothermal conditions. Higuchi et al. *(32,33)* have investigated isochronal conditions, by undertaking a series of stress relaxation measurements in glassy polycarbonate (Lexan) over a range of aging times and temperatures. They then examined the relation between the mean relaxation time, derived from a stretched exponential, and free volumes h_p, from equation 10, or h. The general conclusion is that equation 11 is valid under isothermal conditions, but systematically departs under isochronal conditions. Thus the temperature dependence of the free volume quantity is not the sole contributing factor, since the parameters in equation 11 are temperature dependent. However, on approaching the glass transition range from below, the viscoelastic properties are increasingly controlled by free volume, and equation 11 tends to become valid. Similar conclusions follow for PVAc. If we accept Cohen-Turnbull rationalization of equation 11, *(34)* as applied to stress relaxation, the implication is a minimum free volume requirement for the occurrence of the process, which increases with increasing temperature. *(33)*

Relaxation Theory.

So far the h-function has served to correlate thermodynamic with other properties. The next subject is relaxation theory. This subject owes much to the incisive research of the late André Kovacs and his colleagues. The issue for us is the construction of theories with the h-function as the central quantity, and a kinetics depending on the availability of free volume. A strictly molecular theory would have to provide this dependence. Instead we remain at the level reminiscent of the classical theory of Brownian motion, where macroscopic relationships are invoked (see below).

So far, only the average values of the hole fraction have received attention. In the two versions of a theory dealing with a time dependent h-function, dispersions of h are the focus.

Diffusion Theory. *(35)* The diffusion rate is determined by a Doolittle dependence, equation 11, imposed on the local free volume. The input are the eos scaling parameters, the Doolittle constants, derived from viscoelastic data in the melt, and a diffusion length scale as an adjustable parameter. The experimental information consists of volume recovery, following simple up or down temperature steps, or two steps in opposite directinons, separated by a time interval. The integrated h-function, derived from the solution of the diffusion equation, is to be compared with h extracted from the measured volume recovery. The theory is successful for the single step experiments in PVAc. In the double step experiment the location in time of the observed maximum is correctly predicted, but its height is underestimated. One may anticipate that a distribution of diffusion functions could formally correct this deficiency.

Stochastic Theory. *(36)* This version operates with free volume dispersions resulting from thermal fluctuations. The agent for the drive to equilibrium are local motions involving a characteristic number of segments. These encompass a certain volume in which the fluctuations are considered. The mean square fluctuations $<h^2>$, derived from the free energy, equations 1 and 2, and the average $<h>$ derived from the eos are assumed to determine the distribution of free volume levels. The kinetics of free volume change is described by transition probabilities. These in turn are Doolittle or WLF functions of the particular free volume level plus a contribution from the surrounding regions, which is approximated by the mean free volume. The result is a Fokker-Planck equation for the distribution function w (h,t)

$$\partial w \,/\, \partial t = \partial \,/\, \partial h \{ D(h) \partial w \,/\, \partial h - D(h) w \, d\ell n w_\infty \,/\, dh \} \qquad (12)$$

This describes the "diffusion" of w to the equilibrium function $w_\infty(h)$, governed by the diffusion (Doolittle) function D(h). The first term inside the brace represents a diffusion current. The second term arises from a conservative force as the gradient in h of a potential - $kT \,\ell \, nw_\infty$. This formulation can be extended to a study of the effect of external forces on w and hence on the macroscopic volume relaxation. So far, solutions of equation 12 have not been obtained. In the original formulation *(36)* solutions for the case of discrete free volume states were developed by Robertson, which yield series of coupled differential equations for the transition probabilities.

The first applications of the theory were concerned with volume recovery following single and multiple temperature steps. We refer for details to Ref. 36. Good agreement with Kovacs' experimental data *(37)* on PVAc is observed for the former. In the latter the agreement is semi-quantitative in reproducing the locations and heights of the experimental maxima obtained by diverse thermal histories. This situation is significantly remedied by allowing for an Arrhenius thermal activation process in addition to the structural (h) contribution to the kinetics. *(36)* We note finally that this theory leads in a natural way to a distribution of retardation times.

Further Applications. Detailed evaluations and numerical simulations of the theory are found in Vleeshouwers' thesis. *(31,38)* One example concerns the frequency and temperature dependence of the complex bulk modulus. Another is the simulation of cooling the melt over a transition region into the glassy state. To proceed, the continuous temperature profile is approximated by a series of discrete steps, depending on the cooling rate. The relaxing free volume and volume following each T-jump are then computed. In this manner the theory provides information about the effect of cooling rate on T_g. This is illustrated in Figure 8 for the case of PVAc. Most importantly, it describes the influence of the eos parameters, see equation 3, in particular of the attraction energy ε^* and of c, on T_g.

Conclusions

The structure function h obtained at equilibrium by a free energy minimization, is extracted in the glass from its equation of state for a specific formation history. This procedure yields satisfactory correlations, when the role of h as an ordering and as a free volume quantity respectively is adopted. The additional assumption is that the difference between equilibrium melt and its glass resides solely in the temperature and pressure dependence of h. This leads to a partial freeze-in of h at T_g, which is a function of T_g and approaches completeness for low T_g. The assumption however fails in accounting for the change ΔC_p at T_g. It also fails in describing the temperature dependence of the positron cavity fraction h_p in the glass, although it does so in the melt. Both discrepancies indicate that further attention should be paid to the behavior of c, the number of volume dependent, external degrees of freedom. In the melt, the equation of state has been quantitatively successful with a constant c. The results of a formal reconciliation of h derived from the equation of state and the positron h_p suggest however a modification of the

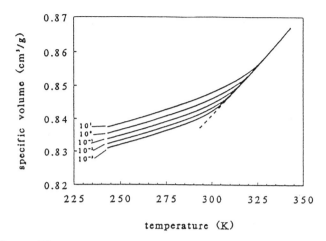

Figure 8.   Theoretical volume-temperature curves of a poly (vinyl acetate) glass as a function of cooling rate. Adapted from ref. 31.

theory which allows for a pressure and temperature dependent c in melt and glass. A coupling of c and the free volume h remains to be investigated. On the other hand, there is the possibility that the differences between h_p and h involve radiation induced rate effects, and depending on experimental conditions. The introduction of h as a free volume function in melt flow and viscoelasticity shows simple relationships over limited ranges of temperature and pressure. Connections between degrees of freeze-in and activation energies at T_g remain to be explored. Kinetic theories, based on relaxation of the h-function as the determinant, capture the essentials of the volume relaxation process following temperature or pressure changes. However the comparisons with multi-step temperature experiments suggest the existence of an activation process in addition to the structural (h) contribution. A noteworthy result of the theory, to be further exploited, is the relation between T_g and the characteristic molecular interaction parameters and the c-factor, and thus the physical properties of the melt.

Literature Cited

1.   Simha, R.; Somcynsky, T., *Macromolecules* **1969**, 2, 342.
2.   Rodgers, P. A., *J. Appl. Polym. Sci.* **1993**, 48, 1061.
3.   Simha, R.; Wilson, P. S., *Macromolecules* **1973**, 6, 908.
4.   Simha, R.; Carri, G. A., *J. Polym. Sci. Part B: Polym. Phys.* **1994**, 32, 2645.
5.   Goldstein, M., *J. Phys. Chem.* **1973**, 77, 667.
6.   McKinney, J. E., *Ann. N.Y. Acad. Sci.* **1976**, 279, 88.
7.   McKinney, J. E.; Simha, R., *J. Res. Nat. Bur. Stand. (U.S.)* **1977**, 81A, 283.
8.   McKinney, J. E.; Simha, R., *Macromolecules* **1976**, 9, 430.
9.   Berg, J. L.; Simha, R., *J. Non-Crystalline Solids* **1976**, 22, 1.
10.   McKinney, J. E.; Simha, R., *Macromolecules* **1974**, 7, 894.
11.   Utracki, L. A., personal communication.

12. Rehage, G., *J. Macromol. Sci. Phys.* **1980**, B18(3), 423.
13. Fischer, E. W.; Wendorff, J. H., *Kolloid Z. u. Z. Polymere* **1973**, 251, 876.
14. Simha, R.; Jain, S. C.; Jamieson, A. M., *Macromolecules* **1982**, 15, 1517.
15. Wiegand, W., *Inaugural Dissertation*, Marburg/Lahn **1977**; Ruland, W., *Pure Appl. Chem.* **1977**, 49, 905.
16. Papazoglou, E.; Simha, R., *Macromolecules* **1988**, 21, 1670.
17. Simha, R.; Roe, J. M.; Nanda, V. S., *J. Appl. Phys.* **1972**, 43, 4312.
18. Yee, A. F.; Maxwell, M. A., cited in Hong et. al., *J. Polym. Sci. Part B: Polym. Phys.* **1983**, 21, 1647.
19. Olabisi, O.; Simha, R., *Macromolecules* **1975**, 8, 211.
20. Yu, Z.; Yahsi, U.; McGervey, J. D.; Jamieson, A. M.; Simha, R., *J. Polym. Sci. Part B: Polym. Phys.* **1994**, 32, 2637.
21. Carri, G. A., Thesis, Case Western Reserve University, Cleveland, OH, **1995**.
22. Zoller, P., *J. Polym. Sci. Polym. Phys. Ed.* **1982**, 20, 1453.
23. Yu, Z., Thesis, Case Western Reserve University, Cleveland, OH, **1995**.
24. Brandrup, J.; Immergut, E. J., eds. in Polymer Handbook, 3rd Edition, **1989**, John Wiley, New York, 413.
25. Papazoglou, E.; Simha, R., unpublished.
26. Batschinski A. J., *Z. Phys. Chem.* **1913**, 84, 644.
27. Doolittle, A. K.; Doolittle, D. B., *J. Appl. Phys.* **1957**, 28, 901.
28. Utracki, L. A., *Can. J. Chem. Eng.* **1983**, 61, 753.
29. Curro, J. G.; Lagasse, R. R.; Simha, R., *J. Appl. Phys.* **1981**, 52, 5892.
30. Curro, J. G.; Lagasse, R. R., *Macromolecules* **1982**, 15, 1559.
31. Vleeshouwers, S., Thesis, Eindhoven Technical University, Eindhoven, Netherlands, **1993**.
32. Higuchi, H.; Yu, Z.; Jamieson, A. M.; Simha, R.; McGervey, J. D., *J. Polym. Sci. Part B: Polym. Phys.* **1995**, 33, 2295.
33. Higuchi, H.; Jamieson, A. M.; Simha, R., *ibid.* **1996**, 34, 1423.
34. Cohen, M. H.; Turnbull, D., *J. Chem. Phys.* **1959**, 31, 1164.
35. Curro, J. G.; Lagasse, R. R.; Simha, R., *Macromolecules* **1982**, 15, 1621.
36. For a detailed discussion of the Robertson, Simha, Curro (RSC) theory see Robertson, R. E. "Free Volume Theory and its Application to Polymer Relaxation in the Glassy State" in Computational Modeling of Polymers, Bicerano, J. Ed., Marcel Dekker Inc., New York, **1992**, pp. 297-361.
37. Kovacs, A. J., *Fortsch. Hochpolym. Forsch.* **1963**, 3, 394.
38. Vleeshouwers, S.; Nies, E., *Macromolecules* **1992**, 25, 6921.

Chapter 9

Structural Relaxation and Fragility of Glass-Forming Miscible Blends Composed of Atactic Polystyrene and Poly(2,6-dimethyl-1,4-phenylene oxide)

Christopher G. Robertson and Garth L. Wilkes

Department of Chemical Engineering, Polymer Materials and Interfaces Laboratory, Virginia Polytechnic Institute and State University, Blacksburg, VA 24061–0211

Structural relaxation rates for miscible polymer blends of atactic polystyrene (a-PS) and poly(2,6-dimethyl-1,4-phenylene oxide) (PPO) were assessed for isothermal aging in the glassy state and compared to the pure polymer relaxation rates. Specifically, volume relaxation rates were measured via dilatometry at several undercoolings from 15 to 60°C below the inflection glass transition temperature (T_g), and enthalpy relaxation rates were determined for aging at T_g-30°C. The compositional dependence was qualitatively similar for volume relaxation compared to enthalpy relaxation for aging at 30°C below T_g, and the trend featured the blend relaxation rates falling below weighted contributions from the structural relaxation rates of the pure components. The blends exhibited greater fragility values than expected based on the values of the neat materials. This suggests that the decreased structural relaxation rates for the blends aged at T_g-30°C were a consequence of an increased degree of required cooperativity between relaxing segments in the blend compared to the pure polymers.

As the temperature of an amorphous polymer is decreased from an initial state of thermodynamic equilibrium in the liquid state, a glass transition temperature (T_g) is reached where the molecular mobility is reduced to an extent which disallows the equilibrium state to be attained for a finite quench rate. During cooling a glass-forming material at constant pressure, equilibrium relaxation times typically increase in a non-Arrhenius manner toward a temperature asymptote at the Vogel temperature (T_0) where the equilibrium relaxation times diverge to infinity. If the kinetic temperature limit, T_0, is considered to be a true thermodynamic transition temperature or Kauzmann temperature (T_K), then fragility (F) can be considered a measure of how close the kinetic glass transition approaches the thermodynamic transition ($F = T_0/T_g$)

according to Angell (*1*). Cooling a polymer into the glassy state inherently results in molecules which are initially "frozen" into a non-equilibrium state but which can slowly relax in a local segmental manner, in accordance with the thermodynamic driving force, towards an equilibrium state of lower volume, enthalpy, and entropy. This temporal change in the thermodynamic state of a glassy material, known as structural relaxation, is accompanied by changes in application properties, common examples of which include mechanical embrittlement and a decrease in permeability (*2,3*). These changes in the macroscopic properties are referred to as physical aging, although the term physical aging is also used in a more general manner to describe all thermoreversible changes associated with the non-equilibrium glassy state, regardless of whether the changes are associated with the thermodynamic state (structural relaxation) or bulk properties.

Polymer blends can often be tailored to provide unique combinations of desirable properties, thereby making blends candidates for novel applications. The importance of mechanical and barrier characteristics to the successful utilization of polymeric materials and the fact that physical aging can significantly affect these properties in glassy polymers, therefore, justifies research efforts aimed at understanding the physical aging of polymer blends. Miscible amorphous blends represent potentially interesting systems to study with regards to physical aging. These single-phase systems can display largely homogeneous behavior which in some instances is intermediate to the pure component responses (e.g. compositional dependence of T_g) and in other cases is not anticipated based upon the expected contributions of the pure components comprising the blend (e.g. mechanical property synergism).

This communication presents volumetric and enthalpic structural relaxation rates as a function of composition for miscible blends of atactic polystyrene (a-PS) and poly(2,6-dimethyl-1,4-phenylene oxide) (PPO). This preliminary investigation is part of a larger study aimed at understanding the influence of specific interactions on the structural relaxation behavior and corresponding mechanical response changes for miscible blend systems. This ongoing research effort will aid in obtaining a greater understanding of the physical aging of miscible blends, a research area previously investigated in only a limited number of studies (*4-16*). From a broader perspective, the research to be discussed is also designed to probe the ability of using glass-forming behavior upon cooling through the glass transition region (i.e. fragility) to predict structural relaxation rates in the glassy state.

Experimental Details

Blend Preparation and Characterization. Blends of a-PS (Dow Chemical 685-D) and PPO (Polysciences, Cat.# 08974) were prepared by mixing at 265°C for 15 minutes in a Brabender (Model 5501) melt blender. Films were compression molded from the neat materials and blends, and the resulting films had an approximate thickness of 0.2 mm. All materials were stored in a dessicator cabinet prior to testing. The inflection glass transition temperature (T_g) was investigated as a function of blend composition in a differential scanning calorimeter (Perkin Elmer DSC 7) for samples weighing 8 to 11 mg at a heating rate of 10°C/minute following a quench from the equilibrium liquid state (T_g+50°C) at 200°C/min (see *Enthalpy Relaxation*

Measurements section for further DSC details). The onset to the glass transition (T_{onset}) and the end of the transition region (T_{end}) were also assessed for the blend system in order to determine values of glass transition breadth ($T_{end}-T_{onset}$). Density measurements were made at 23°C using a pycnometer manufactured by Micromeritics (Model AccuPyc 1330).

Enthalpy Relaxation Measurements. Prior to aging, samples weighing approximately 10 mg were loaded in aluminum pans and quenched into the glassy state at 200°C/min in the DSC after annealing at T_g+50°C for 10 minutes. Samples were aged isothermally at T_g-30°C (±0.5°C) in ovens under nitrogen purge for various amounts of time ranging from 1 to 100 hours. Each sample was then scanned in the Perkin Elmer DSC 7 from T_g-50°C to T_g+50°C using a heating rate of 10°C/minute (first heat). In order to provide an unaged reference with which to compare an aged DSC trace, each sample was then annealed in the DSC at T_g+50°C for 10 minutes, quenched at 200°C/min, and scanned from T_g-50°C to T_g+50°C at 10°C/minute (second heat). Extent of enthalpy relaxation was determined from the first and second heating scans using a method to be described later. In some cases, a third heat was performed after an additional 10 minute hold at T_g+50°C to verify, by comparison with the second heat, that the isothermal hold at T_g+50°C did not result in chemical degradation. All DSC testing utilized a nitrogen purge. An instrument baseline was generated every two hours of testing at a heating rate of 10°C/minute using empty pans with lids in the reference and sample cells. The ice content in the ice/water bath was maintained at approximately 30-50% by volume during all testing. The DSC temperature was calibrated using the melting points of indium and tin, and the heat flow was calibrated using the heat of fusion of indium.

Volume Relaxation Measurements. Isothermal volume relaxation was monitored for compression molded bar samples (13 mm x 13 mm x 38 mm) using a precision mercury dilatometry apparatus described in detail elsewhere (*17*), and just prior to volume relaxation measurements, the materials were annealed in the dilatometers for 10 minutes at T_g+50°C and then quenched using an ice bath. The thermodynamic state of the quenched dilatometer samples was essentially identical to that of the quenched DSC samples used to assess enthalpy relaxation as was verified in the following manner. Films of a-PS were alternately stacked with Teflon® films to form a composite bundle with the approximate dimensions of a typical dilatometer sample, and this film bundle was encapsulated in a dilatometer bulb with mercury. The encased material was then annealed in the bulb at T_g+50°C for 10 minutes and quenched to room temperature by immersion of the dilatometer bulb in an ice bath. Following this quench, the a-PS films were extracted from the bulb and then the outer and center layers were used to generate ~10 mg DSC samples. These samples were subsequently scanned in the DSC from T_g-50°C to T_g+50°C at 10°C/minute, annealed at T_g+50°C for 10 minutes, quenched at 200°C/min, and scanned a second time from T_g-50°C to T_g+50°C at 10°C/minute. Fictive temperature (T_f) calculations were performed on both heating scans for the two film samples using the Perkin Elmer analysis software provided with the instrument. The first heats of the outer and center film samples provided T_f values of 103.5°C and 103.0°C, respectively. The second heats for both samples yielded essentially equivalent fictive temperatures equal to 103.4°C, a value also obtained for a-PS samples which were freshly quenched (DSC quench at 200°C/min) and which were not previously subjected to enclosure in the dilatometer bulb. Because the fictive temperatures were essentially identical for both the DSC and dilatometer quenching conditions, the initial structural states of the samples were equivalent for the volume and enthalpy relaxation experiments. For reference, the use of a typical apparent activation energy for the glass transition temperature region provides an expected T_f reduction of approximately 3°C for a ten-fold decrease in the cooling rate.

Dynamic Mechanical Testing. Dynamic mechanical measurements (tensile) were made with a Seiko DMS 210 using samples (0.2 mm thick, 20 mm long, 5 mm wide) which were freshly quenched after free-annealing at T_g+50°C for 10 minutes. This testing was performed in the α relaxation temperature region using a heating rate of 0.3°C/min and a frequency range of 0.01 to 20 Hz. The slow heating rate essentially allowed isothermal data to be obtained as a function of frequency at temperatures ranging from approximately 5°C below to 30°C above the calorimetric T_g.

Results and Discussion

The structural relaxation of a-PS, PPO, and miscible blends thereof containing 25, 50, 75, and 87.5 wt.% PPO was investigated by assessing reductions in enthalpy and specific volume during isothermal annealing. For comparative purposes, the aging was performed at fixed temperature distances below the inflection glass transition temperature (T_g) values. The compositional dependence of T_g for the miscible blend system was determined via DSC using a heating rate of 10°C/min, and the results are indicated in Figure 1. Using the widely accepted definitions of the onset and end of the glass transition region, the breadth of the glass transition did vary slightly with composition, and the maximum breadth was 12.9°C for the 50/50 wt. a-PS/PPO blend in contrast to values of 6.0°C and 7.6°C for a-PS and PPO, respectively. This certainly raises questions regarding whether scaling of the aging data, which is necessary to allow compositional comparisons, should use the inflection in the glass transition temperature (T_g) as a reference point or the onset of the transition (T_{onset}), and what the possible influence of microheterogeneities, rich in the pure components, is on the physical aging process. This is even more of an issue for a blend system such as a-PS and poly(vinyl methyl ether) (PVME) where the 50/50 wt. blend has a glass transition breadth of approximately 50°C, far greater than the breadth values for pure a-PS and PVME which are both under 10°C (*18*).

Enthalpy Relaxation. Although enthalpy relaxation cannot be directly measured, the reduction in enthalpy occurring during isothermal aging below T_g is recovered when the aged sample is heated back through the glass transition region in a differential scanning calorimeter (DSC). Subtracting the DSC scan for the freshly quenched

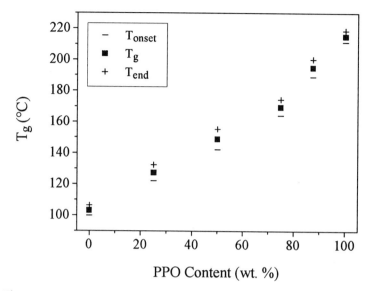

Figure 1: Glass transition temperatures for a-PS/PPO blend system measured using DSC during heating at 10°C/min following quench from T_g+50°C at 200°C/min.

sample (second heat after annealing at T_g+50°C for 10 minutes) from the scan for the aged sample (first heat) and then integrating the result provides an enthalpy difference value essentially equivalent to the reduction in enthalpy which took place during the actual aging period. If samples are aged for different amounts of time (spaced logarithmically), then an isothermal enthalpy relaxation rate can be assessed as follows (19): $\beta_H = d(\Delta H)/dlog(t_a)$ where t_a is the isothermal aging time at T_a, the aging temperature, and ΔH is the endothermic area determined from the aforementioned subtraction procedure, keeping in mind that a positive ΔH really reflects a negative change in enthalpy which occurred during physical aging. Samples of a-PS, PPO, and intermediate blends were isothermally aged at $T_a = T_g$-30°C for periods of time ranging from 1 to 100 hours and the resulting enthalpy relaxation rates are presented in Figure 2 as a function of composition. For this aging temperature, pure PPO relaxes faster than the a-PS homopolymer and the blend rates appear to deviate negatively from what might be anticipated assuming additive responses from the two components in the blend. The compositional dependence of β_H is quite similar to the variation of T_g with PPO content, an interesting feature which does not necessarily have any physical meaning.

Volume Relaxation. Isothermal volume relaxation was assessed for the a-PS/PPO blend system using a precision mercury dilatometer up to a final aging time of approximately 2 days. Four aging temperatures of T_g-15°C, T_g-30°C, T_g-45°C, and T_g-60°C were used, and the resulting relaxation rates ($\beta_V = -(1/V)dV/dlog(t_a)$) (19) are presented in Figure 3 (initial curvature in volume relaxation plots was not used in

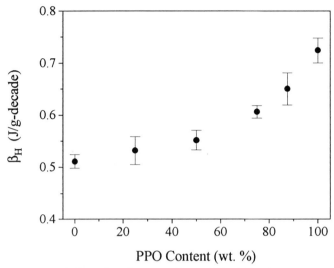

Figure 2: Enthalpy relaxation rates for a-PS/PPO blend system aged 30°C below T_g. Error bars indicate the error in the slope assessed from a plot of ΔH versus $\log(t_a)$.

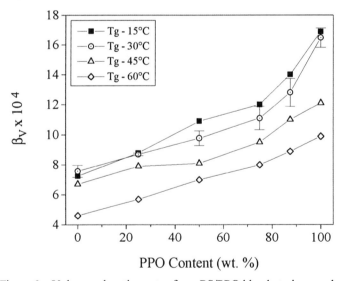

Figure 3: Volume relaxation rates for a-PS/PPO blend at given undercoolings. Data and error bars for $T_g-30°C$ represent three volume relaxation experiments while data at other undercoolings represent one experiment.

the assessment of the β_V values). It is evident from these data that PPO relaxes at a greater rate than a-PS, as was also indicated by the enthalpy relaxation investigation performed isothermally 30°C below Tg. Additionally, the volume and enthalpy relaxation data obtained at T_g-30°C exhibit very similar compositional trends, with blend rates lower than weighted contributions from the pure component rates. Current research on the a-PS/PPO blend system not detailed in this communication involves a comparison of β_V/β_H with the ratio of the step increases in the thermal expansion coefficient and heat capacity to see if the expression $\beta_V/\beta_H = \Delta\alpha/\Delta C_p$ holds as it did for atactic polystyrene according to Oleinik.(20)

An interesting feature of the volume relaxation rate data is that the blend aging rates deviate negatively from a linear interpolation of the pure component rates for aging performed at aging temperatures of 15 and 30°C below T_g, while the blend volume relaxation rates at the lowest aging temperature of T_g-60°C appear to fall directly between the rates of the pure components in a linear fashion. This change in the compositional trend as aging is performed deeper in the glassy state was initially thought to be due to the influence of a secondary relaxation in PPO, appearing as a shoulder to the α transition and manifesting itself in the blends at the same frequency/temperature as in pure PPO (21). This PPO relaxation was later discovered to be the direct result of moisture introduced while tightening the samples in the tensile grips of the dynamic mechanical analyzer at subambient temperatures. Using testing procedures which eliminated this subambient tightening operation, initially dry samples did not exhibit this relaxation and samples exposed to water prior to testing did. This unique dynamic relaxation behavior in PPO, which has not been described in the literature to the knowledge of the authors, will be detailed in a publication which is in preparation. This knowledge, therefore, invalidates the speculative explanation for the aforementioned volume relaxation rate trend. The development of an understanding of the changing compositional dependence of the volume relaxation rates as aging is performed deeper in the glassy state is currently being undertaken.

Fragility. In addition to sub-T_g structural relaxation behavior, general glass-forming behavior was investigated for the blend system as a function of composition. The generation of dynamic mechanical master curves in the α relaxation region (from approximately T_g-5°C to T_g+30°C) allowed Williams-Landel-Ferry (WLF) (22) parameters to be determined for the a-PS/PPO blends, and these parameters are given in Table I. Angell's (1) recent definition of fragility (F), $F = 1 - (C_{2,g}/T_g) = T_0/T_g$, was used to assess fragility as a function of blend content and the resulting values are evident in Figure 4. The blends are clearly more fragile than a weighted average of the fragility values of pure a-PS and PPO indicated by the dashed line in Figure 4. Consistent with this observation, estimates of the number of segments in a cooperatively relaxing domain at T_g (z_g) using the Adam-Gibbs approach (23,24) resulted in greater values for the blends in comparison to the pure components. Based on the WLF equation, the activation energy at T_g is $\Delta E = 2.303 R C_{1,g}(T_g)^2 / C_{2,g}$ and the activation energy in the limit of $1/T \rightarrow 0$ ($T \rightarrow \infty$) is $\Delta\mu = 2.303 R C_{1,g} C_{2,g}$. The barrier height $\Delta\mu$, also known as the Vogel-Fulcher activation energy, can be viewed as representing the activation energy for the independent relaxation of a segment while ΔE is the apparent activation energy at T_g which involves cooperative segmental

Table I: WLF Parameters at Calorimetric T_g for a-PS/PPO Blend System

PPO Content (wt. %)	$C_{1,g}$	$C_{2,g}$ (K)
0	15.1 (± 0.1)	47.8 (± 0.2)
25	15.5 (± 0.6)	50.2 (± 3.3)
50	17.5 (± 1.1)	54.8 ± 6.2
75	17.4 (± 0.3)	62.2 (± 1.8)
87.5	18.0 (± 0.3)	68.9 (± 1.5)
100	20.4 (± 1.3)	83.8 (± 6.5)

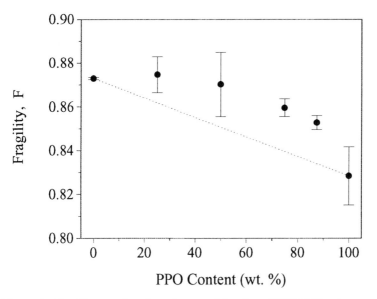

Figure 4: Fragility as a function of composition for a-PS/PPO blend system.

relaxation. Therefore, the number of segments (z) which must relax cooperatively at T_g can be expressed as: $z_g = \Delta E/\Delta \mu = (T_g / C_{2,g})^2$. Angell's expression for fragility given above was developed based on the assumption that $C_{1,g}$ does not vary for different materials and should be on the order of 16 to 17(*1*), which does not appear to be the case for the a-PS/PPO system which exhibits compositional dependence of the $C_{1,g}$ parameter. Inspection of the expression for z_g indicates that this quantity is independent of $C_{1,g}$ although no assumption regarding the constancy of this parameter is invoked. Therefore, fragility defined as $F = 1 - (C_{2,g}/T_g)$ can be used as a measure of cooperativity, independent of arguments concerning whether $C_{1,g}$ should be invariable, since expressions for both F and z_g involve only the parameters $C_{2,g}$ and T_g, although the exact functional specifics vary. Returning to a discussion of the experimental data, a probable cause for the increased relaxing domain size for the blends in comparison to the neat polymers is that the presence of specific interactions between a-PS and PPO in the blends during relaxation requires enhanced cooperativity. However, as can be seen in Figure 5, the formation of these specific interactions also results in a negative change in volume upon mixing, as is often observed for miscible polymer blend systems. At this point, therefore, it is difficult to separate the enthalpic constraints provided by the specific interactions from the geometric constraints imposed by a decrease in free volume when the specific interactions are formed upon blending.

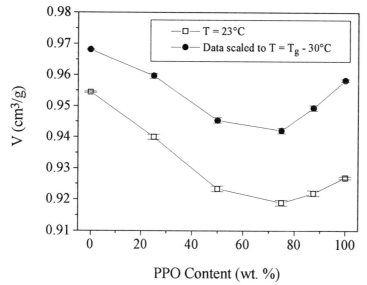

Figure 5: Specific volume for freshly quenched a-PS/PPO blends. Data was obtained at 23°C using pycnometer and scaled to $T = T_g$-30°C using thermal expansion coefficients given by Zoller and Hoehn (*25*).

It is interesting to see if a correlation exists between structural relaxation rates in the glassy state and fragility, which is an indicator of glass-forming behavior. In comparison to weighted contributions from the pure component responses, the blends were more fragile and exhibited lower volume and enthalpy relaxation rates for aging performed at the lowest undercoolings investigated (15 and 30°C below T_g), suggesting that decreases in structural relaxation correspond to increases in fragility. This correlation is clearly evident in Figure 6 which illustrates linear decreases in both enthalpy and volume relaxation rates at $T_a = T_g$ - 30°C with increases in fragility for the a-PS/PPO blend system. The implication is that the same molecular features which govern the build-up of relaxation times during glass formation upon cooling also influence the temporal change in the thermodynamic state during annealing of the non-equilibrium glass.

142

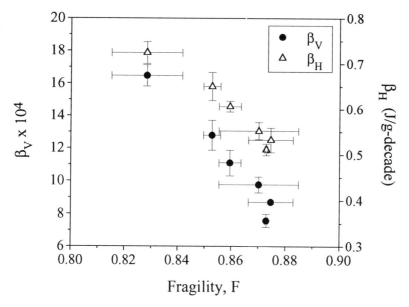

Figure 6: Apparent correlation between structural relaxation rates and fragility for the a-PS/PPO blend system aged at 30°C below T_g

Conclusions

Structural relaxation was investigated as a function of composition for blends of a-PS and PPO. For aging at temperatures 15 and 30°C below T_g, the blend rates followed by both volume and enthalpy were clearly less than expected based upon weighted rates of the pure components. The blends were more fragile and possessed larger cooperatively relaxing domain sizes than expected based on the behavior of a-PS and PPO, an observation believed to be a result of the specific interactions in the blends. The convincing correlation between diminishing structural relaxation rates with increasing fragility for the blend system suggests that sub-T_g relaxation was also affected by the interactions.

Acknowledgments

The financial support of C. G. Robertson by Phillips Petroleum and Eastman Chemical Company is gratefully acknowledged. The authors also extend their appreciation to Dr. C. A. Angell and Dr. Shiro Matsuoka for helpful research discussions.

Literature Cited

1. Angell, C. A. *J. Res. Natl. Inst. Stand. Technol.* **1997**, *102*, 171.
2. Tant, M. R.; Wilkes, G. L. *Polym. Eng. Sci.* **1981**, *21*, 874.
3. Struik, L. C. E. *Physical Aging in Amorphous Polymers and Other Materials*; Elsevier: New York, 1985.
4. Prest, W. M. Jr.; Luca, D. J.; Roberts, F. J. Jr. In *Thermal Analysis in Polymer Characterization*; Turi, E. A., Ed.; Heyden: Philadelphia, 1981; pp. 24-42.
5. Prest, W. M. Jr.; Roberts, F. J. Jr. In *Thermal Analysis*, Proceedings *of the Seventh International Conference on Thermal Analysis*; Miller, B., Ed.; John Wiley and Sons: New York, 1982; Vol 2; pp.973-8.
6. Cavaille, J. Y.; Etienne, S.; Perez, J.; Monnerie, L.; Johari, G. P. *Polymer* **1986**, *27*, 686.
7. Pathmanathan, K.; Johari, G. P.; Faivre, J. P.; Monnerie, L. *J. Polym. Sci., Part B: Polym. Phys.* **1986**, *24*, 1587.
8. Bosma, M.; ten Brinke, G.; Ellis, T. E. *Macromolecules* **1988**, *21*, 1465.
9. Cowie, J. M. G.; Ferguson, R. *Macromolecules* **1989**, *22*, 2312.
10. Mijovic, J.; Ho, T.; Kwei, T. K. *Polym. Eng. Sci.* **1989**, *29*, 1604.
11. Ho, T.; Mijovic, J. *Macromolecules* **1990**, *23*, 1411.
12. Elliot, S. *Ph.D. Dissertation;* Heriot-Watt University (U.K.), 1990
13. Ho, T.; Mijovic, J.; Lee, C. *Polymer* **1991**, *32*, 619.
14. Oudhuis, A. A. C. M.; ten Brinke, G. *Macromolecules* **1992**, *25*, 698.
15. Pauly, S.; Kammer, H. W. *Poly. Networks Blends* **1994**, *4*, 93.
16. Chang, G.-W.; Jamieson, A. M.; Yu, Z.; McGervey, J. D.; *J. Appl. Polym. Sci.* **1997**, *63*, 483.
17. Shelby, M. D. *Ph.D. Thesis*; Virginia Polytechnic Institute and State University, 1996.
18. Schneider, H. A.; Cantow, H.-J.; Wendland, C.; Leikauf, B. *Makromol. Chem.* **1990**, *191*, 2377.
19. Hutchinson, J. M. *Prog. Polym. Sci.* **1995**, *20*, 703.
20. Oleinik, E. F. *Polymer Journal* **1987**, *19*, 105.
21. Robertson, C. G.; Wilkes, G. L. *Proc. Am. Chem. Soc. Div. Polym. Mat. Sci. Eng. (PMSE Preprints)*, **1997**, *76*, 354.
22. Williams, M. L.; Landel, R. F.; Ferry, J. D. *J. Am. Chem. Soc.* **1955**, *77*, 3701.
23. Adam, G.; Gibbs, J. H. *J. Chem. Phys.* **1965**, *43*, 139.
24. Matsuoka, S.; Quan, X. *Macromolecules* **1991**, *24*, 2770.
25. Zoller, P.; Hoehn, H. H. *J. Polym. Sci.: Polym. Phys. Ed.* **1982**, *20*, 1385.

Chapter 10

Dielectric Spectroscopy of Bisphenol-A Polycarbonate and Some of Its Blends

G. J. Pratt[1] and M. J. A. Smith[2]

[1]Department of Mechanical and Manufacturing Engineering, University of Melbourne, Parkville 3052, Australia
[2]Department of Physics, University of Warwick, Coventry CV4 7AL, United Kingdom

Dielectric spectroscopy has been applied to the study of commercial bisphenol-A polycarbonate and some of its blends and to their susceptibility to environmental factors. The multiplicity of overlapping absorptions is consistent with the high impact strength of polycarbonate over an extended temperature range. Two intermediate-temperature losses are differentiated and delineated; their origins and reason for diminution by annealing are examined. Dielectric spectroscopy is seen as a particularly sensitive means for detecting and possibly characterizing radiation-induced changes in polymers before significant deterioration of other properties has occurred. For impact-modified PC/PBT and PC/PET blends evidence is provided for partial miscibility of the component polymers and for a two-phase morphology with a polyester-rich dispersed phase in a continuous matrix rich in polycarbonate.

The high impact strength, dimensional stability and optical clarity (low crystallinity) of bisphenol-A polycarbonate (PC) together with its low dielectric loss have led to a range of applications embracing optical components, CD-ROMs, film capacitors and safety-related products. Subsequent market demands for enhanced physical properties has stimulated the development of a range of commercial blends of which rubber-modified bisphenol-A polycarbonate (PC) with polybutylene terephthalate (PBT) or polyethylene terephthalate (PET) are amongst the more successful.

Dielectric spectroscopy is concerned with the dependence of complex permittivity on temperature and frequency. The relatively low level of d.c. conduction in polycarbonate ensures that the principal relaxations associated with polycarbonate's active C=O dipole can be observed over a useful range of frequency and temperature. In multi-phase or multi-component polymers charge accumulation at the sub-structure interfaces leads to Maxwell-Wagner-Sillars (MWS) contributions to the overall polarization.

The sensitivity of dielectric spectrometry is such that it may be used with polymers to study how molecular motion is influenced by low molecular weight additives, crosslinking, branching, inter- and intra- molecular interactions, blending, copolymerization, and degradation. Dielectric data can complement that obtained using dynamic mechanical analysis, differential scanning calorimetry, spectrophotometry, electron spin resonance, or nuclear magnetic resonance. Much of the useful dielectric information on polymeric materials is provided by observations within the frequency range 10^{-3} Hz to 10^8 Hz.

The present paper reviews the application of dielectric spectroscopy to an integrated study of commercial bisphenol-A polycarbonate and some of its blends [1-3] including its susceptibility to environmental factors such as u.v. radiation [4,5] and humidity [6].

Experimental

The clear polycarbonate materials investigated were General Electric LEXAN 141, a medium viscosity grade containing a small amount of phosphite processing stabilizer, the corresponding u.v.-resistant compound (143), and an additive-free grade (145). The blends investigated were a PC/PET blend (XL1339) from GE Plastics Europe, a 5:4 melt blend of LEXAN 145 PC and VALOX 315 PBT containing a transesterification inhibitor (as used in the production of commercial XENOY materials), and a similar composition containing 10% of added rubber (impact modifier). XL1339 contains 68.5% PC, 21.0% PET and 7.0% impact modifier.

Sample preparation and measurement techniques have been described previously [2,7]. Apart from some samples of LEXAN 141 which were deliberately exposed to the laboratory atmosphere for at least one month, each sample was equilibrated in dry nitrogen at 30 to 35°C for at least 24 hours prior to investigation. The dielectric data was obtained over the ranges 0.1 Hz to 3 MHz and -190°C to +195°C using the wide-band capacitative-T permittivity bridge of Pratt and Smith [7]. For each material a single sample of approximate dimensions 25 x 20 x 0.4 mm was sufficient to cover the experimental range without a change of apparatus or technique and without resort to Fourier transform analysis. Temperature control was achieved using a baffled flow of dry nitrogen gas at the desired temperature.

For the investigation into the influence of u.v. radiation on the dielectric properties of polycarbonate, pressed sheets of the material were irradiated at room temperature in air by a standard Hanovia mercury lamp. The lamp provided a spectrum of wavelengths within the range 254 to 546 nm and sample intensities of 2.47 mW cm^{-2} and 0.91 mW cm^{-2} respectively at the principal wavelengths of 365 nm and 254 nm. A voltage-stabilized power supply assisted consistency of spectral output.

Immediately prior to a dielectric measurement, a Keithley Instruments 602 electrometer was used to determine the quasi-d.c. conductivity of the sample after allowing time, typically up to an hour, for the decay of current transients. The equivalent imaginary permittivity caused by the quasi-d.c. conductivity was subtracted from the measured total permittivity.

Results and Discussion

Bisphenol-A carbonate has been widely studied by dielectric [8-26], dynamic mechanical [27-31] and thermally stimulated depolarization (TSD) [10-13,32-35] techniques. However, differences in the compositions of the materials studied, and in their thermal history and pretreatment, have led to apparently conflicting results being reported in the literature, as discussed in detail in a recent paper [6]. In the present study contour maps of complex relative permittivity for both basic and u.v.-resistant grades of LEXAN have been obtained over an extended range of experimental conditions using a single apparatus, with each grade of material subject to the same thermal history.

Unirradiated Polycarbonate Homopolymers. Dielectric spectroscopy has identified eight distinct absorption and dispersion regions in nitrogen-equilibrated commercial bisphenol-A polycarbonate [1]. A clear correspondence is found between the temperature-frequency regimes in which changes occur in ε' and ε'' as expected from the Kramers-Kronig relations. Increases in ε' coincide with peak maxima for ε'' as seen, for example, in Figures 1 and 2.

High- and Low-Temperature Relaxations. A particularly sharp and prominent loss peak in the ε'' contour map for unirradiated LEXAN 141, 143 or 145 is attributed to the primary (α) relaxation which is associated with gross motions of the polymer chain in the region of Tg. The widths of the α peaks for each of the three materials are comparable with that shown in Figure 3 for unequilibrated LEXAN 141 and suggest a similar, low degree of crystallinity [17].

The most extensive feature in each ε'' map is a broad β-peak in a region where ε' changes gradually. Deconvolution

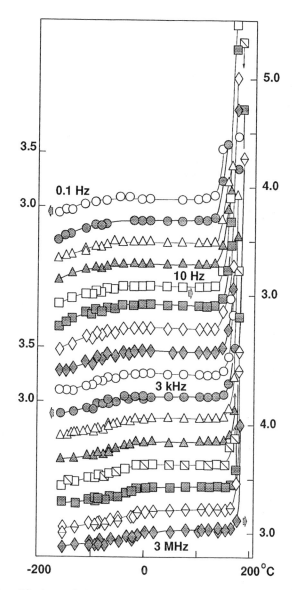

Figure 1: Plots of the temperature-variation of real relative permittivity ε′ of unequilibrated LEXAN 141 polycarbonate at half-decade frequencies from 0.1 Hz to 3 MHz. The scale markings on the vertical axes refer to ε′ values for 3 MHz, 10 kHz, 30 Hz and 0.1 Hz respectively, as indicated.

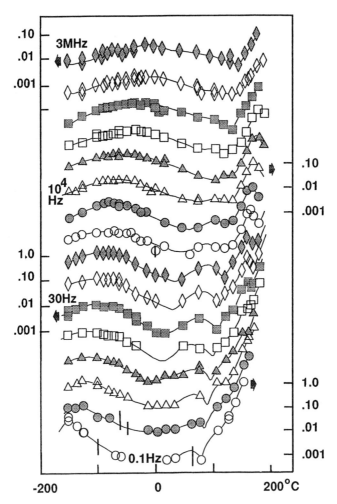

Figure 2: Plots of the temperature-variation of imaginary relative permittivity ε" of unequilibrated LEXAN 141 polycarbonate at half-decade frequencies from 0.1 Hz to 3 MHz. The scale markings on the vertical axes refer to ε" values for 3 MHz, 10 kHz, 30 Hz and 0.1 Hz respectively, as indicated.

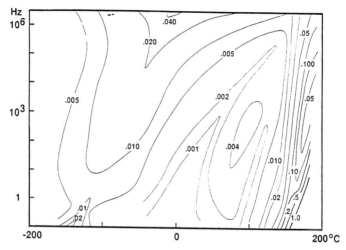

Figure 3 : Contour map showing the temperature-frequency variation of imaginary relative permittivity ε'' for unequilibrated GE LEXAN 141 polycarbonate.

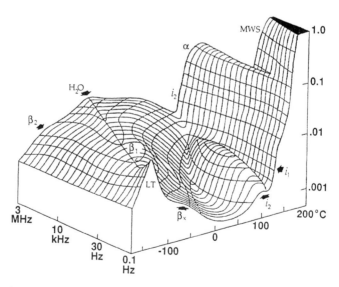

Figure 4: 3-D representation showing the temperature-frequency variation of imaginary relative permittivity ε'' for unequilibrated LEXAN 141 polycarbonate.

of the peak indicates that a range of segmental motions contribute to its breadth and shape [36]. At lower temperatures the phenyl groups are immobile and localized segmental motion is hindered by adjacent phenyl rings despite the open packing of the polymer chains in glassy PC. At higher temperatures the mobility of the phenyl groups facilitates tightly-coupled combined motion of carbonate and phenyl groups.

In unequilibrated LEXAN 141 the presence of an additional loss region above -40°C is indicated by the increase of slope above 30 kHz in the Arrhenius plot (Figure 5) for the overall β-peak which does not occur for the material equilibrated in dry nitrogen. The higher activation energy of this water-related relaxation relative to the fundamantal β-process is consistent with the existence of hydrogen-bonding, or some other form of inter-molecular interaction which hinders the cooperative movements of adjacent carbonyl and phenyl groups.

Both LEXAN 141 and 143 [1] show a distinctive loss of similar intensity in the range -120°C to -130°C below 3 Hz which is not observed in LEXAN 145 [3]. It is concluded that the low temperature process below 3 Hz is a consequence of the presence of the phosphite processing stabilizer in the former which is absent in LEXAN 145.

Exceedingly large losses at low frequencies above 150°C are attributed to Maxwell-Wagner-Sillars (MWS) polarizations arising from conduction mismatches at the structural interfaces between a continuous matrix of amorphous polycarbonate and a crystalline or densified second phase. Provided that the discontinuous phase tends towards a two-dimensional aspect and has a conductivity less than that of the matrix, theory predicts substantial MWS losses even with a low concentration of the discontinous phase [37].

Intermediate - Temperature Relaxations. Secondary relaxations in the glassy state at temperatures intermediate between those of the α- and β- relaxations have been reported, but workers disagree as to their nature, location and origin. Confusion arises in part from a failure to recognize the existence of two separate processes. Krum and Müller [19] observed an intermediate relaxation only for injection-moulded or cold-drawn polycarbonate samples. Since the magnitude was diminished by annealing and the loss was not detected in fully annealed samples, they concluded that the intermediate process is a non-equilibrium effect associated with residual stresses.

In the present work two separate intermediate relaxations are observed and defined in the glassy state at temperatures above ambient in all LEXAN grades which have been investigated, whether equilibrated [1,3] or unequilibrated [6], u.v.-irradiated [4] or unirradiated [1,3,6]. The first (i_1) appears in Figures 3 and 4, for example, as a shoulder from about 105°C at 0.1 Hz to

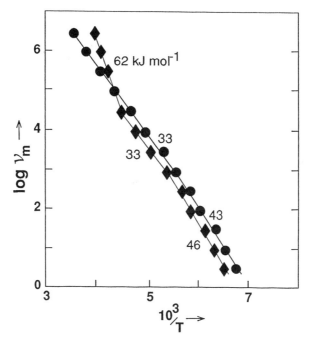

Figure 5: Arrhenius plots of the logarithm of the frequency location of the β-peak vs. reciprocal temperature for equilibrated (●) and unequilibrated (♦) LEXAN 141. Equilibrated samples were stored in dry N_2 at 30-35°C for 24 hours prior to testing.

140°C at 300 Hz. This absorption cannot be discerned above 1 kHz, and corresponds closely with that reported by Watts and Perry [26] for LEXAN 101. The second (i_2) persists throughout the frequency range covered by the present work, and most probably corresponds with that reported by Müller and co-workers [18-19], Sacher [23-25] and others [8-9,28-29,38-39]. At low frequencies the i_2 absorption exists as a separate peak at moderate temperatures (Figure 4) which at higher frequencies gradually merges with the α-peak to become a shoulder at about 160°C at 3 MHz.

On balance, the i_1 relaxation process is now regarded as involving partially-correlated motion of a moderate number of molecular segments or repeat units as precursor to the cooperative but uncorrelated micro-Brownian motion which occurs above Tg. It is postulated that the origin of the i_2-absorption is a relaxation process in which the phenyl groups undergo 180° flips as might arise when adjacent carbonyl groups interchange between *cis-trans* and *trans-trans* conformations [40-42]. Annealing reduces the available free-volume, thereby inhibiting the ring-flip process. Rather than the relief of frozen-in stresses, this reduction in free volume is considered to be the underlying cause of the diminution by annealing of the intermediate relaxations.

Influence of U.V. Radiation. Above 180°C the values of ε′, ε″ and tan δ rise significantly more rapidly for the u.v.-resistant LEXAN 143 than for either 141 or 145 and in a manner which seems to be independent of frequency [1]. It is postulated that u.v.-protection in the 143 material is conferred by intimate association of the benzotriazole additive with the polymer molecules, and that above 180°C this additive is liberated from the polymer molecules, freeing additional polarizable groups which contribute to the observed increase in ε′, ε″ and tan δ.

For the unstabilized LEXAN 141 exposed to u.v. radiation, additional dispersions appear as a result of new polar groups created by the irradiation (Figure 6b). Continuation of irradiation causes the additional dispersions to move to lower temperature and frequency (Figure 6c) in a manner which is broadly consistent with a sequence of photo-Fries reactions and subsequent photo-oxidation [4]. Using polycarbonate samples containing identifiable photo-products, in particular phenyl salicylate, some progress has been made towards establishing specific correlations between radiation-induced absorptions and expected products of photo-degradation or photo-oxidation [5]. Moreover, an indication of photo-degradation can be obtained at an early stage before significant deterioration of properties has occurred.

For the stabilized polycarbonate there is evidence that the benzotriazole imparts partial u.v. stabilization

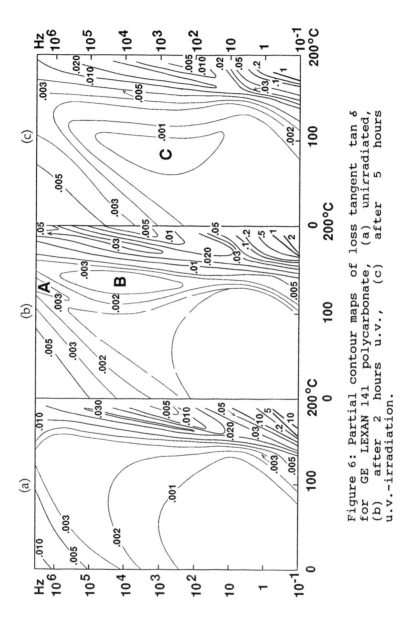

Figure 6: Partial contour maps of loss tangent tan δ for GE LEXAN 141 polycarbonate, (a) unirradiated, (b) after 2 hours u.v., (c) after 5 hours u.v.-irradiation.

by a mechanism of preferential u.v. absorption until the additive is itself consumed [5].

Polycarbonate / Polyester Blends. Contour maps of the temperature-frequency variation of complex relative permittivity have been obtained for impact-modified PC/PBT and PC/PET blends [3,43] and for their constituent polycarbonate [3], PBT [44] and PET [45] homopolymers. In the impact-modified PC/PBT blend (Figure 7), as in each of the blends, a single broad β-absorption is observed but the separate α-absorptions of the constituents persist.

The persistence of the glass transition temperatures of the component polymers in each of the blends, and the lack of a third Tg peak at an intermediate temperature (see, for example, Figure 8), confirms that the blends contain two phases. The temperature shift in each Tg is a consequence of the partial miscibility of the component polymers in the melt. The very slight increase in the Tg of PBT and PET in the PC/PBT and PC/PET blends respectively indicates that the polyester-rich phase contains only a small amount of polycarbonate, whereas the significant decrease in Tg in the PC component of the blends suggests the presence of substantial proportions of PBT or PET respectively in the phase consisting predominantly of polycarbonate.

A reason for blending PBT with PC is to overcome the comparatively poor solvent resistance of polycarbonate. The enhanced solvent resistance of the blend would be expected if PBT was the continuous matrix phase in the blend, as suggested by Hobbs et al [46-47]. However, the observation that the d.c. conductivity of polycarbonate and its blends is orders of magnitude less than that of PBT and PET over a substantial temperature range indicates unambiguously that the PC-rich phase forms the continuous matrix in each of the blends investigated here. It would seem that the enhanced solvent resistance in PC/PBT blends can only arise from the substantial proportion of evenly-distributed PBT which is contained in the matrix.

Conclusions.

Eight distinct absorption regions are identified in nitrogen-equilibrated polycarbonate, with bound water causing a further absorption in the unequilibrated material. The multiplicity of overlapping absorptions is consistent with the high impact strength of polycarbonate over an extended temperature range. Two intermediate-temperature losses are differentiated and delineated; their origins and reason for diminution by annealing are examined. Rather than the relief of frozen-in stresses, the associated reduction in free volume is considered to be the underlying cause of the diminution by annealing of the intermediate relaxations.

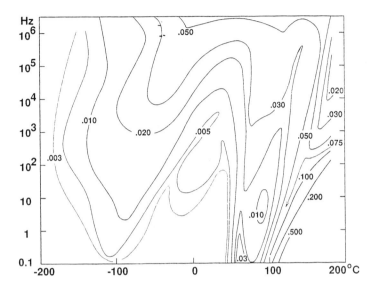

Figure 7: Contour map of imaginary relative permittivity ε" for an impact-modified polycarbonate /PBT blend. [*adapted from ref. 3*]

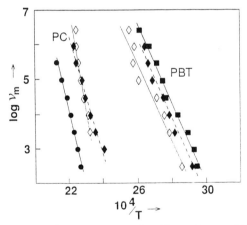

Figure 8: Arrhenius plots of the logarithm of the frequency of maximum loss vs. reciprocal temperature for the dielectric α-relaxations in GE LEXAN 145 PC (●), VALOX 315 PBT (■), a 5:4 PC/PBT blend (◆) and a 5:4 PC/PBT blend containing 10% impact modifier (◇). [*adapted from ref. 48*]

Dielectric spectroscopy is a particularly sensitive means for detecting radiation-induced changes in polymers before significant deterioration of other properties has occurred. For u.v.-irradiated polycarbonate the dielectric data is broadly consistent with a sequence of photo-Fries reactions and subsequent photo-oxidation. Changes in the additional region observed at high temperatures in the u.v.-resistant polycarbonate imply that the benzotriazole imparts partial u.v. stabilization by a mechanism of preferential u.v. absorption. Some progress has been made towards establishing specific correlations between radiation-induced absorptions and expected products of photodegradation or photo-oxidation.

For impact-modified PC/PBT and PC/PET blends evidence has been presented for partial miscibility of the component polymers and for a two-phase blend morphology with a polyester-rich dispersed phase in a continuous matrix rich in polycarbonate. Other absorptions are attributed respectively to MWS interfacial polarization, to the presence of the impact modifier and to a phosphite processing stabilizer.

Literature Cited

1. Pratt,G.J.; Smith,M.J.A. *Brit.Polym.J.* **1986**, *18*, 105
2. Pratt,G.J.; Smith,M.J.A. *Polymer* **1989**, *30*, 1113
3. Pratt,G.J.; Smith,M.J.A. *Plastics, Rubber and Composites Processing and Applications* **1991**, *16*, 67
4. Pratt,G.J.; Smith,M.J.A. *Polym.Degr.Stab.* **1989**, *25*, 2671
5. Pratt,G.J.; Smith,M.J.A. *Polym.Degr.Stab.* **1997**, *56*, 197
6. Pratt,G.J.; Smith,M.J.A. *Polym.Internatl.* **1996**, *40*, 239
7. Pratt,G.J.; Smith,M.J.A. *J.Phys.E:Sci.Instrum.* **1982**, *15*, 927
8. Allen,G.; McAinsh,J.; Jeffs,G.M. *Polymer* **1971**, *12*, 85
9. Allen,G.; Morley,D.C.W.; Williams,T. *J.Mater.Sci.* **1973**, *8*, 1449
10. Aoki,Y.; Brittain,J.O. *J.Polym.Sci.;Polym.Phys.Ed.* **1976**, *14*, 1297
11. Aoki,Y.; Brittain,J.O. *J.Appl.Polym.Sci.* **1976**, *20*, 2879
12. Aoki,Y.; Brittain,J.O. *J.Appl.Polym.Sci.* **1977**, *21*, 199
13. Hong,J.; Brittain,J.O. *J.Appl.Polym.Sci.* **1981**, *26*, 2459 & 2471
14. Bair,H.E.; Johnson,G.E.; Merriweather,R. *J.Appl.Phys.* **1978**, *49*, 4976
15. Ito,E.; Hatakeyama,T. *J.Polym.Sci.;Polym.Phys.Ed.* **1974**, *12*, 1477
16. Ito,E.; Hatakeyama,T. *J.Polym.Sci.;Polym.Phys.Ed.* **1975**, *13*, 2313
17. Matsuoka,S.; Ishida,Y. *J. Polym.Sci.,C* **1966**, *14*, 247
18. Müller,F.H.; Huff,K. *Kolloid Z.* **1959**, *164*, 34
19. Krum,F.; Müller,F.H. *Kolloid Z.* **1959**, *164*, 81

20. Pochan,J.M.; Hinman,D.F.; Turner,S.R. *J.Appl.Phys.* **1976**, *47*, 4245
21. Pochan,J.M.; Gibson,H.W.; Froix,M.F.; Hinman,D.F. *Macromolecules* **1978**, *11*, 165
22. Pochan,J.M.; Gibson,H.W.; Pochan,D.L.F. *Macromolecules* **1982**, *15*, 1368
23. Sacher,E. *J.Macromol.Sci.-Phys.* **1974**, *B9*, 163
24. Sacher,E. *J.Macromol.Sci.-Phys.* **1974**, *B10*, 319
25. Sacher,E. *J.Macromol.Sci.-Phys.* **1975**, *B11*, 403
26. Watts,D.C.; Perry,E.P. *Polymer* **1978**, *19*, 248
27. Chung,C.I.; Sauer,J.A. *J.Polym.Sci.,A-2* **1971**, *9*, 1097
28. Illers,K.H.; Breuer,H. *Kolloid Z.* **1961**, *176*, 11
29. Illers,K.H.; Breuer,H. *J.Colloid Sci.* **1963**, *18*, 1
30. Locati,G.; Tobolsky,A.V. *Adv.Mol.Relax.Proc.* **1970**, *1*, 375
31. Reding,F.P.; Faucher,J.A.; Whitman,R.D. *J.Polym.Sci.* **1961**, *54*, S56
32. Jain,K.; Agarwal,J.P.; Mehendru,P.C. *Il Nouva Cimento* **1980**, *55B*, 123
33. Mehendru,P.C.; Jain,K.; Agarwal,J.P. *J.Phys.D;Appl. Phys.* **1980**, *13*, 1497
34. Linkens,A.; Vanderschueren,J. *Polym.Letters* **1977**, *15*, 41
35. Vanderschueren,J.; Linkens,A.; Haas,B.; Dellicour,J. *J.Macromol.Sci.-Phys.* **1978**, *B15*, 449
36. Pratt,G.J.; Smith,M.J.A. *Polymers at Low Temperatures (Proceedings)*, PRI: London, 1987; pp. 9/1-10
37. van Beek,L.K.H. in *Progress in Dielectrics*; Birks,J.B.; Hart,J.,Eds.; Heywood: London, 1967; pp.7 & 69
38. Heijboer,J. *J.Polym.Sci.,C* **1968**, *16*, 3755
39. Bussink,J.; Heijboer,J. *Proc.Intern.Conf.IUPAC*, Delft, 1964
40. Jones,A.A. *Macromolecules* **1985**, *18*, 902
41. O'Gara,J.F.; Jones,A.A.; Hung C.-C.; Inglefield,P.T. *Macromolecules* **1985**, *18*, 1117
42. Connolly,J.J.; Inglefield,P.T.; Jones,A.A. *J.Chem. Phys.* **1987**, *86*, 6602
43. Pratt,G.J.; Smith,M.J.A. *Electrical, Optical and Acoustic Properties of Polymers - EOA III*, PRI: London, 1992; pp.13/1-8
44. Pratt,G.J.; Smith,M.J.A. *J.Mater.Sci.* **1990**, *25*, 477
45. Pratt,G.J.; Smith,M.J.A. *Electrical, Optical and Acoustic Properties of Polymers - EOA II*, PRI: London, 1990; pp.P9/1-4
46. Hobbs,S.Y.; Dekkers,M.E.J.; Watkins,V.H. *J.Mater.Sci.* **1988**, *23*, 1219
47. Hobbs,S.Y.; Dekkers,M.E.J.; Watkins,V.H. *Polym.Bull.* **1987**, *17*, 341
48. Pratt,G.J.; Smith,M.J.A. *Polym.Internatl.*, **1997**, *43*, 137

Chapter 11

Physical Aging near the Beta Transition
of a Poly(amide–imide)

George J. Dallas[1], Thomas C. Ward[2], James Rancourt[3], and Mia Siochi[4]

[1]Beloit Manhattan Research and Development, 155 Ivy Park Clarks Summit, PA 18411
[2]NSF Science and Technology Center, Department of Chemistry, Virginia Polytechnic Institute and State University, Blacksburg, VA 24061
[3]Polymer Solutions, 1872 Pratt Drive, Blacksburg, VA 24060
[4]NASA Langley Research Center, Hampton, VA 23681

Water and low temperature (20-35°C) aging influence the dynamic mechanical properties of a poly(amide-imide). At concentrations below 2 weight percent, water contributes to a low temperature relaxation between -120 and -50°C. Above 2 weight percent the water influences the beta transition. The enthalpy of activation for the beta relaxation is dependent upon aging temperature and time. Aging temperatures closer to the beta transition temperature result in higher activation enthalpies for that dispersion.

Physical aging is the reversible densification of an amorphous material below its glass transition temperature. Experiments traditionally have focused on volume and mechanical property variations with thermal history. Later experiments have focused on understanding structure-property relationships and measuring the evolution of the distributions of relaxations. Early investigations of physical aging include the work of Tool (1) on inorganic glasses and Fuoss (2) on the dielectric properties of PVC. The influence of thermal history on creep and volume were extensively examined by Struik (3) and Kovacs (4) respectively, along with many others (5-19). Although the alpha transition was the dispersion found to be most affected by aging, the contribution of the beta process was also explored. The relationship between alpha and beta molecular motions and mechanical properties has been reviewed by Boyer (20-22) and Hartman (23). A more current discussion of physical aging in the vicinity of the beta transition temperature has been conducted by McKenna (24-28) with contributions by Struik (29), Read (30), Hougham (31) and Dallas (32). Dilatometric measurements by Lee (33) in atactic polystyrene indicate that volume relaxation occurs below the beta transition. These data support mechanical experiments done by Read and Dean (34) on poly(methylmethacrylate). The effect of physical aging on dynamic mechanical properties of a polyimide has been investigated by others (35-39).

Prior research has shown that the amount of physical aging is dependent upon the aging temperature and time below the glass transition temperature T_g. The smaller the difference between the glass transition temperature and the aging temperature, the faster the aging. Exceeding the T_g reversibly erases the aging process. Thermal history below the T_g strongly influences the resulting distribution of relaxation times.

In the present research, variables associated with physical aging near the alpha transition were used to examine the beta transition of a poly(amide-imide). After sample drying, the relationship between the beta transition's activation enthalpy and thermal history was explored.

The effects of water in polymers have been reviewed in a symposium (40-47) and by various researchers (48-64). The effect of diluents on the dynamic mechanical properties of polymers has also been examined extensively (65-68). In Gillham's work on water ingression into a polyimide, the maximum of the damping peak for water at 1 Hz was examined. The peak ranged from -120 to -90°C depending upon the structure of the polymer. From these data it was shown that water ingressed into a polymer in stages, first forming hydrogen bonds, then water clusters. At higher percentages of water there were distortions of the polymers' alpha and beta relaxations.

The concepts derived from physical aging and water ingression into polymers will be used to interpret the dynamic mechanical results for the poly(amide-imide) in the present study. It is beneficial to examine the effects of water and aging simultaneously in order to decouple their influence on the dynamic mechanical loss properties of the polymer.

Experimental

The poly(amide-imide) samples were obtained from Amoco as Torlon 4203L. This poly(amide-imide) was derived from 4 - 4' oxydianiline and trimellitic anhydride. Samples were dried in a Varco oven at 190°C under vacuum for periods of 3 to 12 hours. A minimum of six samples was cycled through all temperature profiles. The samples were quenched in liquid nitrogen for less than 30 seconds. They were stored in a sealed Ziploc® bag purged with nitrogen. The samples were either stored in a desiccator at 20°C or a Fisher variable temperature oven at 35 and 60°C. They were aged for 15 days, 4 days and 23 hours, respectively. The aging time was chosen to be greater than the experimental time of seven hours.

The dynamic mechanical properties were measured on a Polymer Labs dynamic mechanical thermal analyzer connected to a Hewlett-Packard controller. They were analyzed by a multifrequency temperature scan at 0.03, 0.3, 0.1, 3.0 and 30 Hz. The samples were heated at 0.5°C/min from -150 to 280°C in the ʋater study, while a temperature range of 10 to 220°C was used for analysis of the beta dispersion. The samples had three different weight percentages of water: 1) dried with 0%, 2) as-received with 1.5 to 2.5%, and 3) saturated with approximately 4 to 5%. The as-received sample was exposed to the atmosphere for more than 150 days. This sample contained both water and 1-methyl-2-pyrrolidinone (NMP). The samples saturated with water were prepared by placing them in boiling water overnight. The water content was determined either by weighing on a Mettler analytical balance or by thermogravimetry. Thermogravimetry was done on a Perkin Elmer Model 7 thermogravimetric analyzer.

Thermogravimetry-mass spectrometry was done with a Perkin Elmer TAS 2 TGA connected by a capillary transfer line to a Hewlett Packard mass selective detector. A heating rate of 40°C/min was used so that the volatiles would rapidly desorb from the sample. Helium was used as a carrier gas at a pressure of 40 psi. The time traces of water and NMP were done at mass to charge ratios of 18 and 99 respectively.

Dielectric thermal analysis was done with a Polymer Laboratory DETA from -140 to 230°C at a heating rate of 1°C/min using frequencies of 1, 3, 10, 50, and 100 kHz. The 20 mm diameter electrodes were used. The 0.57 mm thick poly(amide-imide) samples were tested either as-received or after drying at 190°C for 7 hours. The samples were tested under a nitrogen atmosphere to reduce the chance of ambient water condensing on the surface of the samples.

Results and Discussion

A set of tests was done to characterize the weight percent of volatiles in the as-received sample, the effect of saturating the polymer with water and the effect of thermal aging below the beta transition. They included TGA-mass spectrometry to identify the amount of water and NMP in the as-received sample. Dynamic mechanical thermal analysis was used to determine the effect of water and physical aging on the loss modulus of the polymer. Differential scanning calorimetry was used on the saturated sample to determine if the absorbed water was in a clustered form. Also, dielectric thermal analysis was used to determine how the dielectric storage ε' and dielectric loss ε'' varied between the as-received and dried samples. These results will show that as the concentration of water increased, the form of the water in the polymer changed. Also, thermal aging below the beta transition changed the dynamic mechanical loss properties of this polymer in a fashion similar to physical aging of the glass transition.

Figures 1a and 1b show the TGA-mass spectrometry data for water and NMP on the as-received poly(amide-imide), indicating that neither liquid was removed until five minutes into the run, which corresponded to approximately 220°C. The thermogravimetry showed that there was a total of approximately 2 weight percent of water and NMP in the polymer. Isothermal thermogravimetric experiments were conducted on the as-received sample between 145 and 165°C. Temperatures above 155°C were enough to dry the sample in 3 hours. Although, further drying of the samples were done at 190°C, an intermediate value between 155 and 220°C. Even this simple experiment showed that there were strong associations of the water and NMP to the poly(amide-imide).

Figure 2 is a plot comparing the log of the dynamic mechanical loss modulus (E") vs. temperature from -150 to 280°C at 1 Hz for the poly(amide-imide) having different water content: 1) as-received (1.5%), 2) dried (0%), and 3) saturated (4%) water. The as-received sample was aged more than 150 days at room temperature, the dried sample had no aging, while the sample saturated with water was aged overnight in boiling water at 100°C. The three distinguishing regions in this log E" vs temperature comparison are from -120 to -50°C, from 0 to 175°C, the beta transition region and from 250 to 300°C, the glass transition temperature. The data will be discussed from the lower to the upper temperature dispersion.

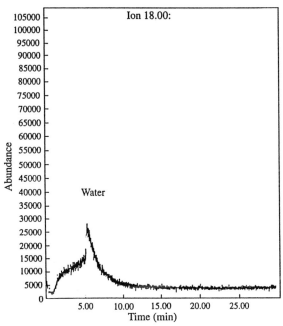

Figure 1a. TGA-mass spectrometry data for water on the as-received poly(amide-imide).

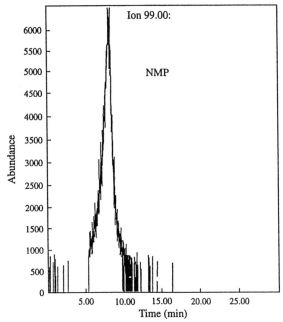

Figure 1b. TGA-mass spectrometry data for NMP on the as-received poly(amide-imide).

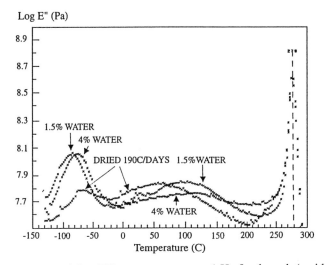

Figure 2. Loss modulus (E″) vs. temperature at 1 Hz for the poly(amide-imide) with different water contents: as-received (1.5%), dried (0%), and saturated (4.5%).

The dispersion peak from -120 to -50°C was small for the dried sample. Increasing the water content caused the peak height to reach a maximum at 1.5% water and remain unchanged with the higher water content of 4%. Above approximately 2% water, the peak height did not increase in proportion to the increase in water. Only the peak maximum shifted from -80 to -70°C. This peak has been associated with water that is hydrogen bonded to its local environment or is in clusters (44, 53). This concept is supported by an activation energy for this relaxation of ~30 kJ/(mol of hydrogen bonds) which is within the accepted range for water (69, 70). This activation energy was determined by an Arrhenius plot of the DMTA peak maxima for this relaxation. The effect of aging on this low temperature dispersion was never explored. Since this low temperature peak did not increase in proportion to the weight percent of water, this implied that above 2 weight percent water, the water existed in some other environment within the polymer.

The dispersion region from 0 to 175°C showed that the dried sample with zero aging had a lower peak maximum temperature than the as-received sample with 1.5% water/NMP aged >150 days. The sample with 4% water showed a lower peak height than the 1.5% and 0% water values. This implied that at higher percentages of water, the water associated with functional groups that contributed to the beta transition and reduced its loss values at 1 Hz. This peak was identified by Arnold (39) to be due to amine or imide groups which have an excellent affinity for water.

The glass transition region for all three samples was similar in magnitude and location. It is postulated that any effects that might be due to water were eliminated at the higher temperatures after the water desorbed from the polymer, before the glass transition temperature was reached.

Differential scanning calorimetry was used on the 4% water sample to determine if the water was in a clustered form in the polymer. No melt transition for water at 0°C was observed; thus, the water was not shown to be in a detectable amount of the clustered form, as was shown by Johnson (44) for polyvinylacetate.

Analyses of the data with increased water content led to the speculation that water ingressed in stages, from 0 to 1.5% water ingressed into the interstices of the polymer, from 1.5 to 4% the water associated with functional groups that contributed to the beta transition. Since the as-received sample had greater than 150 days of aging at room temperature, the exact effect of aging on the beta transition could not be decoupled from the effects of water and NMP. This led to the following isothermal aging experiments below the beta transition temperature on dried samples.

A series of isothermal aging experiments below the beta transition was conducted in order to determine the changes in the activation enthalpy of the beta relaxation with aging time and temperature. The as-received sample was used for comparison. The samples were: 1) as-received, 2) dried 190°C-quenched-no aging, 3) dried 190°C-quenched-aged 15 days at 20°C, 4) dried 190°C-quenched-aged 4.5 days at 35°C and 5) dried 190°C-quenched-aged 23 hours at 60°C. Figures 3a - 3d show the DMTA plots of the beta relaxation for these samples. Comparing the as-received, Figure 3a, to the dried no aging, Figure 3b, showed two distinct differences. One, the low frequency peaks shifted to lower temperatures after drying and two, the small hump in the 100 to 150°C region was removed. Most notably, the 0.03 Hz peak decreased from 70 to 42°C. It is not clear whether the shift of the peaks was due to absorbed

164

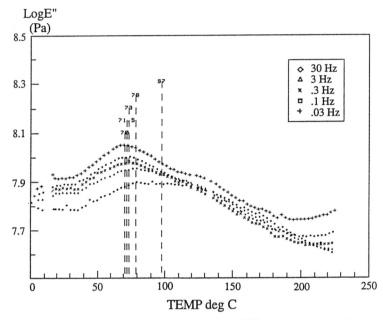

Figure 3a. Loss modulus (E″) vs. temperature at different frequencies for the as-received poly(amide-imide).

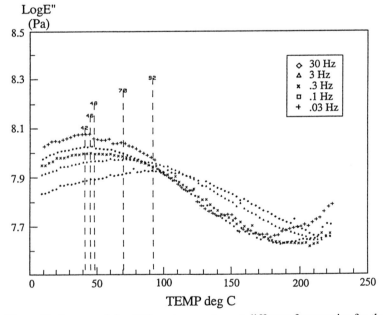

Figure 3b. Loss modulus (E″) vs. temperature at different frequencies for the poly(amide-imide) dried at 190°C-quenched-no aging.

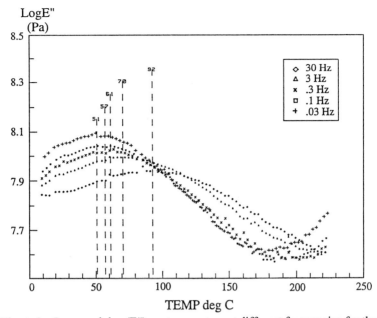

Figure 3c. Loss modulus (E″) vs. temperature at different frequencies for the poly(amide-imide) dried at 190°C-quenched-aged 15 days at 20°C.

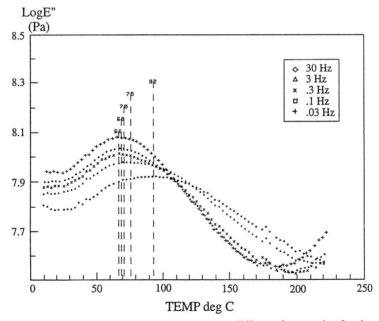

Figure 3d. Loss modulus (E″) vs. temperature at different frequencies for the poly(amide-imide) dried at 190°C-quenched-aged 4.5 days at 35°C.

molecules (water or NMP) modifying the motions related to the beta transition or if it was due to aging at room temperature. Figures 3c-3d were for samples dried and aged at different times and temperatures. Comparing the dried sample with no aging, Figure 3b, to the dried sample aged for 4.5 days at 35°, Figure 3d, showed that the low frequency peaks increased to higher temperatures. Most notably, the 0.03 Hz peak increased from 42 to 66°C. Comparing Figure 3c and 3d showed that the amount the peaks shifted was also dependent upon aging temperature and time. The 20°C aging resulted in the peaks shifting to temperatures intermediate to the unaged and 35°C aged values. Aging at 60°C resulted in a beta relaxation that resembled the unaged sample Figure 3b. These experiments demonstrated that the dynamic mechanical properties of the beta transition were affected by aging below that dispersion. Also, aging at 35°C for 4.5 days resulted in a beta dispersion that was similar to the as-received sample.

The calculated activation enthalpy values are shown in Table 1 for a typical sample aged with the four thermal cycles. These were not the results of four separate samples, but one sample cycled through the four aging profiles. Several samples were analyzed with the four aging experiments to test for repeatability and all resulted in similar loss modulus profiles. Between each experiment the samples were dried at 190°C for more than 7 hours and quench cooled to "reverse" the previous thermal histories. This is analogous to the rejuvenation procedures used with the glass transition.

The data in Table 1 were obtained from the peaks of the loss modulus vs. temperature plots. The E_a values in the first column were derived using Arrhenius assumptions. Comparing the values for the as-received sample to the dried sample with zero aging, showed a decrease in the activation energy from 316 to 125 kJ/M. Aging at room temperature for 15 days increased the activation energy to 170 kJ/M, while aging at 5 days at 35°C increased the activation energy to ~ 260 kJ/M. The $\Delta G^{\ddagger}$, $\Delta H^{\ddagger}$ and $\Delta S^{\ddagger}$ values were calculated from transition state assumptions using a graphical method similar to that of Kauzmann (71). The $\Delta S^{\ddagger}$ values behaved similarly to the $\Delta H^{\ddagger}$ values with aging. This means that not only were the enthalpic interactions increasing with aging time, but a structure was also being developed. These data showed that enthalpic interactions can increase more rapidly with a higher aging temperature, Ta (smaller temperature difference between T_β and Ta) and shorter aging times. Therefore, the rate the activation enthalpy changed occurred faster at a temperature closer to the beta transition temperature maximum. This is similar to physical aging of the glass transition, where aging occurs faster the smaller the difference between the aging temperature and the glass transition temperature.

These data indicate that the activation enthalpy for the beta transition was dependent upon the samples' thermal history below the beta transition. This increased activation enthalpy with aging was anticipated by Read (34), while the peak shift of the beta transition with aging below the beta transition was observed with dielectric analysis by Hougham (31). Other experiments that could be used to identify or infer transient mechanical and water effects might be residual stress or sorption experiments like those conducted by Ree (72) or Silverman (63) on polyimides. To completely reveal how a dispersion changes shape and location with physical aging, we suggest a continued focus on imaging the spectrum of mechanical relaxations associated with T_g

and T$_\beta$ and identification of the functional groups associated with these relaxations. The latter could possibly be done by 2D NMR (73).

Table 1. Activation Energies vs. Aging time for the Beta Transition
of a Poly(amide-imide)

Arrhenius	E$_a$ (kJ/M)		
Kauzmann	$\Delta G^{\ddagger}$ (kJ/M)	$\Delta H^{\ddagger}$ kJ/M	$\Delta S^{\ddagger}$ J/M•K
As Received	316	302	626
No aging Rejuvenated	125	117	123
Aged 15 days @ 20 - 25°C	170	163	245
Aged 4.5 days @ 35°C	260	251	490

The influence of water on the poly(amide-imide) was also determined with dielectric thermal analysis. Two samples were run: one was as-received and the other was dried for 7 hours at 190°C. They are shown in Figures 4a and 4b. Figure 4a illustrates the effects of water on the low temperature loss between -100 and 0°C. There were also ionic conductivity losses between 0 and 70°C. The ionic conductive losses were determined from the slope (-1) of the loss maximum peak height versus log frequency plots. These were attributed to mobile ions. This ionic mobility was dependent on water.

Based on the dielectric and dynamic mechanical data, it appears that water and small polar molecules contribute to three dispersions in this poly(amide-imide). One is the low temperature relaxation between -100 and 0°C. This may be a hydrogen bonded relaxation since the activation enthalpy was 30 kJ/mol. This occurs at concentrations of water ranging between 0 to 4 weight percent. Two, the dielectric relaxation between 0 and 70°C can probably be attributed to conductive contaminants whose mobility is dependent upon a minimum amount of water. Three, at high water concentrations, greater than 2 weight percent, the water/NMP contributes to the beta relaxations observed between 50 and 150°C.

Conclusions

The as-received sample contained a total of 2 weight percent of water and NMP. To remove the residual solvent and water, the Torlon needed to be dried above 190°C for several hours. For the dried and as-received samples, increasing the volatiles from 0 to 2 weight percent increased the low temperature dynamic mechanical loss peak. Also, the dielectric loss showed that there were conductive losses between 0 and 70°C for the

168

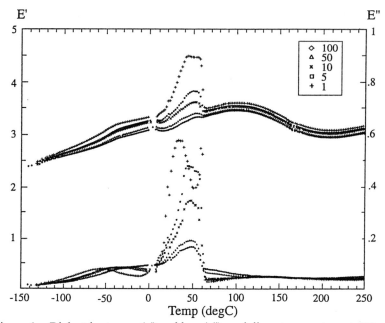

Figure 4a. Dielectric storage (ε') and loss (ε'') moduli vs. temperature at different frequencies (1-100 Hz) for the as-received poly(amide-imide) with 2.5% water.

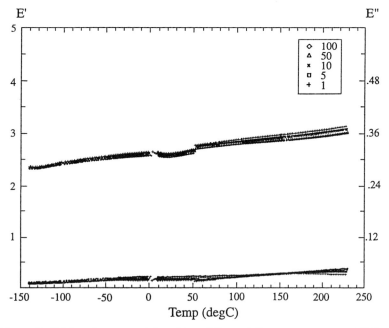

Figure 4b. Dielectric storage (ε') and loss (ε'') moduli vs. temperature at different frequencies (1-100 Hz) for the dried poly(amide-imide) with 0% water.

as-received sample. From 2 to 4 weight percent water the beta transition decreased in size, while the height of the low temperature peak remained unchanged. The location and activation enthalpy for motion of the beta relaxation was shown to be a function of the temperature and time of low temperature aging between 20 and 35°C. The increase in activation enthalpy with aging was reversible. It was only necessary to heat above the beta transition to a temperature of 190°C to reverse aging near the beta relaxation and not above the glass transition temperature of 280°C. This rejuvenation of the beta transition could not have been observed without proper drying of the sample to decouple the effects of aging and water.

Acknowledgments

George John Dallas would like to acknowledge the advice and guidance from Dr. T. C. Ward, Dr. Garth Wilkes and Dr. Herve Marand. I also appreciate the TGA-MS work from Dr. J. Rancourt and the editorial advice from Dr. Mia Siochi.

Literature Cited

1. Tool, A.Q. and D.B. Lloyd, J. Research, **1930**. *5*(National Bureau of Standards): p. 627.
2. Fuoss, R.M., J. Amer. Chem. Soc., **1939**. *61*(9): p. 2255-2334.
3. Struik, L.C.E., *Physical Aging in Amorphous Polymers and Other Materials.* **1978**, Amsterdam: Elsevier.
4. Kovacs, A.J., R.A. Stratton, and J.D. Ferry, J. Phys. Chem., **1963**. *67*(152).
5. Algeria, A., L. Goitiandia, I. Telleria, and J. Colmenero, J. Non-Crys. Solids, **1991**. *131-133*: p. 457-461.
6. Bauwens-Crowet, C. and J.C. Bauwens, J. Non-Crys. Solids, **1991**. *131-133*: p. 505-508.
7. Hadley, D.W. and I.M. Ward, *Mechanical Properties*, in *Encyclopedia of Polymer Science and Engineering*. 1985, John Wiley and Sons: New York. p. 379-466.
8. Hunston, D., W. Carter, and J.L. Rushford, *Linear Viscoelastic Properties of Solid Polymers as Modeled by a Simple Epoxy*, in *Developments in Adhesion*, A.J. Kinloch, Editor. 1980, Applied Science. p. 125-174.
9. Hutchinson, J., Prog. Polym. Sci., **1995**. *20*: p. 703-760.
10. Lee, A. and G.B. McKenna, Polymer, **1988**. *29*: p. 1812-1817.
11. Lee, A. and G.B. McKenna, Polymer, **1990**. *31*(March): p. 423-429.
12. Mijovic, J., L. Nicolais, A. D'Amore, and J.M. Kenny, Polym. Eng. Sci., **1994**. *34*(Mid-March No. 5): p. 381-389.
13. Ngai, K.L. and R.W. Rendell, J. Non-Crys. Solids, **1991**. *131-133*: p. 942-948.
14. Santore, M.M., R.S. Duran, and G.B. McKenna, Polymer, **1991**. *32*(13): p. 2377-2381.
15. Starkweather, H., Macromolecules, **1981**. *14*: p. 1277-1281.
16. Van Krevelen, D.W., *Properties of Polymers*. 3rd ed. **1990**, Amsterdam: Elsevier.

17. Wimberger-Friedl, R. and J.g.d. Bruin, Macromolecules, **1996**. *29*: p. 4992-4997.

18. G'Sell, C. and G.B. McKenna, Polymer, **1992**. *33*: p. 2103-2113.

19. Duran, R.S. and G.B. McKenna, J. Rheo., **1990**. *34*(6): p. 813-839.

20. Boyer, R.F., Polym. Eng. Sci., **1968**. *8*(3): p. 161-185.

21. Boyer, R. *Introductory Remarks: Symposium on Thermomechanical Anaysis*. in *Symposium on Thermomechanical Anaysis*. **1974**: Mercel Dekker.

22. Boyer, R., Polymers, **1976**. *17*(November): p. 996-1008.

23. Hartmann, B. and G.F. Lee, J. Appl. Polym. Sci., **1979**. *23*: p. 3639-3650.

24. McKenna, G.B. and R.W. Penn, Polymer, **1980**. *21*(February): p. 213-220.

25. McKenna, G.B., *Glass Formation and Glassy Behavior*, in *Comprehensive Polymer Science*. 1990, Pergamon: Oxford. p. 311-362.

26. McKenna, G.B., M.M. Santore, A. Lee, and R.S. Duran, J. Non-Crys. Solids, **1991**. *131-133*: p. 497-504.

27. McKenna, G.B. and C.A. Angell, J. Non-Crys. Solids, **1991**. *131-133*: p. 528.

28. McKenna, G.B. and C.A. Angell, J. Non-Crys. Solids, **1991**. *131-133*: p. 528-536.

29. Struik, L.C.E., J. Non-Crys. Solids, **1991**. *131-133*: p. 395.

30. Read, B.E., J. Non-Crys. Solids, **1991**. *131-133*: p. 408-419.

31. Hougham, G., G. Tesoro, and J. Shaw, Macromolecules, **1994**. *27*(3): p. 3642-3648.

32. Dallas, G., *Fundamental Experiments on the Response of Solutions, Polymers and Modified Polymeric Materials to Electromagnetic Radiation*, in *Materials Engineering and Science*. 1992, Virginia Polytechnic Institute and State University: Blacksburg, VA. p. 186.

33. Lee, H. and F.J.M. Garry, J. Mole. Sci. Phy. Ed., **1990**. *B29*(2 & 3): p. 185-202.

34. Read, B.E. and G.D. Dean, Polymer, **1984**. *25*: p. 1679-1686.

35. Venditti, R.A. and J.K. Gillham, ACS Div. Polym. Mat. Sci. and Eng. San Fran., **1992**. *66*: p. 196.

36. Feller III, F.A. and A.B. Brennan. *Physical Aging Behavior of Poly(etherimide)*. in *NATAS*. **1991**. Minneapolis, MN.

37. Feller III, F. and A.B. Brennan. *Relaxation Behavior of a Poly(imide)*. in *NATAS*. **1992**.

38. Chung, T.-S. and E.R. Kafchinski, J. Appl. Polym. Sci., **1996**. *59*: p. 77-82.

39. Arnold, F., K. Bruno, D. Shen, M. Eashoo, C. Lee, F. Harris, and S. Cheng, Polym. Eng. Sci., **1993**. *33*(Mid-November No. 21): p. 1373-1380.

40. Bretz, P.E., *Effect of Moisture on Fatigue Crack Propogation in Nylon 66*, in *Water in Polymers*, S.P. Rowland, Editor. 1980, American Chemical Society: Washington, D. C. p. 531-554.

41. Brown, G.L., *Clustering of Water in Polymers*, in *Water in Polymers*, S.P. Rowland, Editor. 1980, American Chemical Society: Washington, D. C. p. 441-450.

42. Fuzek, J.F., *Glass Transition Temperature of Wet Fibers*, in *Water in Polymers*, S.P. Rowland, Editor. 1980, American Chemical Society: Washington, D.C. p. 515-530.

43. Illinger, J.L. and N.S. Schneider, *Water-Epoxy Interactions in Three Epoxy Resins and Their Composites*, in *Water in Polymers*, S.P. Rowland, Editor. 1980, American Chemical Society: Washington, D. C. p. 571-584.

44. Johnson, G.E., *Water Sorption and Its Effect on a Polymer's Dielectric Behavior*, in *Water in Polymers*, S.P. Rowland, Editor. 1980, American Chemical Society: Washington, D. C. p. 451-468.

45. Moy, P. and F.E. Karasz, *The Interactions of Water with Epoxy Resins*, in *Water in Polymers*, S.P. Rowland, Editor. 1980, American Chemical Society: Washington, D. C. p. 505-514.

46. Starkweather, H.W., *Water in Nylon*, in *Water in Polymers*, S.P. Rowland, Editor. 1980, American Chemical Society: Washington, D.C. p. 433-440.

47. Moore, R.S. and J.R. Flick, *The Influence of Water Concentration on the Mechanical and Rheo-Optical Properties on Poly(methyl methacrylate)*, in *Water in Polymers*, S.P. Rowland, Editor. 1980, American Chemical Society: Washington, D. C. p. 555-570.

48. Apicella, A. and C. Carfagna, J. Appl. Polym. Sci., **1983**. *28*: p. 2881-2885.

49. Apicella, A., J. Appl. Polym. Sci., **1984**. *29*: p. 2083-2096.

50. Apicella, A. and N. Nicolais, *Effect of Water on the Properties of Epoxy Matrix and Composite*, in *Advance in Polymer Science*, K. Dusek, Editor. 1985, Springer-Verlag: New York. p. 69-76.

51. Best, M., J.W. Halley, B. Johnson, and J.L. Valles, J. Appl. Polym. Sci., **1993**. *48*: p. 319-334.

52. Child, T.F., Polymer, **1972**. *13*: p. 259-264.

53. Dagani, R., *Water Cluster Cradles H30+ Ion In Stable Cagelike Structure*, in *Chemical and Engineering News*. 1991. p. 47-48.

54. Derbyshire, W. and I.D. Duff, Disc. Far. Soc., **1974**. *57*: p. 243-253.

55. Dounis, D.V., J.C. Moreland, G.L. Wilkes, and D.A. Dillard, J. Appl. Polym. Sci., **1993**. *50*: p. 293-301.

56. Ellis, B. and H.U. Rashid, J. Appl. Polym. Sci., **1984**. *29*: p. 2021-2038.

57. Errede, L.A., J. Polym. Sci.: Polym. Chem. Ed., **1988**. *26*: p. 3375.

58. Froix, M.F. and R. Nelson, Macromolecules, **1975**. *8*(6): p. 726-730.

59. MacQueen, R.C. and R. Granata, J. Polym. Sci. Part B: Polym. Phy., **1993**. *31*: p. 971-982.

60. McCrum, N.G., B.E. Read, and G. Williams, *Anelastic and Dielectric Effects in Polymers Solids*. **1967**, London: John Wiley and Sons.

61. Schneider, N.S., J.L. Illinger, and F.E. Karasz, J. Appl. Polym. Sci., **1993**. *47*: p. 1419-1425.

62. Seferis, J.C., J.D. Keenan, and J.T. Quinlivan, Journal of Applied Polymer Science, **1979**. *24*: p. 2375-2387.

63. Silverman, R.D., J. Appl. Polym. Sci., **1993**. *47*: p. 1013-1018.

64. Wright, W.W., Brit. Polym. J., **1983**. *15*: p. 224-242.

65. Gillham, J.K., Y. Ozari, and R.H. Chow, J. Appl. Polym. Sci., **1979**. *23*: p. 1189-1201.

66. Mohajer, Y., G.L. Wilkes, and J.E. McGrath, J. Appl. Polym. Sci., **1981**. *26*: p. 2827-2839.

67. Robeson, L.M., A.G. Farnham, and J.E. McGrath, *Dynamic Mechanical Characteristics of Polysulfone and Other Polyarylethers*, in *Midland Macromolecular Institute Monographs*, D.J. Meier, Editor. 1978, Gordon Breach. p. 405-426.
68. Plazek, D.J., J. Non-Crys. Solids, **1991**. *131-133*: p. 836-851.
69. Pimentel, G. and A.L. McClellan, *The Hydrogen Bond*, ed. L. Pauling. **1960**, San Francisco: W. H. Freeman.
70. von Hippel, A., IEEE Tran. Ele. Ins., **1988**. *23*(5): p. 817-840.
71. Kauzmann, W., Rev. Mod. Phys., **1942**. *14*: p. 12-44.
72. Ree, M. and T. Nunes, Polymer, **1992**. *33*(6): p. 1228.
73. Spiess, H.W., J. Non-Crys. Solids, **1991**. *131-133*: p. 766-772.

Chapter 12

Low-Temperature Relaxation of Polymers Probed by Photochemical Hole Burning

Shinjiro Machida[1], Satomi Tanaka[1], Kazuyuki Horie[1], and Binyao Li[2]

[1]Graduate School of Engineering, The University of Tokyo, 7-3-1 Hongo, Bunkyo-ku, Tokyo 113, Japan
[2]Changchun Institute of Applied Chemistry, Chinese Academy of Sciences, Stalin Street 109, Changchun 130022 Jilin, Peoples Republic of China

Photochemical hole burning spectroscopy is applied to estimate the low-temperature relaxation properties of polymers. Methacrylate polymers, polyethylene derivatives, and polymers with aromatic groups were studied. Both slight and large changes of microenvironments around doped dye molecules are detected by temperature cycling experiment. Relaxation behavior of the polymers is discussed in relation to their chemical structures. The low-energy excitation mode of each polymer is also estimated and discussed.

There are many methods for studying the structure and properties of glassy polymers. Photochemical hole burning (PHB), or more generally, persistent spectral hole burning (PSHB) is one of the sensitive ways to study structure-property behavior at low temperatures. PHB is a phenomenon where a narrow dip (or a "hole") is created in the inhomogeneous absorption band of a dye embedded in a polymer matrix by photoexcitation. From the PHB measurement we can obtain various types of information about structure and properties of glassy systems such as low-energy excitation, structural relaxation, and electron-phonon interaction (*1,2*).

One of the typical experimental procedures in PHB measurements is temperature cycling. The temperature cycle experiment gives irreversible changes in the microenvironments of dye molecules caused by thermal activation (*3*). Here we present the results of irreversible hole broadening and hole filling in temperature cycling experiments for various types of polymers. By comparing the results with other relaxation data of the polymers, we attribute the origin of the irreversible changes of hole profiles. We also discuss factors which determine the relaxation properties of polymers in relation to their chemical structures.

In addition to the temperature cycle measurement, we estimate the low-energy excitation mode or phonon frequency of the polymers. The low-energy excitation modes are believed to relate to the excess heat capacity of amorphous materials (*4*). We have reported that the low-energy excitation modes of glassy polymers can be estimated from pseudo-phonon side holes in PHB spectra (*5*). The low-energy excitation modes are very sensitive to the microscopic ordered structure (*6*) and agree well with those measured by heat capacity and neutron inelastic scattering.

We discuss the obtained low-energy excitation modes of the polymers from the aspect of their chemical structures and relaxation properties.

Experimental

As host polymer matrices we used poly(alkyl methacrylate)s with various ester groups (methyl: PMMA, ethyl: PEMA, isopropyl: PiPMA, normalpropyl: PnPMA, isobutyl: PiBMA, normalbutyl: PnBMA), poly(methyl acrylate), polyethylene (PE), poly(vinyl chloride) (PVC), poly(vinylidene chloride) (PVDC), poly(vinyl alcohol) (PVOH), polystyrene (PS), polycarbonate (PC), phenolphthalein polyether-ketone (PEK-C), and phenolphthalein polyether-sulfone (PES-C).

X = CO Phenolphthalein polyether-ketone (PEK-C)
X = SO$_2$ Phenolphthalein polyether-sulfone (PES-C)

Scheme. Structures of PEK-C and PES-C.

Tetraphenylporphine (TPP) was used as a guest dye molecule except for PVOH. Tetra(4-sulfonatophenyl)porphine (TPPS) was used for the PVOH sample. Free-base porphyrins such as TPP and TPPS undergo a hydrogen tautomerization reaction by photoexcitation at low temperatures. The hydrogen tautomerization is the mechanism of hole burning for the measured systems. When the TPP undergoes the hydrogen tautomerization, the chemical structure of the TPP itself does not change at all. However, the interaction between the TPP and the surrounding matrix changes due to the tautomerization reaction giving rise to the frequency shift of the dye.

Most samples were prepared by solvent casting, drying under vacuum at T_g, and molding under 1 kg/cm^2 at T_g. The PVOH sample was prepared only by solvent casting, and the PE sample was prepared by melting and quenching. The samples are 0.5-1.0 mm thick and contain about 1×10^{-3} mol/kg of the dye molecules.

Samples were placed in a cryostat with a cryogenic refrigerator and were cooled down to about 20 K. Hole burning was performed by irradiating an Ar$^+$-laser-pumped dye laser with the intensity of 1-10 mW/cm^2 to the lowest S$_0$-S$_1$ transition of TPP around 640 nm for 1-10 min. Burned holes were detected by a photomultiplier and a lock-in amplifier through the change in intensity of the transmitted light of a halogen lamp by scanning a 1-m monochromator.

Results and Discussion

Methacrylate Esters. Figure 1 shows typical hole profiles of TPP/PMMA during the temperature cycle experiment and a schematic diagram of the temperature cycling. At first a hole is burned and measured on the profile at 20 K. Then the temperature of the sample is elevated to a certain excursion temperature. The sample is kept at the excursion temperature for 5 min. Then we cool the sample again to 20 K and measure the hole profile. We repeat the temperature cycle with increasing elevated temperature.

The irreversible change in the hole profile includes two parts. One is an increase in hole width and the other is a decrease in hole area. The increase in hole width, called spectral diffusion, means small frequency shifts (or small microenvironments change) of each dye molecule. Slight changes of matrix conformation are the cause of this phenomenon. The decrease in the hole area, which is called hole filling, corresponds to relatively large shifts of resonant frequencies of dye molecules. The backward reaction of hole burning processes or large changes of microenvironments of the dye molecules can induce the hole filling phenomenon. Because the thermally induced hydrogen tautomerization of TPP does not occur below 100 K, the decrease in hole area below 100 K reflects large changes of microenvironments caused by some relaxation of the matrix polymers.

Figure 2 shows the increase in hole width measured at 20 K after cooling from excursion temperatures for TPP in PMMA, PEMA, PiPMA, PnPMA, PiBMA, and PnBMA. The values in parentheses correspond to the initial hole width before temperature cycling. The hole broadening profiles can be classified into three groups, namely, ①PMMA, ②PEMA and PiPMA, and ③PnPMA, PiBMA, and PnBMA. This classification corresponds to the number of serial carbon atoms in the ester unit. PMMA has only one carbon atom in the ester and shows no marked relaxation below 100 K. In PEMA and PiPMA, which have two serial carbon atoms in the sidechain, the hole width increases gradually from 20 K and there is a plateau region around 60 K. The holes burned in PnPMA, PiBMA, and PnBMA, which contain more than three serial carbon atoms in their ester, begin to broaden suddenly from ~50 K.

Low temperature relaxation properties of methacrylate polymers with various ester groups have been studied by dielectric loss (7) and mechanical loss (8) measurements. The relaxation curves of hole width are quite similar to those observed dielectrically and mechanically. The relaxation peaks of dielectric loss at low temperatures in PiPMA, PiBMA, and PnBMA have been attributed to the rotation of the end ester groups from the results of potential energy calculation (7). Therefore, we also attribute the cause of the hole broadening for these polymers to the rotation of the end ester groups. In PMMA although the rotation of the end methyl group must begin below 20 K, such rotation should have no influence on the microenvironments of the TPP.

The change in hole area during the temperature cycling experiments for TPP in methacrylate polymers is shown in Figure 3. The hole area is normalized by the initial value at 20 K. The hole filling behavior of these systems can be classified into the same groups as the hole broadening result shown in Figure 2. In other words, the hole area begins to decrease when the hole width begins to increase in each system. As is mentioned above, the decrease in hole area reflects relatively larger changes of microenvironments than the increase in hole width does. The TPP molecule is so large that there are many sidechain units around the TPP dye. The rotation of many sidechain esters is likely to induce the large changes of microenvironments. Therefore, we conclude that the rotation of end ester groups is also the mechanism of the thermally induced hole filling.

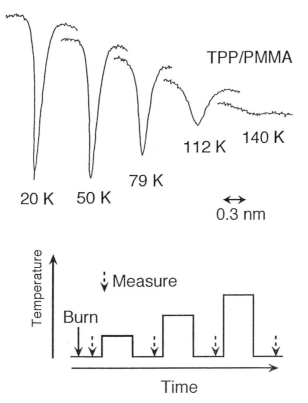

Figure 1. Upper half: Typical hole profiles of TPP/PMMA during temperature cycling. The values in ( ) correspond to the initial hole width before temperature cycling. Lower half: Schematic diagram of temperature cycle experiment.

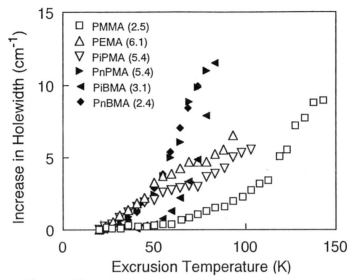

Figure 2. Excursion temperature dependence of decrease in hole area measured for methacylate polymers.

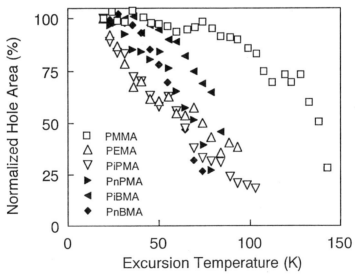

Figure 3. Excursion temperature dependence of normalized hole area measured for methacrylate polymers.

Comparison of Acrylate with Methacrylate.   The excursion temperature dependence of the increase in hole width for PMA and PMMA is shown in Figure 4. In these polymers the rotation of the end methyl group has no influence on the microenvironments of the dyes and the rotation of the whole sidechain ester group (β-relaxation) does not proceed below 150 K. Therefore, we attribute the cause of the hole broadening and hole filling in these polymers to some segmental motion of the mainchain. In PMA the increase in hole width is a little larger than that in PMMA between 50 K and 120 K. This is likely because the α-methyl group enhances the steric hindrance for slight segmental motion of the mainchain. This is consistent with the fact that PMA has a lower glass transition temperature than PMMA.

Figure 5 shows the normalized hole area as a function of excursion temperature for TPP/PMA and TPP/PMMA. The hole area of PMA system decreases at much lower temperatures than that of PMMA. In PMA large segmental motions which induce hole filling begin at much lower temperatures than in PMMA. This can also be explained by the smaller steric hindrance for segmental motion in PMA as is explained in the case of hole width. However, the difference in the results of hole filling between the two polymers is larger than the difference in hole broadening behavior. The segmental motion of the mainchain is likely to need larger conformational change of the polymer compared to the sidechain rotation. Therefore, the absence of α-methyl groups seems to be reflected more sensitively by the large changes of microenvironments leading to hole filling than by the small ones leading to spectral diffusion.

Polyethylene and Its Derivatives.   Figure 6 shows the increase in hole width measured at 20 K after cooling from excursion temperatures in PE, PVC, and PVDC. Because these polymers have no sidechains, the origin of the irreversible hole broadening in these polymers is attributed to some segmental motion of the mainchain. PE has no relaxation below 40 K, but it shows a sudden increase in hole width above 50 K. This hole broadening profile of PE is quite similar to that observed in PnPMA, PiBMA, and PnBMA. Polymer systems with more than three serial carbon atoms seem to show sudden relaxation beginning at ~50 K whether these relaxations occur in the sidechain or the mainchain. In PVC and PVDC the hole width increases almost linearly from 20 K. The extent of chain packing for these polymers would be smaller than in PE due to the existence of chlorine atom, which might be the reason for the hole broadening below 40 K. In Figure 6 PVC shows a larger increase in hole width than does PVDC. This is likely because PVC has a smaller steric hindrance for slight segmental motion than PVDC due to the absence of a chlorine atom.

In Figure 7 the decrease in normalized hole area of TPP/PE, PVC, and PVDC are shown. In PE the hole area decreases suddenly from ~50 K. This is similar relaxation profile with the increase in hole width of the PE. The hole area observed in PVDC changes almost linearly with the hole width in the system. However, PVC shows a smaller decrease in hole area than PVDC above 50 K in spite of the larger increase in hole width in PVC than in PVDC. In these polymers the large segmental motion of the mainchain leading to hole filling might proceed in different ways from the small one which induces hole broadening.

In Figure 8 the hole broadening behavior of PE and PVOH are shown. PVOH has no marked relaxation below 100 K and shows the best thermal durability of the burned hole among the polymer systems. PVOH has hydroxyl groups which undergo hydrogen bonding. We conclude that the interchain hydrogen bonds in PVOH suppress the segmental motion of the mainchain. The PVOH system shows no marked relaxation also in the decrease in hole area.

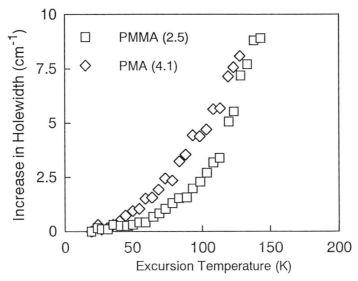

Figure 4. Excursion temperature dependence of increase in hole width of PMMA and PMA. The values in ( ) correspond to the initial hole width before temperature cycling.

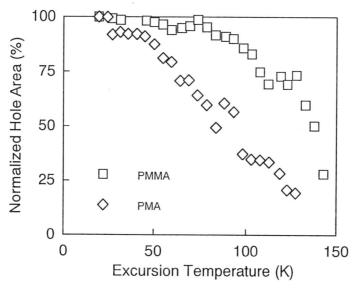

Figure 5. Excursion temperature dependence of normalized hole area of PMMA and PMA.

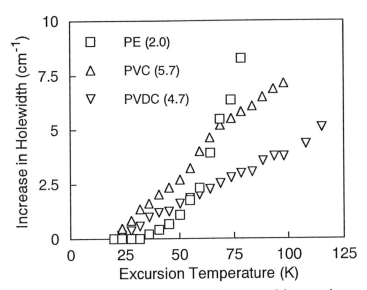

Figure 6. Excursion temperature dependence of increase in hole width of PE, PVC, and PVDC. The values in ( ) correspond to the initial hole width before temperature cycling.

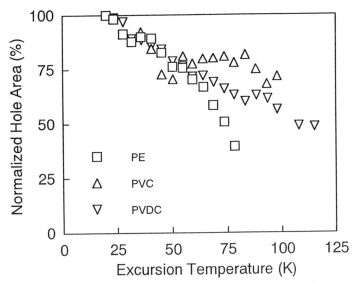

Figure 7. Excursion temperature dependence of normalized hole area of PE, PVC, and PVDC.

Polymers with Aromatic Rings. Figure 9 shows the excursion temperature dependence of the increase in hole width for PEK-C, PES-C, PC, and PS. The extent of hole broadening is greater in PES-C than that in PEK-C from 40 K to 80 K. This is because the O=S=O group is bigger than the C=O group and there is more free volume for slight segmental motions in PES-C than in PEK-C. This interpretation is consistent with the result of PC which has a C=O group in the mainchain. PC shows a quite similar hole broadening curve with PEK-C below 75 K. The sudden increase in hole width of PC above 75 K might be due to the fluctuation around O-CO-O groups in the mainchain. In Figure 9 PS shows relaxation from much lower temperature than other polymers with aromatic mainchain. The phenyl group attached to the sidechain can fluctuate more easily than the phenyl group in the mainchain and thus PS showed relaxation from lower temperature than polymers with an aromatic mainchain.

The excursion temperature dependence of the hole area for these aromatic polymers (PEK-C, PES-C, PC, and PS) are shown in Figure 10. The curves of the hole area do not show so clear a difference between each polymer as the hole width profiles. However, PC shows a small decrease in hole area among the polymers. We attribute this to the absence of bulky units like phenyl rings in the sidechain of PC. The presence of bulky units in the sidechain would make some local free volume for relatively large segmental motions at low temperatures.

Estimation of Low-Energy Excitation Modes. An absorption line profile of each dye molecule at low temperature consists of two components: a sharp zero-phonon line and a broad phonon side band. The energy difference between the zero-phonon line and the phonon side band coincides with the low-energy excitation mode or phonon frequency of the matrix when multi-phonon processes can be ignored.

Figure 11 shows a saturated hole profile burned to estimate the low-energy excitation mode, E_s, of TPP/PnPMA. A deep zero-phonon hole is created at 644.8 nm. In addition, there are two broad holes at both sides of the zero-phonon hole. The broad hole at the shorter wavelength is called a "phonon side hole" and the one at the longer wavelength is called a "pseudo-phonon side hole". The phonon side hole consists of a phonon side band of the zero-phonon hole, and the pseudo-phonon side hole is made of the zero-phonon line of the reacted molecules which have been excited via phonon side band.

Table 1 lists the values of E_s estimated from the energy difference between zero-phonon holes and pseudo-phonon side holes of the measured polymers. In Table 1 the E_s of other several polymer systems are also shown. The E_s is large in polymers with hydrogen bonds such as PVOH and poly(2-hydroxyethyl methacrylate). The interchain hydrogen bonds form some rigid microscopic structure leading to a high phonon frequency. Methacrylate and acrylate polymers have similar values of E_s which are in the range of 13-15 cm^{-1}. This result is in contrast to the relaxation behavior observed in temperature cycle experiments which markedly depends on the kind of ester unit. Therefore, the low frequency phonon of matrix polymers has quite different origin from the thermal relaxation observed from irreversible changes in hole profiles. In Table 1 polymers with a phenyl ring in the sidechain show relatively low E_s (10-11 cm^{-1}). This is probably due to the low frequency vibration mode of the sidechain phenyl groups.

Conclusion

We measured the irreversible changes in hole profiles and the low-energy excitation modes for several porphyrin-doped polymer systems. The cause of the hole broadening

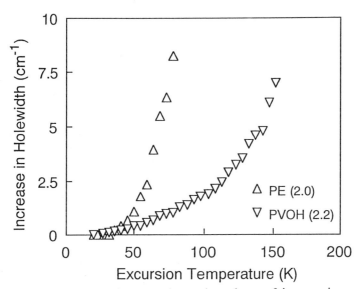

Figure 8. Excursion temperature dependence of increase in hole width of PE and PVOH. The values in ( ) correspond to the initial hole width before temperature cycling.

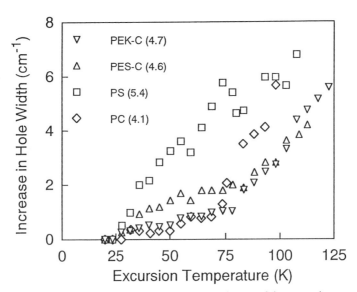

Figure 9. Excursion temperature dependence of increase in hole width of PEK-C, PES-C, PS, and PC. The values in ( ) correspond to the initial hole width before temperature cycling.

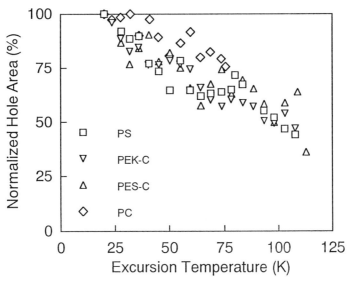

Figure 10. Excursion temperature dependence of normalized hole area of PEK-C, PES-C, PS, and PC.

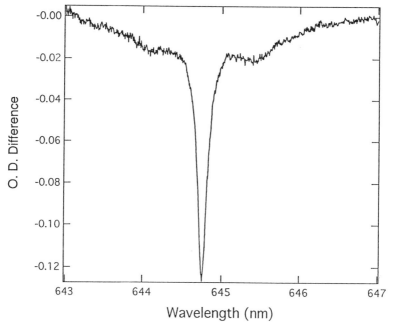

Figure 11. A typical hole profile burned for TPP/PnPMA.

Table I. Low-Energy Excitation Modes of Measured Polymers.

Polymer	Es (cm^{-1})
PMMA	13
PiPMA	13
PnPMA	15
PiBMA	13
PMA	14
PE	17
PVOH	24
PEK-C	11
PES-C	11
PS	10
PHEMA (*9*)	23
PET (undrawn) (*10*)	11
PET (5-times drawn) (*10*)	12
phenoxy resin (*11*)	15
epoxy resin (DGEBA/EDA) (*12*)	15
epoxy resin (DGEBA/HMDA) (*12*)	17

PHEMA: poly(2-hydroxyethyl methacrylate)
PET: poly(ethylene terephthalate)
DGEBA: diglycidyl ether of bisphenol A, EDA: ethlene diamine
HMDA: hexamethylene diamine.

in methacrylate polymers was attributed to the rotation of the end ester groups. For other polymers, some local segmental motion of the mainchain is concluded to induce an irreversible increase in hole width and an irreversible decrease in hole area. Steric hindrance induced by substituents attached to the mainchain lead to a smaller increase in hole width, which should be due to the enhanced steric hindrance for segmental motion. Polymers which are expected to have larger free volume showed a greater extent of relaxation. Interchain hydrogen bonds suppress the relaxation of the mainchain dramatically. In some polymers decrease in hole area showed different behavior from increase in hole width. This fact indicates that relatively large segmental motion occurs in a different way from slight one in such polymers. The low-energy excitation modes are larger in the polymers with hydrogen bonds and smaller in the polymers with aromatic rings.

The changes of microenvironments around the doped dye molecules have a strong correlation with the characters of matrix polymers such as steric hindrance for segmental motion, amount of free volume, and interchain hydrogen bonds. By the PHB measurements, we can detect both slight and large environmental change around the dyes doped in polymers at low temperatures. The PHB spectroscopy is expected to be extended to other polymer systems to study low-temperature relaxation properties of them.

Literature Cited

1. *Persistent Spectral Hole Burning; Science and Applications;* Moerner W. E. ed., Springer-Verlag, Berlin, Heidelberg, New York, **1988**.
2. Horie K.; Machida S., In *Polymers as Electrooptical and Photooptical Active Media;* Shibaev V. ed., Springer-Verlag, Berlin, Heidelberg, New York, **1996**, pp1-36.
3. Tanaka S.; Machida S.; Horie K., *Macromol. Chem. Phys.*, **1996**, *vol.197*, pp 4095-4103.
4. Horie K.; Furusawa A., In *Macromolecules 1992;* Kahovec J., Ed.; VSP BV, Utrecht, **1993**, pp 313-321
5. Du Y.; Horie K.; Ikemoto M., *Polym. Eng. Sci.*, **1994**, *34*, pp 1362-1367.
6. Furusawa A.; Horie K.; Mita I., *Chem. Phys. Lett.*, **1989**, *161*, pp 227-231.
7. Shimizu K.; Yano O.; Wada Y., *J. Polym. Sci.*, **1975**, 13, pp 1959-1974.
8. Heijboer J.; Pineri M.; In *Nonmetallic Materials and Composites at Low Temperatures 2;* Hartwig G.; Evans D., Eds., Plenum Press, New York, **1982**, pp 89-116.
9. Saikan S.; Kishida T.; Kanematsu Y.; Aota H.; Harada A.; Kamachi M., *Chem. Phys. Lett.*, **1990**, *171*, pp358-36.
10. Horie K.; Kuroki K.; Itaru M.; Ono H.; Okumura S.; Furusawa A., *Polymer*, **1991**, *32*, pp 851-855.
11. Furusawa A.; Horie K.; Kuroki K.; Mita I., *J. Appl. Phys.*, **1989**, *66*, pp 6041-6047.
12. Furusawa A.; Horie K.; Suzuki T.; Machida S.; Mita I., *Appl. Phys. Lett.*, **1990**, *57*, pp 141-143.

PHYSICAL AGING

Chapter 13

Fitting Differential Scanning Calorimetry Heating Curves for Polyetherimide Using a Model of Structural Recovery

Sindee L. Simon

Department of Chemical Engineering, University of Pittsburgh, Pittsburgh, PA 15261

Differential scanning calorimetry (DSC) heating curves obtained at 10°C/min for polyetherimide cooled at various rates are fit using the Tool-Narayanaswamy-Moynihan model of structural recovery coupled with Scherer's equation which describes the dependence of the characteristic relaxation time on structure and temperature. The parameters for Scherer's equation are obtained from the temperature dependence of shift factors needed to form reduced curves for isothermal enthalpy recovery experiments and isothermal creep data. Hence the equation for the relaxation time correctly describes the limiting behavior in both glass and equilibrium regimes over large ranges of temperature. Still, the model parameter β which describes the nonexponentiality is found to decrease as the cooling rate decreases. Thermal gradients in the DSC sample have been postulated to be responsible but incorporation of the effect of a thermal gradient in the model calculation did not impact the dependence of β on thermal history.

Models of structural recovery [1,2,3,4,5,6,7] account for both the nonlinearity and the nonexponentiality of the structural recovery process. The Tool-Narayanaswamy-Moynihan formulation, in particular, has been used to successfully describe the results of differential scanning calorimetry (DSC) enthalpy recovery experiments.[8,9,10] However, there are unresolved issues pertaining to the models [4,7,11,12], one of which is the apparent dependence of model parameters on thermal history. It has been postulated that this may be due to the presence of thermal gradients within the sample during DSC experiments. [13,14,15] Another possibility may be that the Tool-Narayanaswamy equation generally used for the characteristic relaxation time τ is Arrhenius and hence, only describes the structure and temperature dependence over a relatively small temperature range with a single set of parameters. In this work, we will show that using a more complicated equation capable of describing the limiting temperature dependence of τ in both the glass and equilibrium regimes over a wide temperature range with a single set of parameters still results in other model parameters being dependent on thermal history. We then show that incorporation of the effects of the thermal gradient in the sample itself cannot account for the problem. The results

188

indicate that other explanations for the dependence of model parameters on thermal history need to be explored. Various other explanations have been proposed [12] (e.g., the assumption of thermorheological and thermostructural simplicity, the use of reduced time in the generalized stretched exponential Kohlrausch-William-Watts (KWW) [16,17], or the assumption that tau depends on the average and instantaneous temperature and fictive temperature of the system).

This work is organized as follows. We first review the Tool-Narayanaswamy model of structural recovery and discuss equations for the relaxation time τ_o and how the equation parameters were obtained from isothermal data. We then fit the experimental DSC heating curves using the model and show that the stretched exponential parameter β depends on thermal history despite an equation for the relaxation time which describes the limiting temperature dependence in the glass and equilibrium regimes. Next, we discuss how the effects of the thermal gradient were incorporated into the model of structural recovery. Finally, we compare the fit of the modified and unmodified models and show that incorporating thermal gradients does not impact the dependence of the parameter β on the thermal history.

Tool-Narayanaswamy Model of Structural Recovery
Moynihan's formulation [3] of the Tool-Narayanaswamy [1] model is used in this work. In Moynihan's equations, the fictive temperature, T_f, originally defined by Tool [18], is used as a measure of the structure of the glass. The evolution of fictive temperature is represented by the generalized stretched exponential Kohlrausch-William-Watts (KWW) function [16,17]:

$$\frac{dT_f}{dT} = 1 - \exp\left[-\left\{\int_0^t (dt/\tau_0)\right\}^\beta\right] \tag{1}$$

The nonexponentiality of the process is described by β. The nonlinearity is incorporated into the model by allowing the characteristic relaxation time τ_0 to be a function of both temperature and structure (T_f). Given an equation for τ_0, equation 1 can be solved numerically for a given thermal history which begins at a temperature above T_g where the material is in equilibrium, as described by Moynihan et al.[3] In order to compare the calculated evolution of the fictive temperature during a DSC scan to experimental data, the heat capacity is calculated from the temperature derivative of the fictive temperature [19]:

$$C_p(T) = C_{pg}(T) + \Delta C_p(T_f)\frac{dT_f}{dT} \tag{2}$$

where $C_{pg}(T)$ is the glass heat capacity and $\Delta C_p(T_f)$ is the difference between the liquid and glass heat capacities. Both heat capacities are extrapolated to temperatures T and T_f, respectively. We have found [20] that for the polyetherimide examined in this work,

$$C_{pg}(T) = 0.94 + 0.0034 \, T(°C) \quad \text{from 125 to 165°C} \quad (J/g \, °C) \tag{3}$$
$$\Delta C_{pl}(T) = 0.57 - 0.0017 \, T(°C) \quad (J/g°C) \tag{4}$$

The equation for ΔC_p is valid only in the vicinity of the glass temperature since it was obtained by extrapolating the values of glass and liquid heat capacities into the transition region. The reported values for C_p and ΔC_p are comparable to those obtained by Hay et al. [21].

Phenomenological Equation for the Relaxation Time

Many phenomenological equations have been proposed for relating the relaxation time τ_0 to temperature and structure (T_f) [1,22,23,24,25]. The most-widely used is the Tool-Narayanaswamy equation [1], an Arrhenius-like equation which does not describe the observed Vogel-Tammann-Fulcher (VTF) [26] (or Williams-Landell-Ferry WLF [27]) temperature dependence of polymers in the equilibrium state. However, since the annealing peak observed in DSC experiments occurs over a narrow temperature range, an Arrhenius approximation is adequate for describing the data. In fact, Moynihan obtained identical fits using the Tool-Narayanswamy equation and an equation based on Adam and Gibb's theory (see below).[12] A problem may arise, however, when fitting multiple DSC curves which have annealing peaks at significantly different temperatures – in such a case, the apparent activation energy may need to be varied in order to fit all of the data.

To avoid the need to vary the apparent activation energy with thermal history, an equation could be used which gives the observed VTF dependence at equilibrium. Two such equations have been derived based on the Adam and Gibbs' theory for the configurational entropy [28]. The equation of Hodge [22] was derived assuming the configurational change in heat capacity at T_g varies as C/T, where C is a constant:

$$\ln \tau_o \;=\; \ln A \;+\; \frac{D/RT}{1-\dfrac{T_2}{T_f}} \tag{5}$$

whereas the equation by Scherer [23] was derived assuming the configurational change in heat capacity at T_g varies as linearly with temperature (i.e., a + bT):

$$\ln \tau_o = \ln A + \frac{D/RT}{\left[a\ln\!\left(\dfrac{T_f}{T_2}\right) + b\!\left(T_f - T_2\right) \right]} \tag{6}$$

where D/R, T_2, ln A, a, and b are taken to be constants. Equation 5 has the same number of constants as the Tool-Narayanaswamy equation, whereas equation 6 introduces one more fitting parameter. (The constant a/b is taken as an additional fitting parameter because although the temperature dependence of ΔC_p can be measured, the temperature dependence of the configurational contribution to ΔC_p is required in the model). Both equations 5 and 6 are capable of describing the VTF behavior in the equilibrium regime, as was shown many years ago for equation 5 [29]. Both equations also describe an Arrhenius-like temperature dependence in the glass. The primary difference between the two equations that is relevant to this work is that for a given VTF temperature dependence in the equilibrium state, the Arrhenius temperature dependence in the glass is fixed for equation 5, whereas due to the additional parameter in equation 6, the apparent activation energy in the glass can vary independently.

One approach to find the parameters for equations 5 or 6 is by a best fit of the DSC data. A more satisfying approach may be to obtain the parameters from the temperature dependence of shift factors used to reduce isothermal data, where the shift factor is simply the difference between the $\ln \tau_o$ at the temperature of interest and the $\ln \tau_o$ at a reference temperature, taken in this work to be the nominal glass temperature of 207.5°C:

$$\ln a_T = \ln \frac{\tau_o}{\tau_{o,\,ref}} \tag{7}$$

We now introduce the data and data reduction procedures used to obtain the parameters in equations 5 and 6.

The VTF behavior in the equilibrium regime is determined from the temperature dependence of isothermal creep recovery data at temperatures above T_g. Although it may not be obvious that creep and enthalpy should have the same temperature dependence, previous results from our laboratory [30] indicate that this is the case for polyetherimide over the narrow temperature range studied near T_g. An additional reason for using the creep data to determine the equilibrium temperature dependence is that the data was obtained over a broad temperature range. Figure 1 shows isothermal creep recovery data at temperatures ranging from 210 to 298°C for the polyetherimide studied (unpublished data courtesy of B. Lander and D. J. Plazek). Also shown in Figure 1 is the reduced curve obtained by time-temperature superposition of the data to a reference temperature of 207.5°C. Data at the two highest temperatures (272 and 298°C) do not superpose perfectly indicating degradation may have occurred at these temperatures. A comparison of the reduced curve to other data on polyetherimide previously published from our laboratories [30] shows a discrepancy of one and a half decades at the nominal T_g of 207.5°C. This corresponds to a 4°C temperature difference and may be attributable to plasticization of the previous specimen. The temperature dependence of the equilibrium shift factors obtained from the reduction of the recoverable compliance curves follows the VTF equation and can be fit by either equation 5 or 6. However, to obtain the additional parameter in equation 6, we need shift factor data in the glassy regime where $T_f \neq T$.

Figure 2 shows the change in the fictive temperature during isothermal enthalpy recovery experiments for the polyetherimide at several temperatures. The change in the fictive temperature during aging increases logarithmically with aging time and then levels off once equilibrium is achieved. The values presented were obtained by dividing the enthalpy change measured by DSC [30] by the change in heat capacity extrapolated to the aging temperature:

$$\Delta T_f = T_{f,unaged} - T_f = \frac{\Delta H_a}{\Delta C_p(T_a)} \tag{8}$$

where $T_{f,unaged}$ is the initial fictive temperature of the glass (= 215.3°C for all but the highest aging temperature). A horizontal shift of this data (time-temperature superposition) to the aging curve at 207.5°C is shown in Figure 3 and results in a curve with a common asymptote in the glassy region. Data deviate from the common asymptote as they approach equilibrium. The shift represents an iso-T_f shift for the case where $T_f > T_a + 1K$. The shift factors can thus be used to obtain the temperature dependence of the relaxation time τ_o along the glass line. The shift factors differ slightly from those obtained from a simple shift of ΔH_a versus log time data [30]; that shift was performed prior to our obtaining the temperature dependence of ΔC_p and assumed a constant ΔC_p which is equivalent to assuming that ΔH_a is directly proportional to T_f which is not strictly valid if ΔC_p is a function of temperature.

Figure 4 shows the temperature dependence of the shift factors resulting from the reduction of the equilibrium creep data and the isothermal enthalpy recovery data shown in Figures 1, 2, and 3. The fit of equations 5 and 6 are also shown. $T_{f,ref}$ is taken as 207.5°C. The parameters used in the fits are, for equation 5: $D/R = 1920$ and $T_2 = 162°C$; and for equation 6: $D/Ra = 780$, $b/a = -1/560K$ and $T_2 = 151°C$. Equation 6 seems to be better able to describe the temperature dependence in the glass due to the additional parameter. For modeling enthalpy recovery, however, the results are expected to be identical with either equation since the temperature range of interest is

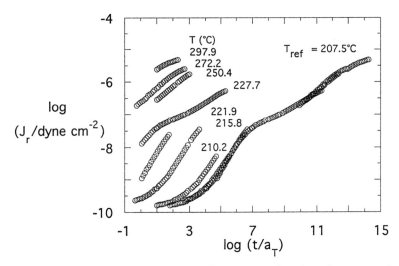

Figure 1: The recoverable creep compliance as a function time at various temperatures and the reduced curve at a reference temperature of 207.5°C. Unpublished data, courtesy of B. Lander and D. J. Plazek.

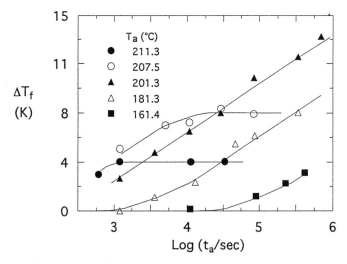

Figure 2: The change in fictive temperature during structural recovery at selected aging temperatures as a function of aging time. Lines indicate the trends in the data.

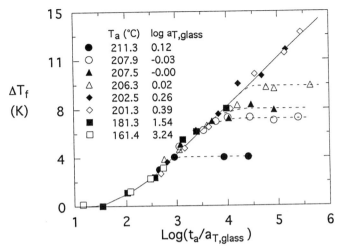

Figure 3: Time-temperature superposition of the change in fictive temperature versus log aging time data to a common asymptote in the glassy state. The reference temperature is 207.5°C. Lines indicate the trends in the data

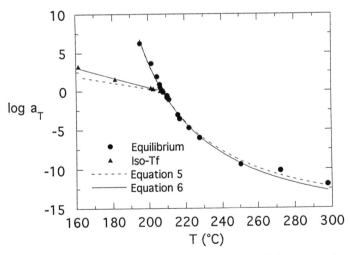

Figure 4: Temperature dependence of the iso-structural shift factors used to obtain the reduced curves shown in Figures 1 and 3. The dashed line indicates the fit of equation 5; the solid line the fit of equation 6.

near and above T_g where the equations overlap. Equation 6 was used to compute the curves shown in this work.

Fitting DSC Cooling Curves

Experimental DSC heating curves are shown by the data points in Figure 5 for cooling rates of 0.1 and 10°C/min. The data were obtained using a Perkin Elmer DSC 7, equipped with a custom-designed water cooling system which yields a stable and reproducible baseline, as described in more detail elsewhere [20]. As the cooling rate decreases, the amount of structural recovery which can occur during the cooling ramp increases and the annealing peaks observed in the DSC heating scans increase in area (as ΔH_a increases and T_f decreases) and move to higher temperatures (as the characteristic relaxation time increases). The behavior is consistent with the results of other researchers.[3,7,31,32]

The fit of the Moynihan-Tool-Narayanaswamy model is also shown in Figure 5 by the solid lines. Since the parameters in the phenomenological equation for τ_o were determined independently as described in the previous section, the curve fits were made using only two adjustable parameters ($\tau_{o,ref}$ and ß). The fits were made such that the height and placement of the calculated annealing peak corresponded to the experimentally observed value (rather than doing a least squares fit of the entire curve). The value of ß decreases as the cooling rate decreases, as shown in Figure 6 for fits to data over a wider range of cooling rates. The trend in ß indicates that physical aging results in a broadening of the relaxation time distribution. This result is consistent with the results of Torkelson et al. [33] who have found that ß decreases with decreasing temperature, as well as with the work of other researchers [34,35,36,37] who have found that thermorheological simplicity and structural-rheological simplicity do not hold. The value of $\ln \tau_{o,ref}$ /sec also changes but the changes are not as significant, ranging 8.7 to 9.5 over the same range of cooling rates, first decreasing then increasing with increasing cooling rate.

Incorporating Thermal Gradients into the Model Calculations

The presence of a thermal gradient in the DSC sample will affect DSC heating curves by broadening the annealing peak and moving the peak to somewhat higher temperatures. Thus, the presence of thermal lag in DSC measurements is a potential explanation for apparent shortcomings in model calculations, as proposed by Hodge and Huvard [13] and by Hutchinson and co-workers [14]. However, recently O'Reilly and Hodge [15] found that model parameters varied with thermal history even for relatively slow heating rates (1.25°C/min) where thermal lag was assumed to be negligible. These last results indicate that thermal lag cannot account for all of the shortcomings of the model. However, we would like to examine the effect of thermal gradients on the shape of calculated enthalpy recovery curves for typical heating rates (10°C/min) and on the values of model parameters needed to fit the data by incorporating the effects of sample thermal gradients into the model. It is noted that we do not incorporate the effects of instrumental thermal lag into the calculations because it is presumed that these effects can be eliminated by proper calibration procedures.

The effects of a thermal gradient can easily be incorporated into the Tool-Narayanaswamy-Moynihan model of structural relaxation by calculating the average response in the sample. To do this, first we calculate the temperature profile and average temperature in the DSC sample as a function of program temperature for a given thermal history, as described elsewhere [20]. Then the fictive temperature and its temperature derivative were calculated at various points in the sample and averaged to give the sample response at the program temperature T_p, $\left(\dfrac{dT_f}{dT}\right)_{T_p}$:

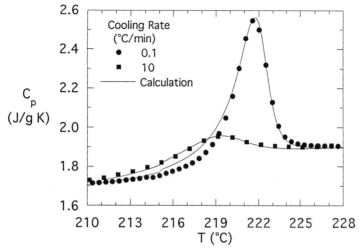

Figure 5: Experimental (data points) and calculated (lines) enthalpy recovery for heating at 10°C/min after cooling at cooling rates of 0.1°C/min and 10°C/min. Temperature gradients were not incorporated into the calculations.

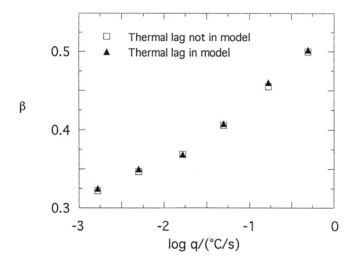

Figure 6: Dependence of model parameter β on thermal history for the calculations where thermal lag is not incorporated (Figure 5) and for calculations where the thermal gradient in the DSC sample is incorporated (Figure 7).

$$\left(\frac{dT_f}{dT}\right)_{T_p} = \int_0^1 \left(\frac{dT_f}{dT}\right) d\xi \qquad (9)$$

where ξ is the normalized distance from the bottom to the top of the sample. It is noted that the temperature profile in the DSC sample was calculated assuming that the top and bottom of the sample pan are maintained at the program temperature (T_p). This will give a conservative estimate of the thermal gradient in the sample; lags of 0.6°C and 1.2°C were calculated at the center of the sample in the glass and liquid regimes, respectively . Gradients of this magnitude (≈ 1°C) have been estimated by various researchers by placing melting standards above and below the sample in the DSC pan and measuring the differences in the observed melting temperatures.[4,14,38,39] Calculations were also performed using different boundary conditions giving lags of 3°C; the conclusions reached below were unchanged.

Figure 7 compares model calculations incorporating the thermal gradient with experimental data for cooling rates of 0.1°C/min and 1.0°C/min. Calculations made without thermal lag (shown in Figure 5) are also displayed. The model parameters β and τ_o were varied in order that the height and placement of the calculated curve match the experimental peak. There is no significant difference between the fit of the calculations assuming no thermal gradients and that assuming an ideal gradient – the two calculated curves are indistinguishable except for a slight broadening when thermal gradients are incorporated. However, in order to match the height and placement of the observed annealing peak for the cooling rate of 0.1°C/min, $\tau_{o,ref}$(207.5°C) was decreased to 5500 s and β was increased slightly to 0.325, changes of 30% and 1%, respectively, compared to the values used in the calculation without a thermal gradient. Similar changes in model parameters were found for the other cooling rate. The values of β have essentially the same temperature dependence as shown in Figure 6 where both values of β are plotted. Hence, the presence of a thermal gradient in the sample itself cannot account for the dependence of model parameters on thermal history; this agrees with the previous conclusions of O'Reilly and Hodge [15].

The results suggest that one of the shortcomings of the Tool-Narayanaswamy-Moynihan model of structural recovery, the dependence of model parameters on thermal history is not due to thermal gradients in the DSC sample itself. Other explanations [12] need to be examined, including the validity of the equations used for the relaxation time, the assumption of thermorheological and thermostructural simplicity, and the way in which the nonlinearity is incorporated into the model.

Conclusions

DSC heating curves for polyetherimide were fit using Moynihan's formulation of the Tool-Narayanaswamy structural recovery model. Only two adjustable parameters, β and $\tau_{o,ref}$ were used to make the fits; the parameters which describe the structure and temperature dependence of τ_o were determined independently from the shift factors used to reduce isothermal creep and enthalpy recovery data. To obtain a good fit of the DSC curves, the stretched exponential parameter β decreased with decreasing cooling rate. Incorporating the effects of the thermal gradient present in the DSC sample was found to not effect the dependence of β on thermal history.

Literature Cited

1.   Narayanaswamy, O. S. *J. Am. Ceram.* Soc. **1971**, 54, 491.
2.   Kovacs, A. J. *Ann. N.Y. Acad. Sci.* **1981**, 371, 38.

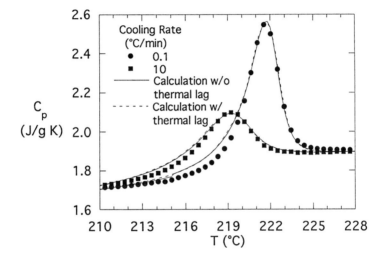

Figure 7: Experimental data (points) for heating at 10°C/min after cooling at 0.1°C/min and 1.0°C/min. Two calculations are shown for each cooling rate: one incorporating no thermal gradient in the DSC sample (solid line) and the other incorporating an ideal thermal gradient (dashed line). No significant difference can be discerned between the fits of the calculations with and without thermal gradients.

3. Moynihan, C. T.; Macedo,P. B.; Montrose, C. J.; Gupta, P. K.; DeBolt, M. A. Dill, J. F.; Dom, B. E.; Drake, P. W.; Esteal, A. J.; Elterman, P. B.; Moeller, R. P.; Sasabe, H.; Wilder, J. A.; *Ann. N.Y. Acad. Sci.* **1976**, 15, 279.
4. Hodge, I. M. *J. Non-Cryst. Solids* **1994**, 169, 211.
5. O'Reilly, J. M. *CRC Crit. Rev. in Solid State and Mat. Sci.* **1987**, 13 (3), 259.
6. Rendell, R. W.; Aklonis, J. J.; Ngai, K. L.; Fong, G. R. *Macromolecules* **1987**, 20, 1070.
7. McKenna, G. B. in *Comprehensive Polymer Science, Volume 12, Polymer Properties*; C. Booth and C. Price, Ed., Pergamon: Oxford, 1989, Vol. 12, pp 311 - 362.
8. Foltz, C. R.; McKinney, P. V. *J. Appl. Polym. Sci.* **1969**, 13, 2235.
9. Petrie, S. E. B. *J. Polym. Sci., Part A-2* **1972**,10, 1255.
10. Struik, L. C. E. *Physical Aging in Amorphous Polymers and Other Materials*; Elsevier: Amsterdam, 1978.
11. Moynihan, C. T.; Crichton, S. N.; Opalka, S. M. *J. Non-Cryst. Sol.* **1991**, 131-133, 420.
12. McKenna, G. B.; Angell, C. A. *J. Non-Cryst. Sol.* **1991**,131-133, 528.
13. Hodge, I. M.; Huvard, G. S. *Macromolecules* **1983**,16, 371.
14. Hutchinson, J. M.; Ruddy, M.; Wilson, M. R. *Polymer* , **1988**, 29, 152.
15. O'Reilly, J. M.; Hodge, I. M. *J. Non-Cryst. Sol.* **1991**,131-133, 451.
16. Kolrausch, F. *Pogg. Ann. Phys.* **1847**, 12, 393.
17. Williams, G.; Watts, D.C. *Trans. Faraday Soc.* **1970**, 66, 80.
18. Tool, A. Q. *J. Am. Ceram. Soc.* **1946**, 29, 240.
19. Moynihan, C. T.; et al. *J. Am. Cer. Soc.* **1976**, 59 (1-2), 12.
20. Simon, S. L. *Macromolecules* **1997**, 30, 4056.
21. Biddlestone, F.; Goodwin, A. A.; Hay, J. N.; Mouledous, G. A. C. *Polymer* **1991**, 32, 3119.
22. Hodge, I. M. *Macromolecules* **1987**, 20, 2897.
23. Sherer, G. W. *J. Am. Cer. Soc.* **1984**, 67 (7), 504.
24. Matsuoka, S.*Relaxation Phenomena in Polymers*, Oxford University Press: New York, 1992.
25. Spathis, G. *Polymer* **1994**, 35 (4), 791.
26. Vogel, H.*Phys. Z.* **1921**, 22, 645.
27. Williams, M. L.; Landell, R. F.; Ferry, J. D. *J. Am . Chem. Soc.* **1955**, 77, 3701.
28. Adam, G.; Gibbs, J. H. *J. Chem. Phys.* **1965**, 43 (1), 139.
29. Magill, J. H. *J. Chem. Phys.* **1967**, 47, 2802.
30. Echeverria, I.; Su, P-C.; Simon, S. L.; Plazek, D. J. *J. Polym. Sci.: Part B: Polym. Phys.* **1995**, 33, 2457.
31. Hodge, I. M.; Beren, A. R. *Macromolecules* **1982**, 15, 762.
32. Hutchinson, J. M. ; Ruddy, M. *J. Polym. Sci.: Part B: Polym. Phys.* **1990**, 28, 2127.
33. Dhinojwala, A.; Wong, G. K.; Torkelson, J. M. *J. Chem. Phys.* **1994**, 100 (8), 6046.
34. Zorn, R.; Mopsik, F.I.; McKenna, G.B.; Willner, L.; Richter, D. *J. Chem. Phys.*, in press; also, McKenna, G.B.; Mopsik, F.I.; Zorn, R.; Willner, L.; Richter, D. *SPE ANTEC* **1997**, II, 1027.
35. Plazek, D. J. *J. Rheol.* **1996**, 40 (6), 987.
36. Wortmann, F.-J.; Schulz, K. V. *Polymer* **1995**, 36 (9), 1611.
37. Matsuoka, S.; Quan, X. *J. Non-Cryst. Solids* **1991**, 131-133, 293.
38. Debolt, M. A., PhD Thesis, Catholic University of America, Washington, D. C., 1976.
39. Donoghue, E.; Ellis, T. S.; Karasz, F. E. in *Analytical Calorimetry*, Vol 5, P. Gioll, Ed., Plenum: NY 1984.

Chapter 14

The Physics of Glassy Polycarbonate: Superposability and Volume Recovery

Paul A. O'Connell, Carl R. Schultheisz, and Gregory B. McKenna[1]

Structure and Mechanics Group, Polymers Division, National Institute of Standards and Technology, Gaithersburg, MD 20899

The mechanical behavior of polymeric materials reflects the underlying physics of the mobility of the polymer chains. We have investigated the response of polycarbonate under three different conditions that will alter the polymer mobility: changes in temperature, strain and aging. We have examined the degree to which we can superpose the effects of these conditions to create master curves that can extend predictions beyond the range of feasible laboratory time scales. It was found that superposition could be successfully applied in each case, but that time-temperature superposition led to a master curve that was significantly different from that found using time-strain superposition. It was also found that curves fit to the [relatively short time] individual relaxation measurements at different temperatures could not be fit to the resulting master curve. Experiments were also performed to investigate the polymer mobility in volume recovery below the glass transition. It was found that volume equilibration required much longer than equilibration of the mechanical response in aging experiments, suggesting that simple free volume models may be inappropriate.

In the vicinity of its glass transition temperature (T_g), the mechanical response of a polymeric material is a strong function of both the rate of deformation and the temperature, with the modulus or compliance changing by several orders of magnitude as the Tg is traversed. These changes reflect the mobility of the polymer molecules, and indicate that the time scale of the experimental measurement is similar to the characteristic times associated with the relaxation mechanisms available to the polymer. Assuming that the same relaxation mechanisms are always active and that the characteristic time of each mechanism is affected in the same way by a change in

[*]Corresponding author.

temperature leads to the concept of time-temperature [or frequency-temperature] superposition (1). The most important result of superposition is that it allows for the creation of a master curve that can predict behavior well beyond typical experimental time scales. The first topic in this paper is an investigation into the time-temperature superposition of torsional stress relaxation data. It was found that the data could be superimposed, and that the stress relaxation data at each temperature could be fit with a [slightly different] stretched exponential function. However, the resulting master curve could not be fit with a stretched exponential function.

The idea that the relaxation process reflects the mobility of the polymer molecules can be investigated using other means to change that mobility, such as the addition of a solvent or by applying a stress or strain. In this case, an applied torsional strain has been used to accelerate the deformation processes. Again, the assumption that each relaxation mechanism is affected identically suggests the possibility of time-strain superposition, and allows for the construction of a master curve. While time-strain superposition is applicable, the master curve is significantly different from the master curve obtained using time-temperature superposition.

Finally, since the polymer relaxation processes are very slow below T_g, one must recognize that the polymer is not generally in an equilibrium state, but is evolving slowly toward thermodynamic equilibrium. This evolution has been examined directly through measurements of the volume, and is also reflected in changes in the mechanical response [modulus or compliance] in a process labeled physical aging (2). The changes in polymer mobility during aging [at constant temperature] again suggests the idea of time-aging time superposition, which allows for some prediction of the behavior at long aging times through a shift rate. However, direct measurement of the equilibration of both the volume and the stress relaxation behavior indicates that these properties equilibrate at different aging times, which suggests that the mechanical response is not directly coupled to the polymer volume.

The material employed in all investigations was General Electric Lexan LS-2, a commercial, UV stabilized, medium viscosity grade Bisphenol-A polycarbonate (3). T_g was measured as 141.3 °C using differential scanning calorimetry heating at 10 °C/min (4). For the torsion experiments, the polycarbonate was obtained in the form of extruded rods of 25 mm diameter, while for the volume recovery measurements, pellets of the polycarbonate intended for injection molding were used. The pellets were approximately 3 mm in diameter by 5 mm long.

Time-Temperature, Time-Strain and Time-Aging Time Superposition

Stress Relaxation Experiments in Torsion. Cylinders of the polycarbonate were machined to a length of 50 mm and diameter 12 mm and a gauge section further machined of 30 mm length and 4 mm diameter. In order to remove the effects of previous thermal and/or mechanical history, the samples were heated to 145 °C [approximately 4 °C above the measured T_g] for 1 hour prior to testing. Residual birefringence was not observed on looking through crossed polars.

The torsion measurements were carried out on a Rheometrics RMS 7200 (3) load frame, modified in our laboratory with a computer controlled servo-motor. The sample and grips were housed within a heater chamber for temperature control, with a measured oven stability [based on the range of measurements] of better than ±

0.1 °C. The torque force relaxations were measured from nominal strains γ [based on the cylinder outer radius] from 0.0025 to 0.08:

$$\gamma = R\Phi/L = R\Psi \qquad (1)$$

where Φ = angle of twist, R the cylinder radius, L the length of the gauge section and Ψ the angle of twist per unit length.

Time-Temperature Superposition. Time-temperature superposition is widely used in the description of polymer behavior at temperatures above the glass transition temperature T_g (*1*), where the viscoelastic response function changes with temperature by a change in the time scale by a horizontal shift a_T and in the intensity by a vertical shift b_T. An often-used representation for the response function in stress relaxation experiments is the stretched exponential of Kohlrausch-Williams-Watts (KWW) (*5,6*):

$$G(t) = G_0 \exp[-(t/\tau_0)^\beta] \qquad (2)$$

where $G(t)$ is the shear modulus response at time t, τ_0 is a characteristic time, β a shape parameter related to the breadth of the relaxation curve, and G_0 is the zero time shear modulus. A change in temperature from T_0 to T results in a change in the characteristic relaxation time leading to a temperature shift factor $a_T = \tau_0(T_0)/\tau_0(T)$. Vertical shifts are seen as a temperature dependent zero time shear modulus G_0; the vertical shift factor is $b_T = G_0(T_0)/G_0(T)$. The modulus G is calculated from the torque by assuming that the stress and strain are both linear functions of the cylinder radius, as in Equation (1).

Time-temperature superposition was applied to data generated from tests carried out at a given strain at temperatures between 30 °C and 135 °C. The data presented here are for a strain of 2%. The master curve was constructed as follows. The highest temperature [say T_1] data were taken as the reference curve and the parameters [G_0, τ and β] to fit to the KWW function determined. Keeping β constant, the KWW fit to the next lowest temperature [T_2] was found, and hence the temperature shift factor determined for the temperature change T_1-T_2 :

$$\log[a_T(T_1\text{-}T_2)] = \log[\tau_0(T_2)/\tau_0(T_1)] \qquad (3)$$

$$\log[b_T(T_1\text{-}T_2)] = \log[G_0(T_2)/G_0(T_1)] \qquad (4)$$

The T_2 data were then re-fitted to the KWW function, this time allowing the β term to vary. Keeping this new β constant, the next lowest temperature (T_3) data were fitted to the KWW function and the temperature shift factor determined for the temperature change T_2-T_3. By repeating this procedure for successively lower temperatures, it was possible to build-up the overall time-temperature master curve from the individual shifts at neighboring temperatures. The individual data sets and resulting master curve constructed using the above procedure are shown in Figure 1, from which it is evident that the data do appear to superimpose to form a master curve.

KWW Analysis of Temperature Data. In the following, we performed the KWW analysis in two ways. We first took the master curve obtained above and asked if it could be described by a KWW function. We then looked at the results from fitting KWW functions to the stress relaxation response at each temperature and asked what the apparent change in KWW parameters would be as a function of temperature.

The time-temperature master curve of Figure 1, determined using the 'semi manual' method outlined above, is replotted in Figure 2 and compared to a KWW fit to these data [solid line]. It is immediately clear that the KWW equation does not adequately describe the relaxation behavior over the whole range of data.

Next the KWW equation was applied to the data at each temperature, with no restrictions on the parameters, and the resulting β parameters are shown in Table I. The important observation here is the variation in the β parameter with temperature, which, if taken literally and on the assumption that the master curve is described by a KWW function, implies that time-temperature superposition does not apply to these data. However, above we have shown that a very good master curve representation of the data can be obtained by using time-temperature superposition. We interpret the result of the present analysis to show that the KWW parameters are not strongly determined using the limited time window [which is typical of mechanical tests] available here. Further, it shows the danger of interpreting limited data in terms of the KWW parameters. We also note that the master relaxation curve obtained from the time-temperature superposition was not well described by a KWW function.

The temperature shift factors, log (a_T), required for superposition are shown in Figure 3. At low temperatures the shift factor is not highly temperature dependent, while at temperatures close to Tg the dependence is clearly very strong. The vertical shift factors required for time-temperature superposition are shown in Table II, referenced to 135 °C. The trend is the same as that observed for the time shift factors, with a relatively rapid change at temperatures a little below T_g followed by a leveling off at temperatures far removed from T_g. Although the vertical shifts between adjacent temperature curves is small, over a broad temperature range the cumulative effect can be significant, in this case leading to a maximum vertical shift of approximately 25% over the temperature range 30 °C to 135 °C. Note though that the vertical shifts are still small in comparison to the orders of magnitude by which the timescale is shifted. The shifts imply that the material is becoming intrinsically stiffer at lower temperatures. This may be reasonably expected, since from thermal expansion effects alone the material will be denser at lower temperatures.

Time-Strain Superposition. The principle of time-strain superposition is essentially the same as that for time-temperature superposition, though now there is a strain induced shift [acceleration] in the time scale of the material response (7-9). Again, within the context of the KWW function one can write the time-strain shift function as $a_\gamma = \tau_0(\gamma_0)/\tau_0(\gamma)$, where γ_0 and γ represent the reference and current strains. Similarly, the vertical strain shifts are $b_\gamma = G_0(\gamma_0)/G_0(\gamma)$. Because the shear strain in the samples is a function of radial position r and at large deformations the shear stress is not linear in the deformation, the modulus of the material is a function of r. Hence, for the time-strain superposition we followed an isochronal analysis developed by McKenna and Zapas (10) based on the elastic scaling analysis of Penn and Kearsley (11,12). In this case we can write :

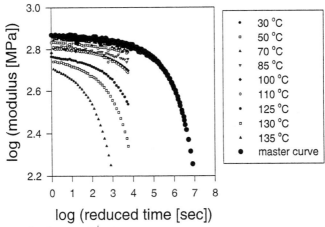

Figure 1. Stress relaxation reponse at a strain of 2% as a function of temperature and the time - temperature master curve.

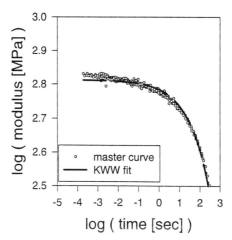

Figure 2. KWW fit to the time - temperature master curve.

Table I KWW fit parameters to relaxation data at each temperature with no restrictions on the parameters

T (°C)	G_o (MPa)	τ (sec)	β
30	780	1.18×10^8	0.20
50	748	4.04×10^9	0.13
70	792	1.35×10^7	0.24
85	700	1.28×10^7	0.22
100	676	2.47×10^5	0.36
110	724	2.47×10^5	0.33
125	609	3.22×10^4	0.37
130	584	7.26×10^4	0.41
135	541	6.68×10^2	0.47

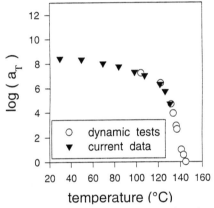

Figure 3. Temperature shift factors, log a_T, as a function of temperature.

Table II. Vertical shift factors for time - temperature and time - strain superposition

Temperature (°C)	time - temperature	Strain	time - strain
30	0	0.0025	0.0
50	0.032	0.005	-0.006
70	0.012	0.01	0.0123
85	0.047	0.02	0.0178
100	0.05	0.035	0.0041
110	0.049	0.045	0.0096
125	0.086	0.055	0.0271
130	0.1	0.07	0.0909
135	0.12		

$$G(t) = \frac{\sigma_{12}(t)}{\gamma} = \frac{2}{4\pi\Psi R^4}\left(3M(t) + \Psi\frac{dM(t)}{d\Psi}\right) \qquad (5)$$

where σ_{12} is the stress at a given value of strain γ and $G(t)$ is the relaxation modulus at that strain, $M(t)$ is the torque response and Ψ is as defined previously.

Time-strain superposition was carried out in the same manner as the time-temperature superposition above. Relaxation data for strains from 0.25 % to 7 % are presented along with the resulting master curve for 30 °C and referenced to a strain of 0.25 % [Figure 4]. The master curve covers approximately 8 decades of time, which is comparable to the range covered with time-temperature superposition. As with time-temperature superposition, small vertical shifts were also required, the magnitudes of which are given Table II.

Comparison of Master Curves. The two master curves obtained above, one from time-temperature superposition [at a strain of 2 % and referenced to 30 °C] and one from time-strain superposition [at 30 °C and referenced to a strain of 2 %] are shown in Figure 5. At short times, that is at low temperatures or small strains, the data coincide. At longer time [>4 decades] the two data sets begin to diverge, with the time-temperature data showing a significantly greater relaxation rate compared to the time-strain data. The implication is that temperature and strain affect the relaxation response in different ways. However, this is problematic in view of the fact that the time-temperature or time-strain superpositions are both based on the concept that there exists an underlying viscoelastic response function that is characteristic of the material (1,13). Hence, shifting the spectrum by applying an increased temperature or an increased strain should only alter the time-scale not the shape of the response function. As a result, it is fair to ask if one or both of the master curves is wrong since two very different long term behaviors are obvious in the data plotted in Figure 5. Because time-temperature superposition has been successful and deviations from it usually reported to be only subtle, our preliminary conclusion here is that the time-strain master curve is not the correct master curve. Further work needs to be done to establish the apparent validity of the time-temperature master curve. As an additional point, the apparent superposability of the time-strain data leads to questions about the use of reduced time concepts in non-linear constitutive equations when data are obtained only isothermally. How does one further test the form of the equation if the apparent memory function constructed by the time-strain superposition is incorrect? A brief discussion of such models is given in (14).

Time-Aging Time Superposition. In order to examine the effects of aging, a loading regime first proposed by Struik (2) was followed. In this, the strains are applied sequentially at aging times t_e that approximately double with each test. The duration of the applied strain, τ_1, was such that the ratio t_1/t_e was constant at 0.10. The applied strains then are essentially 'probes' into the material structure, and are of a sufficiently short duration such that aging effects do not significantly influence the measurements. By allowing the sample to recover for a time t_2 [= $9t_e$], the sample essentially 'forgets' the effect of the previous loading cycle. In order to confirm that this loading regime is

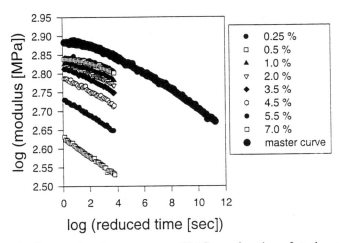

Figure 4. Stress relaxation response at 30 °C as a function of strain and the reduced time - strain master curve.

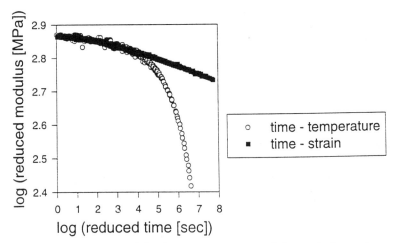

Figure 5. Comparision of the time - temperature and time - strain master curves.

valid and that the sample really does 'forget' the previous loading cycle, a number of tests were carried out where the sample was taken above T_g, cooled and allowed to age for a given time at the appropriate test temperature and then a single probe strain applied. Aging times from 0.5 to 18 hours were examined. The applied strains ranged from 0.0025 to 0.07 over a temperature range from 30 °C to 140 °C

Each stress relaxation curve was again examined using the Kohlrausch-Williams-Watts (KWW) function (5,6) given in Equation 2. For a given temperature and strain, and assuming that β is independent of aging time, it is possible to perform time-aging time superposition of the data by reducing the curves to a reference aging time via a shift along both the time and force axes [an additional proviso here is that the response is sufficiently 'curved' to allow the distinction between the required vertical and horizontal shifts for superposition]. The aging time shift factor a_{te} is then defined from the KWW function as:

$$\log(a_{te}) = \log(\tau_0(t_e)/\tau_0(t_{e,ref}))$$ (6)

where $\tau_0(t_e)$ is the value of τ_0 in Equation 6 at aging time t_e and $\tau_0(t_{e,ref})$ its value at the reference aging time. Having obtained the values of a_{te}, these are then analyzed in the conventional manner of making plots of $\log(a_{te})$ vs $\log(t_e)$, the slope of which has been defined as the shift rate, μ, by Struik (2) as:

$$\mu = d\log(a_{te})/d\log(t_e)$$ (7)

A typical set of stress relaxation curves as a function of aging time is shown in Figure 6 [data for 2% strain at 125 °C]. As can be seen, the data for the longer aging times are shifted to the right on the time axis. That is, the response is shifted to longer times. Because of the greater number of data points for the longest aging time, these data are generally used as the initial reference curve to determine $\tau_0(t_{e,ref})$, $G_0(t_{e,ref})$ and β parameters. The choice of the aging time to use as the reference curve is arbitrary, the only effect being a shift of the final master curve in the time axis. The reference curve is fitted to the KWW function and the parameters determined. The β value then is held constant and the data at shorter aging times fitted to the KWW function by varying $\tau_0(t_e)$ and $G_0(t_e)$. The values of $\log(a_{te})$ and $\log(b_{te})$ can then be determined as discussed above. The assumption of a constant β value can be validated by allowing the G_0, β and τ_0 parameters to vary for each relaxation curve; if this is carried out it is found that the β term is indeed essentially independent of aging time, with the proviso that this is true only over the time scale of these tests [Table III]. Application of the appropriate shift factor to the relaxation data results in a superposition of the data to form the time - aging time 'master curve' [Figure 6]. Note that the data for the master curve have been further off-set by an arbitrary amount in the time direction for clarity.

Figure 7 shows a plot of the $\log(a_{te})$ vs $\log(t_e)$ values at 3.5% strain and 30 °C and 110 °C [note that the data are plotted as referenced to $t_e = 1800$ s]. The data show good linearity, the gradient of which is the double logarithmic shift rate, μ.

Figure 8 illustrates the effect of both temperature and strain on the shift rate, μ, for temperatures up to 110 °C and strains up to 7%. The error bars shown

Table III. β parameter from KWW function as a function of aging time.

log (aging time [sec])	β parameter from KWW function
3.26	0.260
3.58	0.266
3.89	0.266
4.19	0.254
4.50	0.269
4.80	0.273

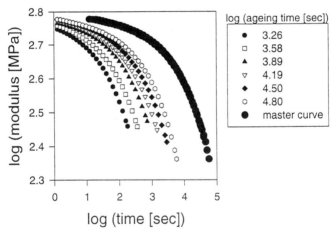

Figure 6. Stress relaxation response at 125 °C and a strain of 2% as a function of aging time and the time - aging time master curve.

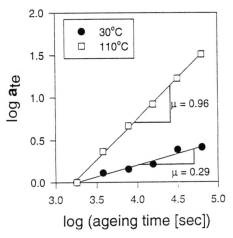

Figure 7. Aging time shift factor as a function of aging time.

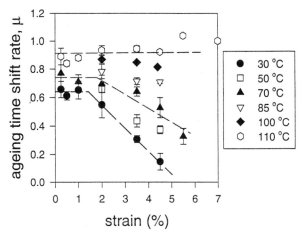

Figure 8. Aging time shift rate as a function of temperature and strain.

are the root mean square deviation from the average value. In the linear viscoelastic range at low strains [<2%] the shift rate is approximately constant at any given temperature. The small strain shift rate can be seen to increase slightly as the temperature increases.

As the strain is increased, that is as the material response moves into the non-linear viscoelastic range, there is a systematic decrease in the shift rate with increasing strain. This strain dependence of the shift rate is seen to become progressively less pronounced at increasing temperatures, as evidenced by a leveling-off of the curves. The temperature dependence of the shift rate then becomes an increasingly strong function of the temperature at higher strains. Significantly, the decreasing dependence of the shift rate on strain continues to such an extent that the shift rate is independent of strain at 110 °C. This was an unexpected result, and to our knowledge is the first time a strain independent shift rate in the large deformation regime has been observed.

The effect of temperature on the shift rate is more clearly seen in Figure 9 where the μ values at three strains are plotted as a function of test temperature. Additional data at higher temperatures [up to 135 °C and omitted for clarity from the previous figure] are also shown. At temperatures near to T_g the shift rate reduces drastically at all strain levels [indeed by definition the shift rate is zero at T_g]. The data show a [strain independent] maximum at approximately 110 °C to 120 °C and then a systematic decrease as the temperature is lowered to 30 °C. As observed above, the temperature dependence of the shift rate is more pronounced at higher strains, shown here by the steeper slope at these strains. Lower strain data have been omitted, again for clarity, as the shift rate was shown to be virtually independent of strain below 2%.

These data are in contradiction to those presented by Struik (2) for polycarbonate where the shift rate remains constant at a value of approximately 1.2 down to temperatures of the order of -50 °C. The reasons for such differences are unknown. We note however that the tests were performed in stress relaxation while those of Struik were performed in creep. Also, Sullivan (12) has reported shifts rates, μ, of approximately 0.75 for polycarbonate in the linear viscoelastic regime from creep experiments; a result in line with the results reported here.

At higher temperatures [approximately T_g-20 °C to T_g-10 °C] the $\log(a_{te})$ vs $\log(t_e)$ data is no longer linear, with the aging rate decreasing at longer aging times [Figure 10]. Eventually the material ceases to age and has thus reached [mechanical] equilibrium. The time scale for this to occur at these temperatures ranges from approximately 10^4 to 10^6 seconds. These timescales are in good agreement with those found in (4) for the same material under dynamic loading conditions. At higher temperatures [above approximately 135 °C] the material is seen to reach equilibrium within 1800 seconds, the shortest aging time examined.

Vertical Shifts. As with the horizontal [time] shifts, a plot of the vertical shifts, $\log(b_{te})$, as a function of $\log(t_e)$ yields a linear relationship, the gradient of which is denoted by b*. The variation of b* with temperature and strain is shown in Table IV. Although there is a relatively large amount of scatter in the b* values, there is no clear trend with either temperature or strain [at least up to the yield point of the material], giving an overall average value of approximately +0.008 over 0.5 to 18 hours of aging. This implies that the zero time or intrinsic modulus is increasing by

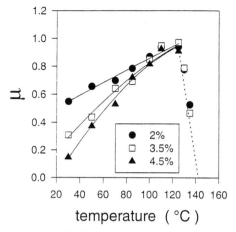

Figure 9. Aging time shift rate as a function of strain and temperature.

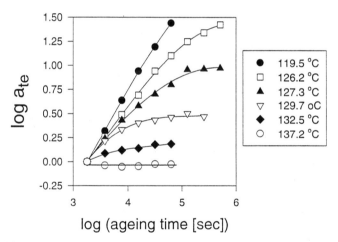

Figure 10. Aging time shift factor as a function of aging time near machanical equilibrium.

Table IV. Vertical shift factor rate as a function of temperature and strain

Strain	30 °C	70 °C	110 °C
0.02	4.3×10^{-3}	6.2×10^{-3}	$9..8 \times 10^{-3}$
0.035	4.6×10^{-3}	8.4×10^{-3}	7.4×10^{-3}
0.045	6.5×10^{-3}	9.5×10^{-3}	5.9×10^{-3}
0.055		7.3×10^{-3}	

approximately 2% per decade of aging. This is perhaps not too surprising in considering that the underlying structure of the material must be changing as the material ages and densifies. The magnitude of the vertical shift is similar to that reported by Struik (2).

Volume Recovery Behavior

Following temperature changes below T_g, the slow evolution of the polymeric structure [as characterized by the volume or enthalpy] towards equilibrium has important kinetics (16-22). These kinetics have been investigated through dilatometric experiments on polycarbonate using temperature-jump histories, which have been fit to a nonlinear model incorporating a material time clock that depends on the current volume.

The evolution of the volume towards equilibrium is also compared to the evolution of the mechanical properties towards an equilibrium state [physical aging (2,4)] as discussed above. Although free volume models (1) would suggest that the volume and mechanical properties should evolve at the same rate, there have been several studies (14,23-29) reporting that the time scales of the evolution of these properties towards equilibrium can be different, and we examine the relative rates of change for the polycarbonate here. In down-jumps close to T_g, comparison of the evolution of the volume and the mechanical response indicates that the mechanical properties cease evolving [reach equilibrium] before the volume. These results are similar to those seen previously in this laboratory with epoxy glasses near T_g (14,25-28). However, this behavior is not universal (30,31) and we suggest that the apparent conflict may be due to differences in materials.

Phenomenological Description of the Volume Recovery. We have employed a hybrid (28) of the Tool-Narayanaswamy-Moynihan-KAHR models of the kinetics of structural recovery (16-20) to obtain relevant phenomenological fits to temperature-jump experiments. The volume departure from equilibrium, δ, is related to the thermal history and material properties as:

$$\delta(z) = (V(z) - V_\infty)/V_\infty = -\Delta\alpha \int_0^z R(z-z')\frac{dT}{dz'}dz' \tag{8}$$

where $V(z)$ is the current volume and V_∞ is the volume of the equilibrium liquid. T is the temperature, and $\Delta\alpha$ is the difference between the liquid and glassy coefficients of thermal expansion. $R(z)$ is a retardation function written in terms of the reduced time of the material denoted by z, which is calculated from the experimental time t through shift factors a_T and a_δ that depend on temperature T and structure δ, respectively:

$$z = \int_0^t \frac{d\xi}{a_T(T(\xi)) \, a_\delta(\delta(\xi))} \tag{9}$$

Assuming the temperature history as a step function from T_1 to T_0 at $t = z = 0$, the convolution integral in Equation 8 reduces to $\delta(z) = -\Delta\alpha \, R(z) \, (T_0 - T_1)$. The volume response function $R(z)$ is expressed as a stretched exponential (5,6) [as was used in (20)],

$$R(z) = \exp[-(z/\tau_r)^\beta] \qquad (10)$$

where τ_r is a characteristic time. The shift factors are exponentials as suggested in (19),

$$a_T = \exp[-\Theta(T-T_r)], \qquad a_\delta = \exp[-(1-x)\Theta\delta/\Delta\alpha] \qquad (11)$$

where T_r is a reference temperature, Θ is a constant, and x is a partitioning parameter ($0 \le x \le 1$) that weights the relative contributions of the temperature and the structure to the shift factor. The parameters $\Delta\alpha$, τ_r, β, Θ and x were then evaluated from the experimental data by adapting a Levenberg-Marquardt nonlinear curve-fitting algorithm (32) as detailed in (28).

Volume Measurements. The volume of the polycarbonate was measured using an automated mercury dilatometer [similar in conception to those used by Kovacs (18)], containing approximately 3.38 g of polycarbonate pellets with a volume of approximately 2.85 cm^3 at 22 °C. The uncertainty in the relative volume measurements [expressed as one standard deviation] is estimated as 5×10^{-5} cm^3. The glassy and liquid coefficients of thermal expansion were measured as $\alpha_G = 2.15 \times 10^{-4}$ $cm^3/(cm^3$ K) and $\alpha_L = 5.94 \times 10^{-4}$ $cm^3/(cm^3$ K), respectively. The uncertainty [expressed as one standard deviation] is estimated as 10^{-5} $cm^3/(cm^3$ K). The difference [used in Equation 8] is $\Delta\alpha = \alpha_L - \alpha_G = 3.79 \times 10^{-4}$ $cm^3/(cm^3$ K).

In order to study the nonequilibrium volume recovery, asymmetry of approach experiments were performed. Up and down-jumps of 5 °C were performed to a final temperature of $T_0 = 140$ °C, and up and down-jumps of 2 °C were performed to a final temperature of $T_0 = 138$ °C. Thermal equilibration took 100 to 200 s.

The results of the curve fit to the temperature jump data are shown in Figures 11 and 12. The points used in the fit are indicated by symbols in the figures, and the lines show the prediction of the model using the calculated parameters given in Table V. Hodge (22) obtained similar parameters of $\beta = 0.46$, $\Theta = 0.87$ K^{-1} and $x = 0.19$ from enthalpy recovery measurements on a Bisphenol-A polycarbonate.

For this limited set of data, the model represents the experiments reasonably well. Simulations using the model indicate that the approximation of the temperature history as a jump is acceptable, because the volume recovery resulting from a finite rate temperature change quickly approaches the volume recovery from a jump once the final temperature has been reached.

Although the temperatures used in the volume recovery experiments were not identical to those used in the torsion experiments, one can readily see that the volume requires a much longer time to reach equilibrium. From Figure 12, it is apparent that the volume takes more than 10^4 s to equilibrate at 138 °C, whereas

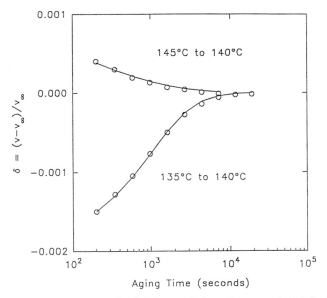

Figure 11. Volume recovery for 5 °C up and down jumps to 140 °C. Symbols are experimental data, and lines are the hybrid model fit to all data in Figures 11 and 12.

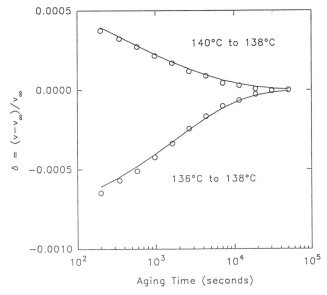

Figure 12. Volume recovery for 2 °C up and down jumps to 138 °C. Symbols are experimental data, and lines are the hybrid model fit to all data in Figures 11 and 12.

Table V. Hybrid model parameters determined from volume recovery data in Figures 11 and 12. Also shown are the parameters evaluated by Hodge [22] from DSC data, and the parameters suggested by KAHR [19].

Parameter	$T_r = 140\ °C$	Hodge [22] (DSC)	KAHR [19]
$\Delta\alpha(10^{-4}\ K^{-1})$	4.4		
$\tau_r\ (s)$	140	280 - 1400	
β	0.4	0.46	
x	0.18	0.19	0.4
$\Theta\ (K^{-1})$	0.88	0.87	1

the mechanical properties have reached equilibrium in less than 1800 s at a lower temperature [137.2 °C], as demonstrated by the flat response of the aging time shift factor at that temperature in Figure 10. This result is consistent with observations from this laboratory in experiments with a model epoxy (14,25-28) using the NIST torsional dilatometer. Note though that other researchers using a range of polymeric systems have found contrasting results in terms of the time scale of the volume, mechanical and enthalpy relaxation times. A summary of these results can be found in (33).

Conclusions

Time-temperature superposition can be successfully applied to polycarbonate at temperatures up to 110 °C below the glass transition temperature. Though the individual relaxation curves at each temperature can be described in terms of the KWW function, the resulting master curve can not. This was attributed to the fact that the KWW parameters are not strongly determined using the limited time window on the individual tests. This in turn leads to a spurious temperature dependence of the KWW β parameter.

Time-strain superposition was also shown to be possible. However, the resulting master curve was significantly different from that obtained using time-temperature superposition. This is problematic if it is assumed that there exists a single underlying response function.

Time-aging time superposition was possible at all strains up to the yield strain and at all temperatures studied. The aging time shift rate. μ, was seen to vary systematically with both temperature and strain. In the linear viscoelastic range, the shift rate remained constant. At increasing strains the shift rate was seen to decrease, though this effect was less evident at higher temperatures. The shift rate increased with increasing temperature up to approximately 110 °C to 120 °C, after which it rapidly decreased to zero at Tg.

The hybrid nonlinear model fits this limited set of volume recovery data quite well, although the calculated difference between the glass and liquid coefficients of thermal expansion is somewhat different from the measured value. The time for volume equilibration following a temperature change was found to be much longer than the time for equilibration of the mechanical response, which indicates that a simple free volume model of the mechanical response may be inappropriate.

References

1. Ferry, J.D., <u>Viscoelastic Properties of Polymers</u>, J. Wiley and Son. 1980
2. Struik, L.C.E., <u>Physical Aging in Amorphous Polymers and Other Materials</u>, Elsevier (Amsterdam) 1978.
3. Certain commercial materials and equipment are identified in this paper to specify adequately the experimental procedure. In no case does such identification imply recommendation or endorsement by the National Institute of Standards and Technology, nor does it imply necessarily that the product is the best available for the purpose.
4. Pesce J.-J., Niemiec J.M., Chiang M.Y., Schutte C.L., Schultheisz C.R., McKenna G.B., *Current Research in the Thermo-Mechanics of Polymers in the Rubbery-Glassy Range AMD-203*, Negahban M.; Ed, ASME (New York) (1995) 77.
5. Kohlrausch F., *Pogg. Ann. Phys*, **12** (1847) 393
6. Williams G. and Watts D.C., *Trans. Faraday Soc.* **66** (1970) 80
7. Shapery R.A.,. *Poly. Eng. Sci.* **9** (1969) 4
8. Matsuoka S., Aloisio C.J. and Bair H.E. *J. Appl. Phys*, **44** No. 10 (1973) 4265
9. Matsuoka S., Bair H.E., Bearder S.S., Kern H.W. and Ryan J.T.. *Poly. Eng. Sci.* **18** No. 14 (1978) 1073
10. McKenna G.B. and Zapas L.J., "Viscoelastic Behavior of Poly(methyl methacrylate): Prediction of Extensional Response from Torsional Data", *Rheology Vol 3 Applications*, Astarita G., Marucci G. and Nicolais L., Ed. Plenum, New York, (1980) pp.299-307
11. Penn R.W. and Kearsley E.A., *Trans. Soc. Rheol.*, **20**, 227 (1976)
12. Nadai A., *Plasticity : A Mechanics of the Plastic State of Matter*, McGraw-Hill, New York (1931)
13. Markovitz H., Superposition in Rheology. *J. Poly. Sci. : Symposium*, **50**, (1975) 431
14. McKenna G.B., "On the Physics Required for the Prediction of Long Term Performance of Polymers and Their Composites", *J. Res. NIST*, **99**, 1994, 169-189.
15. Sullivan J.L., 'Use of Plastics and Plastic Composites : Materials and Mechanisms Issues,' ASME (1993)
16. Tool, A.Q., *J. Res. NBS*, **37** (1946) 73; *J. Am. Ceram. Soc.* **29** (1946) 240.
17. Narayanaswamy, O.S., *J. Am. Ceram. Soc.* **54** (1971) 491.
18. Kovacs, A.J., *Fortschr. Hochpolym.-Forsch.* **3** (1963) 394.

19. Kovacs, A.J., Aklonis, J.J., Hutchinson, J.M., Ramos, A.R., *J. Polym. Sci. Polym. Phys. Ed.* **17** (1979) 1097.

20. Moynihan, C.T., Macedo, P.B., Montrose, C.J., Gupta, P.K., DeBolt, M.A., Dill, J.F., Dom, B.E., Drake, P.W., Esteal, A.J., Elterman, P.B., Moeller, R.P., Sasabe, H., Wilder, J.A., *Ann. N.Y. Acad. Sci.* **279** (1976) 15.

21. McKenna, G.B., *Comprehensive Polym. Sci., Vol. 2: Polymer Properties*, Booth C. and Price C., Eds, Pergamon Press (Oxford) 1989, pp 331 .

22. Hodge, I.M., *J. Non-Crystalline Solids*, **169** (1994) 211.

23. Scherer G.W., *Relaxation in Glass and Composites*, Wiley (New York) 1986.

24. Roe, R.-J., Millman, G.M., *Polym. Eng. Sci*, **23** (1983) 318.

25. Santore, M.M., Duran, R.S., McKenna, G.B., *Polymer*, **32** (1991) 2377.

26. McKenna, G.B., Leterrier, Y., Schultheisz, C.R., *Polym. Eng. Sci*, **35** (1995) 403.

27. Schultheisz, C.R., Colucci, D.M., McKenna, G.B., and Caruthers, J.M., *Mechanics of Plastics and Plastic Composites MD-68/AMD-215*, Boyce M.C., Ed, ASME (New York) (1995) 251.

28. Schultheisz, C.R., McKenna, G.B., Leterrier, Y., Stefanis, E., *Proc. Soc. Exp. Mech.*, Spring Conference, Grand Rapids, MI, June (1995) 329.

29. Delin, M., *Volumetric Analysis in Mechanical Behaviour and Physical Aging of Polymers*, Doctoral Thesis, Chalmers University of Technology, Göteborg, Sweden, 1996.

30. Struik, L.C.E., *Polymer*, **29** (1988) 1348.

31. Echeverria, I., Su, P.C., Simon, S.L., Plazek, D.J., *J. Polym. Sci., Polym. Phys. Ed.*, **33** (1995) 2457.

32. Press, W.H., Flannery, B.P., Teukolsky, S.A., Vetterling, W.T., *Numerical Recipes*, Cambridge University Press (Cambridge) 1986.

33. Simon S., Plazek D.J., Sobieski J.W. and McGregor E.T., J.Poly. Sci, **35** (1997) 929

Chapter 15

Polymer–Polymer Miscibility Investigated by Temperature Modulated Differential Scanning Calorimetry

G. O. R. Alberda van Ekenstein[1], G. ten Brinke[1], and T. S. Ellis[2]

[1]Laboratory of Polymer Chemistry, Materials Science Center, University of Groningen, Nijenbborgh 4, 9747 AG Groningen, The Netherlands
[2]General Motors Research and Development Center, Polymers Department, Warren, MI 48090–9055

In this paper we will review the main features of using both temperature modulated differential scanning calorimetry, T-m.d.s.c., and conventional d.s.c. for investigating phase behavior of polymer mixtures by reference to blends of acrylonitrile-butadiene-styrene copolymers with poly(methacrylate) and chlorinated polymers with aliphatic co-polyesteramides, respectively. We will also elaborate on more recent results concerning the application of T-m.d.s.c. to investigate enthalpy recovery in a blend of polystyrene and a random copolymer of styrene and para-fluorostyrene and whose glass transition temperatures are separated by only 3°C. It will be shown that, although the resolution of T-m.d.s.c. compared to conventional d.s.c is strongly enhanced, unambiguous conclusions about miscibility still requires enthalpy relaxation studies. The advantage of T-m.d.s.c becomes especially clear when overlapping signals such as glass transition and exothermic processes are present.

Relaxation in the amorphous state of glassy and semi-crystalline polymers, often referred to as physical aging, and the subsequent recovery processes on heating through the glass transition, displays individual features that are characteristic of the particular polymer. Several years ago formal procedures were presented (1,2) to illustrate how enthalpy relaxation in the amorphous state of glassy and semi-crystalline polymers can be a valuable tool for elucidating phase phenomena in polymer blends. It has been particularly successful in determining phase behavior in blends whose constituents possess similar glass transition temperatures (T_g) where the detection of a single composition dependent T_g is not possible. Accordingly, when characterized by calorimetric methods, as an enthalpy recovery peak, mixtures of polymers possessing indistinguishable or similar thermal properties can be analyzed quite effectively. The appearance of multiple or asymmetric recovery peaks in polymer mixtures is regarded as conclusive evidence of heterogeneous mixing. Conversely, a single sharp recovery peak is usually indicative of a single homogeneous phase. Although especially suited to these kinds of situations a more recent review (2) has also emphasized the more general utility of the technique to additional multi-component systems. As the emphasis of the determination of polymer-polymer miscibility has progressed to defining the structure-composition relationships involved, particularly by establishing critical miscibility limits through the use of copolymers (3,4), the latter situation is encountered more frequently.

Although specific-heat spectroscopy or AC calorimetry was introduced over thirty years ago (5) and its application to the glass transition of glycerol described in a pioneering paper by Birge and Nagel (6), the adaptation of conventional d.s.c. to include these possibilities is from a much more recent date (7). This technique, using commercially available equipment, is now usually termed temperature modulated differential scanning calorimetry or T-m.d.s.c. The interpretation of the experimental data obtained by it, is a focus of intensive scientific debate (8-10). But letting this aside, it has certainly the potential to provide additional and unique information on multi-component systems. Indeed, recent publications (11) have illustrated that steps have already been taken in this direction. In this paper we will review the main features of the enthalpy recovery technique using both T-m.d.s.c. and conventional d.s.c. and examine the application of both forms of instrumentation to a number of different blend studies.

In a conventional d.s.c., only the total heat flow to the sample is measured. The principal difference between the two types of instrumentation is that the linear heating program applied in T-m.d.s.c. also contains a sinusoidal modulation with frequency ω superimposed over a nominal heating rate; usually considerably lower than that of conventional d.s.c. (1-2°C/min c.f. d.s.c. 10-20°C/min). Although a seemingly minor difference, the advantage of T-m.d.s.c over conventional instrumentation lies in the ability to separate the response of the sample to both the "reversible" and "non-reversing" behavior (7,12). From the modulated temperature and the modulated heat flow signals, the heat capacity is extracted via deconvolution, which, multiplied by the heating rate, gives the reversing heat flow. The average of the modulated heat flow is the total heat flow. The difference between these two heat flows is the non reversing heat flow. This is the more empirical approach. When using deconvolution with phase correction it is also possible to divide both the modulated signals, via heat flow angle and complex heat capacity $C_p^*(\omega)$, into the out of phase "kinetic" heat capacity $C_p''(\omega)$ and the in phase "reversing" heat capacity $C_p'(\omega)$. As usual, a complex response function implies some form of entropy production. In the case of a modulation with amplitude a around a temperature T_o it is rather straightforward to show (6,14) that the minimum entropy production is given by $\Delta S = \pi C_p''(\omega)(a / T_o)^2$. The new element is that the entropy production does not occur inside the system, but rather in the surroundings. In a strict thermodynamic sense, a temperature modulation requires a heat bath for every temperature and the entropy production is due to a redistribution of energy between the heat baths, being transferred from the high temperature heat baths to the low temperature heat baths during every cycle(14).

An extra tool is the possibility to construct Lissajoux figures by plotting, for instance, the modulated heat flow versus derivative modulated temperature. This provides additional information about the validity of the steady state conditions of the experiments, which is visualized by the retrace of the separate ellipses. In these specific Lissajoux figures the width corresponds to the phase lag and the slope to the specific heat. Under the conditions of analysis the enthalpy recovery peak is a non-reversing event and can therefore be separated from the reversible changes of specific heat.

Experimental

Convential d.s.c. was performed using a Perkin Elmer DSC7 with a heating rate of 20°C/min for the poly(methyl methacrylate)/poly(acrylonitrile-*co*-butadiene-*co*-styrene) (PMMA/ABS) systems and the PMMA/poly(styrene-*co*-acrylonitrile) (PMMA/SAN) systems and with 10°C/min for blends of chlorinated polyethylenes/ aliphatic co-polyesteramides. The latter blends were vitrified by quenching from 150 or 250°C into liquid nitrogen before thermal investigations. Preparation, composition and properties can be found in ref.(*3*). SAN copolymers were obtained from Scientific Polymer Products and used as received. ABS materials were supplied by Monsanto as PG298 and Lustran Elite with approximate AN contents in the SAN copolymer, as measured by nmr, of 30 and 25 wt%, respectively. The PMMA was supplied as Plexiglas V502 by Atohaas.

Temperature modulated d.s.c (T-m.d.s.c.) was performed using a TA-Instruments DSC2920. Of the chlorinated polyethylenes/co-polyesteramides, only those blends which are difficult to interpret using conventional d.s.c. curves were investigated by T-m.d.s.c. using an average heating rate of 2°C/min, amplitude 1°C and period 60 sec. For the polystyrene(PS)/fluoropolystyrene blend and its components both T_g and enthalpy recovery studies were performed by T-m.d.s.c. with an average heating rate of 1°C/min, amplitude 1°C and period 60 sec. Polystyrene Dow 666 (M_w=304x10^3, M_n=110x10^3, T_g(midpoint)=104.3°C) and a random copolymer of styrene and para-fluorostyrene P(S-pFS46) (mole fraction of FS equals 0.46, M_w=114x10^3, M_n=64x10^3, T_g(midpoint)=107.3°C) were used. Aging for various amounts of time was performed at 92°C. The 50-50wt% blends were obtained by co-precipitation from a 5 wt% toluene solution into a 20x excess of methanol.

Blend Studies

PMMA/ABS In blends of PMMA and styrene-acrylonitrile (SAN) copolymers, there is a critical composition of the SAN, at approximately 28 wt % acrylonitrile, that marks the transformation from miscibility to immiscibility with PMMA. Confirmation of this behavior is provided by the curves presented in Figure 1a for blends involving SAN containing 25 wt % and 30 wt % AN, respectively, obtained by conventional d.s.c. The matrix of ABS would be expected to behave in a similar manner. However, the presence of the strongly scattering polybutadiene dispersed phase in the ABS excludes optical scattering methods for the determination of phase behavior. Although the close similarity of T_g of the respective materials (PMMA T_g=101°C, ABS T_g=107°C) negates using the criterion of a single composition dependent T_g , enthalpy recovery studies are able to resolve phase behavior (13). The illustrations in Figures 1a and 1b, for miscible and immiscible blends of ABS and SAN, respectively, with PMMA, show the result of subtracting the thermogram of an aged sample from that obtained by quench cooling, recorded using conventional d.s.c., effectively revealing the enthalpy recovery peak(s). ABS resins with AN contents of 28-29 wt% and above, as measured by nmr, exhibit broad multiple recovery peaks, whereas a blend with ABS containing approximately 24 wt% AN in the matrix provides a single sharp peak that is almost twice the height of the former. Measurements of the position of the peak maximum and onset can be used to augment the analysis. An example coming from a ternary blend study is given in Figure 2 showing the peak maxima for PMMA/ABS blends in which a third component, polycarbonate (PC), has been added. Since PC is immiscible with the other components, the presence of the PC has no influence on the enthalpy recovery of the PMMA and ABS. The figure demonstrates that the principle peak is commensurate with that of the pure ABS whereas the onset (not shown) identifies with that of the pure PMMA. Clearly, PMMA and ABS with 31 wt% AN are immiscible (*1,2*). T-m.d.s.c has not been applied to blends such as these.

Chlorinated polymers/Copolyester-amides Recent studies (*3*) of blends of chlorinated polyethylenes with caprolactam(LA)-caprolactone(LO) copolymers have been able to establish a correlation between miscibility and chemical structure within the framework of a binary interaction model. In some of the blends, both components have the ability to crystallize. When one or both of the components can crystallize, the situation becomes rather more complicated. Miscible, cystallizable blends may also undergo segregation as a result of the crystallization with the formation of two separate amorphous phases. Accordingly, it is preferable to investigate thermal properties of vitrified blends. Subsequent thermal analysis also produces exothermic crystallization processes that can obscure transitions and interfere with determination of phase behavior. In these instances T-m.d.s.c has the ability to separate the individual processes and establish phase behavior.

Figure 3 presents thermograms of an immiscible blend of chlorinated polyethylene with 36 wt% chlorine and a co-polyesteramide containing 40.3 mole% lactam(LA). It is obvious that the determination of the second glass transition temperature from the total heat flow is difficult due to the presence of the exothermic crystallization peak. From the reversing heat flow curve this T_g can, however, be established without any difficulty. When the T_g is used as the tool to determine miscibility, conventional d.s.c is clearly not suitable in this case. Hence, the advantages of T-m.d.s.c. become especially clear when the objective is to construct miscibility windows for those systems where overlapping signals such as T_g and exothermic processes are present. A complete miscibility window, as presented in ref (*3*), would not have been possible using conventional d.s.c. only.

Polystyrene/Poly(styrene-*co*-para-fluorostyrene) Systems involving PS and random copolymers of styrene and fluoro-substituted styrenes are among the most

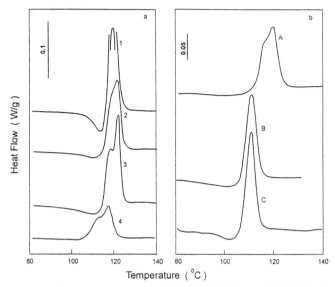

Figure 1a. Enthalpy recovery peaks of PMMA/ABS and PMMA/SAN blends, annealed at $T_a=85°C$: 1) SAN 25 wt% AN, $t_a=1284$min; 2)&3) ABS 31 wt% AN, $t_a=1560$min, blended by coprecipitation and exposed to 150°C and 260°C, respectively; 4) SAN 30 wt% AN, $t_a=265$min.

Figure 1b. Enthalpy recovery peaks of PMMA/ABS blends, annealed at $T_a=85°C$; (A) 50-50 wt% with ABS containing 29 wt% AN, $t_a=2895$min; (B) and (C) 60-40 and 40-60 wt%, respectively, with ABS containing 24 wt% AN, $t_a=390$min.

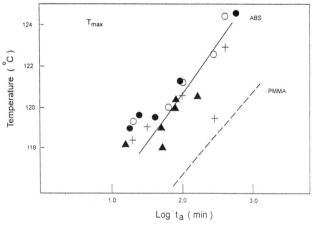

Figure 2 Measured values of maximum of the recovery peak, T_{max}, versus log t_a for PMMA/ABS (31 wt% AN)/PC blends: ▲ 50/50/0; ○ 42/42/15; ● 33/33/33; + 25/25/50. Broken and solid lines represent best fit for T_{max} for pure PMMA and ABS, respectively.

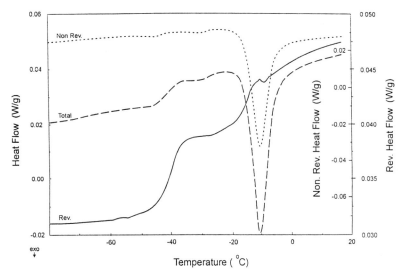

Figure 3. Representative curves, obtained by T-m.d.s.c., of 50-50 wt% blends of chlorinated polyethylene with 36 wt% chlorine and a co-polyesteramide containing 40.3 mole% LA.

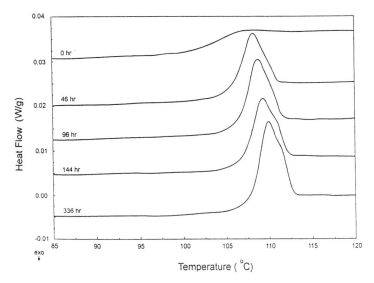

Figure 4. Total heat flow curves of PS/P(S-pSF46) (50-50wt%) aged for indicated times at 92 °C.

obvious examples of blends involving components with nearly identical T_g's. Consequently, they were among the first to be studied by the enthalpy relaxation method (15,16). Although specific heat spectroscopy can in principle provide detailed information about the frequency dependent complex specific heat (5), even this complete spectrum is not sufficient to identify (im)miscibility in blends consisting of components with nearly identical glass transition temperatures (14). Enthalpy recovery studies (1,2) are still required. With the advent of T-m.d.s.c. we decided to reinvestigate a 50/50 wt% blend of PS and a random copolymer of styrene and para-fluorostyrene P(S-pSF46) in order to see what new information might be obtained in this manner. A T-m.d.s.c measurement was taken of the unannealed sample and subsequent measurements involved the same system after being annealed at 92°C for different amounts of time. Figure 4 shows the total heat flow curves. As can be seen, the immiscibility manifests itself as a high temperature shoulder, but only after annealing for 96 hours and more. In Figure 5 the corresponding "reversing" heat flow is presented. The immiscibility is apparent from two T_g's becoming visible already after annealing for 46 hours or more. This figure presents essential new information not available from conventional d.s.c. measurements. With respect to (im)miscibility it only confirms the conclusion which can be drawn from the total heat flow. However, it does provide additional information about the relaxation/recovery behavior of glassy polymers. The total heat flow of physically aged glassy polymers has been described quite successfully by the KAHR model (17) and by the Moynihan approach (18,19) in the past. The fact that in- and out-of-phase data are available now has already stimulated the extension of these approaches to the temperature modulated situation (14,20). Figure 6 presents the "non-reversing" heat flow as the difference between the data presented in Figures 4 and 5.

Finally, Figure 7 presents the derivative of the reversing heat flow corresponding to the data of Figure 5. As noted before, the immiscibility becomes visible after 46 hours or more. The peak corresponds to the midpoint glass transition. For the unaged sample a single peak is observed at 107.1°C. After 46 hours of annealing, still only one T_g is clearly visible, shifted to 108.4°C. An indication of the second T_g can be seen as a high temperature shoulder. After 96 hours of annealing the T_g's are present at 108.5°C and 110.4°C. After 144 hours the peaks have shifted to 109.3°C and 111.2°C. Finally, after 336 hours of annealing the maxima of the two peaks are located at 110.3°C and 112.7°C. The height of the high temperature peak becomes less with increasing annealing time, which might indicate a faster shift of the lower temperature peak of PS gradually overlapping the P(S-pSF46) peak, although the difference between both peaks remains constant (ca. 2 °C).

Conclusions T-m.d.s.c. has already become an indispensable tool for polymer blends studies. Its main advantage is in resolving phase behavior in those situations where additional exothermic processes are present. However, as far as miscibility studies of polymer blends involving components with comparable glass transition temperatures is concerned, we still have to rely on the enthalpy recovery method, that is, assuming that thermal analysis is the experimental technique selected.

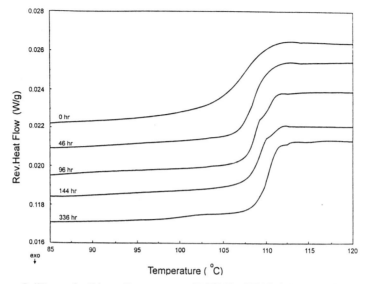

Figure 5. "Reversing" heat flow curves of PS/P(S-pSF46) (50-50wt%) aged for indicated times at 92 °C.

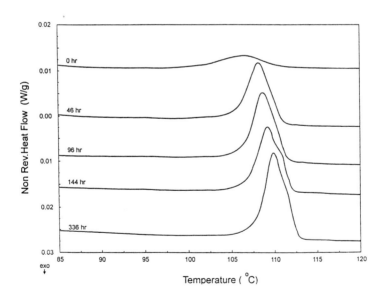

Figure 6. "Non reversing" heat flow curves of PS/P(S-pSF46) (50-50wt%) aged at 92 °C for the indicated amount of time.

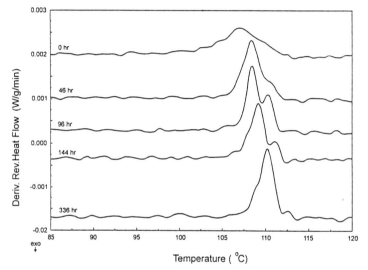

Figure 7. Derivative "reversing" heat flow of PS/p(S-pSF46) (50-50wt%) corresponding to the data in Figure 5.

Literature Cited

1. Bosma, M., ten Brinke, G., Ellis, T. S. *Macromolecules* **1988**, *21*, 1465.
2. ten Brinke, G., Oudhuis, L., Ellis, T. S. *Thermochim. Acta* **1994**, *238*, 75.
3. Alberda van Ekenstein, G. O. R., Deuring, H., ten Brinke, G., Ellis, T. S. *Polymer* **1997**, *38*, 3025.
4. Ellis, T. S. *Macromol. Symposia* **1996**, *112*, 47.
5. Sullivan, P.F., Seidel, G. *Phys. Rev.* **1968**, *173*, 679.
6. Birge, N.O., Nagel, S.R. *Phys. Rev. Lett.* **1985**, *54*, 2674.
7. Reading, M. A., Luget, A., Wilson, R. *Thermochim. Acta* **1994**, *238*, 295.
8. Schawe, J.E.K., Höhne, G.W.H. *Thermochim. Acta* **1996**, *287*, 213.
9. Hutchinson, J.M., Montserrat, S. *Thermochim. Acta* **1996**, *286*, 263.
10. Wunderlich, B., Boller, A., Okazaki, I., Kreitmeier, S. *Thermochim. Acta* **1996**, *282/283*, 143.
11. Hourston, D. J., Song, M., Hammiche, A., Pollock, H. M., Reading, M. *Polymer* **1997**, *38*, 1.
12. Wunderlich, B., Jin, Y., Boller, A. *Thermochim. Acta* **1994**, *238*, 277.
13. Laverty, J. J., Bullach, R. L., Ellis, T. S., McMinn, T. E. *Polymer Recycling*, in press.
14. E. Flikkema, G. Alberda van Ekenstein, G. ten Brinke *Macromolecules*, submitted.
15. Salomons, W., ten Brinke, G., Karasz, F.E. *Polym. Commun.* **1991**, *32*, 185.
16. Oudhuis, A.A.C.M., ten Brinke, G., Karasz, F.E. *Polymer* **1993**, *34*, 1991.
17. Kovacs, A.J., Aklonis, J.J., Hutchinson, J.M., Ramos, R. *J. Polym. Sci., Polym. Phys. Ed.* **1979**, *17*, 1079.
18. Moynihan, C.T., et al. *Ann. N.Y. Acad. Sci.* **1976**, *279*, 15
19. Hodge, I.M., Berens, A.N. *Macromolecules* **1981**, *14*, 1599
20. Hutchinson, J.M., Montserrat, S. *Thermochim. Acta*, **1996**, *286*, 263.

Chapter 16

Physical Aging in PMMA Investigated Using Positron Annihilation, Dielectric Relaxation, and Dynamic Mechanical Thermal Analysis

William J. Davis[1] and Richard A. Pethrick[2,3]

[1]Department of Pure and Applied Chemistry, University of Strathclyde, Thomas Graham Building, 295 Cathedral Street, Glasgow G1 1XL, Scotland

Physical ageing in polymethylmethacrylate has been investigated using a combination of positron annihilation lifetime, dielectric relaxation and dynamic mechanical thermal analysis. No simple theory was capable of fitting all the measurements, and the dielectric data indicate that 'thermorheologically simplicity' is not a valid assumption for PMMA. The free volume distribution appears to change during ageing, however the processes occurring at a molecular level are complex and not simply the result of one single type of conformational rearrangement. An apparent activation energy comparable to that of the glass - rubber transition was observed from an analysis of the data.

The spontaneous, thermally reversible change that occurs in glassy materials when brought close to their glass transition [T_g] is known as *physical ageing (1,2)*. Many of the theories associated with physical ageing are based on free volume as the rate controlling factor. A non-equilibrium thermodynamic conformational distribution arises as the material is cooled through T_g and equilibrium is slowly approached at a rate which depends on the annealing temperature. Below the beta relaxation physical ageing ceases *(1,2)*. The apparent activation energy for the beta relaxation process is typically 30-50 kJ/mol and values of between 200-300 kJ/mole are observed for the glass transition process *(3)*. The former are associated with localised conformational changes requiring motion about one or two bonds whereas the glass transition is a co-operative motion involving between 8-20 bonds. The density, yield stress and elastic modulus of the polymer increase; impact strength, fracture energy, ultimate elongation decrease; creep and stress relaxation rates decrease upon physical ageing.*(1,2)*

*Corresponding author.

Polymethylmethacrylate, [PMMA], a polar polymer exhibits a dielectric relaxation spectrum [DRS] allowing the effects of ageing on the nature of dipole processes to be monitored. Longer range co-operative molecular motions dominate dynamic mechanical thermal analysis [DMTA]. The relaxations observed by these techniques are to a greater or lesser extent influenced by pressure and have associated with them definite volume changes. The T_β process is usually insensitive to pressure whereas T_g is very sensitive and indicates that there is a volume change associated with the process (3). The void structure - free volume at a molecular level can be explored using positron annihilation lifetime spectroscopy [PALS]. Combining data from PALS, DRS and DMTA provides a picture of the way in which free volume and various scales of molecular motion change during ageing.

Below T_g (4,5) co-operative molecular processes are usually assumed to be inactive. However, physical ageing implies that conformational changes may be still able to occur if rather infrequently. Structural relaxation processes are observed to be non-exponential and are represented by a continuous distribution or stretched exponential form (6). Thermorheologically simplicity (TRS) implies that the molecular relaxation process has the same form at different temperatures (7) and the validity of this assumption is addressed in this paper. Isobaric volume recovery (8,9) has been described by a single parameter model, however all free volume models (10,11) have limitations and a distribution of hole sizes and relaxation times leading to a pseudo-linear theory is a more realistic model(12). Comparison of data from various techniques should throw light on the molecular nature of physical ageing.

Experimental

Materials. ICI [Diakon], commercial PMMA, (M_w) - 136,000, T_g - 105°C; was annealed at 120°C. Dog bones - 120 x 10 x 3.25 mm.; square plaques, 60 x 60 x 3.25 mm. and creep bars 15 x 20 x 3.25 mm. were compression molded. Isothermal ageing was performed on samples previously equilibrated at a defined annealing temperature followed by rapid quenching to room temperature, ~298K and reheating to the ageing temperature T_e.

Dielectric Relaxation Spectroscopy - [DRS]. Isothermal and isochronal ageing studies were carried out over a frequency range from 10^{-2} to 6.5×10^5 Hz at 10°C temperature intervals between 30-120°C (13).

Dynamic Mechanical Analysis. A Rheometrics Dynamic Mechanical Thermal Analyser (DMTA) Mk III was used in a single cantilever arrangement. Both E' and tan δ were measured as a function of temperature at a frequency of 1Hz over the temperature range -145 to +150°C and a heating rate of 2°C/min for the isochronal experiments.

Positron Annihilation Lifetime Spectroscopy (PALS). A fast-fast PALS system using cylindrical (40 mm diameter × 15 mm thick) BaF_2 scintillators (14) arranged at 90° to each other to avoid pulse pile up problems. A count rate of 150-300 cps was achieved with a ^{22}Na source and a instrument resolution of 220-240 ps FWHM for a

50 µCi source. Benzophenone and ^{60}Co were used to determine the resolution and source correction. The spectrum was analysed using POSITRONFIT (*15*). A 'fixed analysis'; assumes τ_1 - 0.125 ns; (p-Ps), τ_2 - 0.400 ns (free e^+) and I_3/I_1 at 3:1. Isochronal experiments involved the samples being aged for various times at a defined temperature before being thermally scanned. Isothermal experiments involved the sample being measured continuously as a function of ageing time at a particular temperature. PALS were collected at 10°C intervals from 30-100°C in 60 min corresponding to 500-600 k counts.

Comparison between POSITRONFIT and CONTIN (*16,17*) analysis indicated that only for 10^6 counts or greater could the results of the analysis be differentiated. Spectra with ~ 5 x 10^5 counts were used as optimum in this study.

Results And Discussion

Positron Annihilation Data. The effect of irradiation time and source strength on the o-Ps annihilation (*25*) in PMMA is minimal (*18,19*) and does not have a significant effect on the physical ageing experiments. The possible effects of charging on oPs annihilation will be discussed in detail elsewhere. Samples were annealed for 30 minutes before ageing and then quenched in water for ~ 1 minute. Samples were aged at 30, 50 and 80°C for respectively 0, 4, 8, 16, 32 and 64 hours.

Isothermal Ageing of Poly(methyl methacrylate). The 'equilibrium' liquid line was measured by studying the o-Ps lifetime τ_3, versus temperature between 30-150°C, (Figure 1). Isothermal measurements at 30, 70, 80, 90, 100, 110 and 120°C rather than isochronal experiments were carried out to reduce any errors due to uncertainty in the thermal history. Different models have been used to fit lifetime data (*20*) and include; log, single exponential, double-additive exponential and the Narayanaswamy. Logarithmic regression $\tau_3 = a \times \log(t) + b$; single exponential $\tau_3 = a \times \exp^{\left(-t/\tau\right)} + b$; double additive exponential $\tau_3 = a \times \exp^{\left(-t/\tau\right)}$ $+ b \times \exp^{\left(-t/\tau\right)} + c$; and Narayanaswamy (*7*); $\tau_3 = \exp^{\left(-t/\tau\right)^\beta}$ were explored. The results presented are limited to logarithmic (Figure 2) or simple single exponentials fits, these being the most precise, and more sophisticated models appeared invalid (Table I).

Table I. Summary of fitting parameters for isothermal τ_3 ageing data over temperature range 70-120°C

Ageing Temp (°C)	single exponential fit			log fit
	$\tau_3(t=\infty)$ ns (fixed)	τ_3(hrs)(relaxation time constant)	$\Delta\tau_3$ ns (extent of ageing)	'rate' $d\tau_3/dlogt$ (gradient)
70	1.820	6339	0.165	-0.0050
80	1.888	2478	0.131	-0.0085
90	1.956	1240	0.092	-0.0140
100	2.024	610	0.049	-0.0117
110	2.050	324	0.062	-0.0242
120	/ / / /	/ / / /	/ / / /	-0.0074

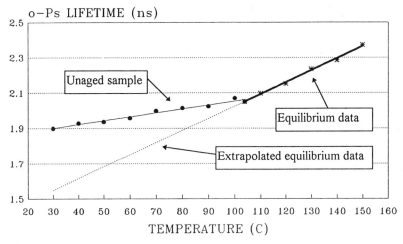

Figure 1.   Plot of τ_3 versus temperature data for PMMA, including extrapolation of equilibrium 'liquid' line below T_g.

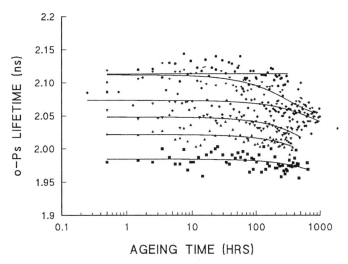

Figure 2.   Exponential fits to isothermal τ_3 data for PMMA over the temperature range 70-120°C. The 120°C data is fit with the equation of a straight line. Data is in order of increasing temperature from the bottom of the plot upwards. ■ 70°C, Δ 80°C, ∇ 90°C, ♦ 100°C, ✷ 110°C, ● 120°C.

Variation of the oPs lifetime relaxation time with ageing temperature gave Arrhenius behaviour with an activation energy of 84±6 kJ/mole. The 'rate' of ageing (Figure 3) obtained from the slope of the log curve, (Figure 4), is similar to Struiks (*1*) ageing rate versus temperature plot and implies a connection between free volume V_f and the mechanical relaxation processes.

Dielectric Relaxation Data. Dielectric permittivity and loss data were measured over the temperature range 30-120°C for unaged and aged PMMA. Comparison of data at 90°C, (Figure 5) indicates the change in the distribution of relaxation processes which occur on ageing (*21,22*). The low frequency tail of the T_g process occurs at approximately 10^{-5} Hz and ageing narrows the β relaxation process and reduces the low frequency amplitude and indicates thermorheological simplicity is not correct.

Mechanical Relaxation Data. Isochronus ageing carried out on unaged samples equilibrated for 100 hours at 90°C in an air circulating oven, (Figure 6). The onset of T_g occurs at a slightly higher temperature in the aged compared to the unaged material. The tan δ plots show two peaks; at around 40°C a β-relaxation and at 120°C an α-relaxation. The largest change in tan δ occurs when samples are aged over the temperature range between the two main relaxation processes.

Isothermal Ageing Experiments. Both E' and tan δ data were collected as a function of the isothermal ageing time over the temperature range 70-100°C at 10°C intervals with a precision of ± 0.1°C at a frequency of 1Hz. The sample heat-up time to the ageing temperature after quenching was ~ 5 minutes, whilst the oven reached this temperature in ~ 2 minutes, a delay in sample equilibration being observed due to the low thermal conductivity of PMMA. Data were collected at 15 minute intervals for periods of up to 200 hrs depending on the ageing temperature. Since tan δ is the ratio E''/E' it does not vary with sample dimensions and is therefore insensitive to changes in geometry during the experiment. Isothermal tan δ data for PMMA over the temperature range 70-120°C weɾe measured, (Figure 7).

Below 110°C tan δ varies approximately linearly with log time over the time range measured (*23*). Above 110°C a rapid decrease in tan δ followed by a stable value indicating equilibration occurs very quickly. Data over the temperature range 70-100°C were fit with both double-exponential and logarithmic functions. The single-exponential fit to the data was inappropriate at lower temperatures and the double-exponential data does not follow the expected pattern shown by both PALS and dielectric data of increasing relaxation time for the relaxation process with decreasing temperature. Attempts were made to mathematically fit the data using a variety of fitting procedures, (Table II). At 110°C tan (delta) varied with log time, Table II and then the ageing 'rate' decreases rapidly to zero. This approach was taken to allow calculation of the ageing 'rate' for 110°C data to allow comparison with both PALS and DRS ageing 'rate' data. DMA experiments over a wide time range (*23*) indicated that a horizontal shift to superimpose materials of different ageing histories was not completely effective and the double-exponential fits over the temperature range 70-100°C, however it cannot be concluded that the relaxation function contains

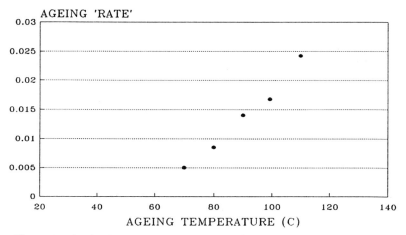

Figure 3. Ageing 'rate' versus temperature for PMMA over the temperature range 70-120°C.

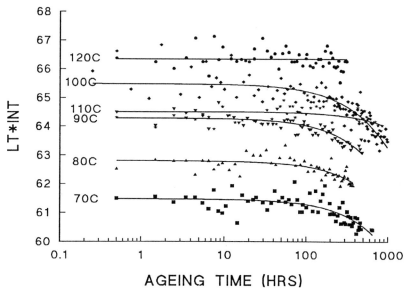

Figure 4. Product of $\tau_3 \times I_3$ versus ageing time for PMMA over the temperature range 70-120°C. Fitted curves are simple single exponentials. ■ 70°C, Δ 80°C, ∇ 90°C, ♦ 100°C, ＊ 110°C, ● 120°C.

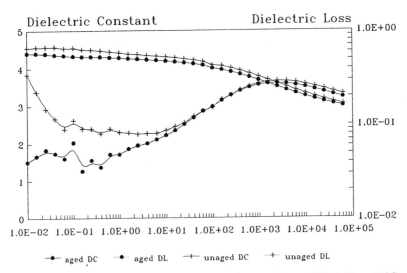

Figure 5. Frequency dependence of ε' and ε'' for aged/unaged PMMA at 90°C.

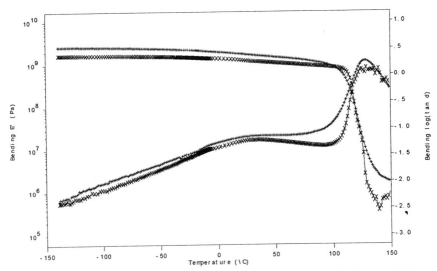

Figure 6. Dynamic mechanical spectra showing both E' and $tan\delta$ of unaged (+) and aged (×) PMMA over the temperature range -145 to +150°C at 1Hz frequency.

two times. This implies that the relaxation process is of a dual molecular origin for the process, however close to T_g 'simplification' is appropriate. The temperature dependence of the rate of ageing, (Figure 8), correlates with the ageing rate measured using PALS. The double-exponential fit represents data over the whole time range but the relaxation parameters do not follow any pattern. A single-exponential fit was appropriate close to T_g, implying only one major process is responsible whereas at lower temperatures two processes are more appropriate.

Table II. Fitting parameters for mechanical data

Ageing temp	1-exp. Fit	2-exp. fit		log fit
(°C)	τ (hrs)	τ_1 (hrs)	τ_2 (hrs)	(gradient)
70	NRS	1.4	40	-0.00528
80	NRS	1.1	30	-0.00767
90	NRS	0.8	62	-0.00841
100	NRS	1.5	33	-0.0276
110	2.3	NRS	NRS	-0.0654
120	0.6	NRS	NRS	NRS

A more detailed presentation of the data and discussion of analysis will appear in subsequent publications.

Discussion

This study shows that none of the various forms of relaxation function used to describe ageing are completely satisfactory and TRS is inappropriate. Correlation between results, (Figure 9) indicates the inherent connectivity between the processes. Curro et al (*24,25*) have studied the change in density fluctuation with temperature and annealing time for PMMA (*26*) and compared it with specific volume data. Positron annihilation data on PMMA (*27,28*) has been interpreted in terms of free volume. For a distribution of hole sizes there will exist many decaying exponentials each with a different characteristic lifetime. The composite of these many exponentials can itself be approximated to an exponential, and it is this decay constant that is used to represent the mean lifetime, and therefore mean hole size.

Conclusions

The following conclusions can be drawn from this study:-
• a correlation exists between the 'free volume' measured by PALS and changes in the DRS and DMTA data clearly indicating a sensitivity of the processes involved to the void structure at a molecular level.
• that 'thermorheological simplicity' is an over-simplification and physical ageing is a redistribution and not just a shift of relaxation processes. The changes in the form of the DRS clearly indicate the way in which certain types of motion are suppressed as physical ageing occurs and other motions are enhanced. In the case

236

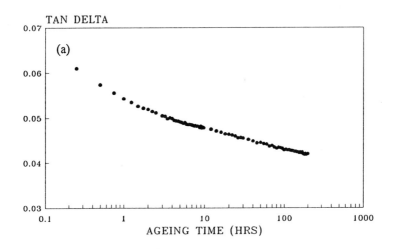

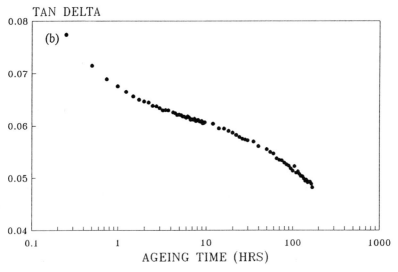

Figure 7. Isothermal *tanδ* data for PMMA aged at 70°C (a) and at 90°C (b).

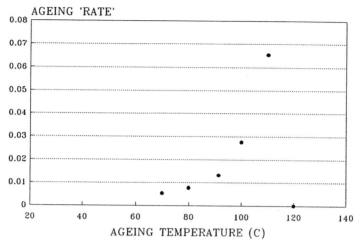

Figure 8. Ageing 'rate' versus temperature for PMMA over the temperature range 70-120°C. The 'rate' was calculated as the slopes of log fits to *tanδ* data.

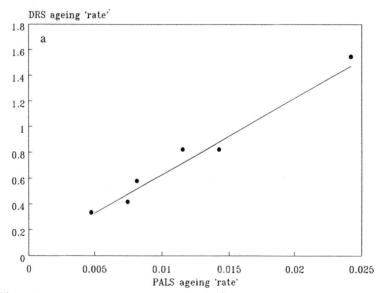

Figure 9. Correlation between ageing 'rates' determined by (a) DRS and PALS experiments, (b) DRS and DMA experiments, (c) DMA and PALS experiments. Ageing 'rates' in each case are determined from gradients of log fits to data.

Continued on next page.

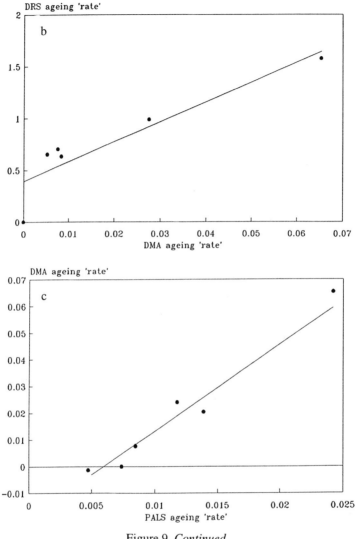

Figure 9. *Continued.*

of PMMA this may reflect the relative contributions of predominantly side chain and backbone motions to the overall relaxation.

- the relaxation distribution does change as the T_g is approached. This observation is consistent with the idea that free volume changes associated with physical ageing occur more rapidly as T_g is approached.

Acknowledgements

WJD wishes to thank the EPSRC and Ford for maintenance during the period of this study.

Literature Cited

1. Struik, L. C. E. *Physical Ageing in Amorphous Polymers and Other Materials*; Elsevier Amsterdam, **1978**.
2. Struik, L. C. E. *Internal Stresses, Dimensional Instabilities and Molecular Orientations in Plastics*; John Wiley, **1990**.
3. Bailey, R. T.; North, A. M.; Pethrick, R. A. *Molecular Motions in High Polymers*; Clarendon Press, Oxford, **1981**.
4. Doolittle, A. K. *J. App. Phys.* **1951**, *22*, 1471.
5. Williams, M. L.; Landel, R. F.; Ferry, J. D. *Journal of the American Chemical Society* **1955**, *77*, 3701.
6. Williams, G. *Adv. Poly. Sci.* **1979**, *33*, 59.
7. Narayanaswamy, O. S. *J. Am. Ceram. Soc.* **1971**, *54*, 491.
8. Kovacs, A. J. *Fortschr. Hochpolym. Forsch* **1963**, *3*, 394.
9. Hutchinson, J. M.; Kovacs, A. J. *J. Poly. Sci.: Poly. Phys Ed.* **1976**, *14*, 1575.
10. Cohen, M. H.; Turnbull, D. *J. Chem. Phys.* **1959**, *31*, 1164.
11. Turnbull, D.; Cohen, M. H. *J. Chem. Phys.* **1961**, *34*, 120.
12. Tool, A. Q. *J. Res.* **1945**, *34*, 199.
13. Hayward, D.; Gawayne, M.; Mahboubian-Jones, B.; Pethrick, R. A. *J. Phys. E. Sci. Instrum.* **1984**, *17*, 683.
14. Chang, T.; Yin, D.; Cao, C.; Wang, S.; Liang, J. *Nuc. Inst. Meth. Phys. Res.* **1987**, *A256*, 398.
15. Kirkegaard, P.; Eldrup, M.; Mogensen, O. E.; Pedersen, N. J. *Computer Physics Communications* **1981**, *23*, 307.
16. Provencher, S. W. *Comput. Phys. Commun.* **1982**, *27*, 229.
17. Gregory, R. B. *Nuc. Inst. Meth. Phys. Res.* **1991**, *A302*, 496.
18. Li, X. S.; Boyce, M. *J. Poly. Sci : Part B : Poly. Physics* **1993**, *31*, 869.
19. Brandt, W.; Wilkenfeld, J. *Physical Review B* **1975**, *12*, 2579.
20. Hill, A. J.; Jones, P. L.; Lind, J. H.; Pearsall, G. W. *J. Poly. Sci.: Part A: Poly. Sci.* **1988**, *26*, 1541.
21. McCrum, N. G.; Read, B. E.; Williams, G. *Anealastic and Dielectric Effects in Polymeric Solids*, Dover Publications, New York, 1991.
22. Malhotra, B. D.; Pethrick, R. A. *Eur. Poly. J.* **1983**, *19*, 457.
23. Venditti, R. A.; Gillham, J. K. *J. App. Poly. Sci.* **1992**, *45*, 1501.
24. Curro, J. J.; Roe, R. *Polymer* **1984**, *25*, 1424.
25. Curro, J. J.; Roe, R. *Macromolecules* **1983**, *16*, 428.
26. Tanaka, K.; Katsube, M.; Okamoto, K.; Kita, H.; Sueoka, O.; Ito, Y. *Bull. Chem. Soc. Jpn.* **1992**, *65*, 1891.
27. Heater, K. J.; Jones, P. L. *Nucl. Inst. Meth. Phys. Research* **1991**, *B56/57*, 610.
28. Hasan, O. A.; Boyce, M. C.; Li, X. S.; Berko, S. *J. Poly. Sci : Part B : Poly. Physics* **1993**, *31*, 185.

MECHANICAL PROPERTIES

Chapter 17

Craze Initiation and Failure in Glassy Poly(ethylene Terephthalate): The Effects of Physical Aging with and Without Exposure to Chemical Environments

Martin R. Tant[1], E. J. Moskala[1], M. K. Jank[1], T. J. Pecorini[1], and Anita J. Hill[2]

[1]**Research Laboratories, Eastman Chemical Company, P.O. Box 1972, Kingsport, TN 37662**
[2]**CSIRO Division of Manufacturing Science and Technology, Private Bag 33, South Clayton MDC, Victoria 3169, Australia**

The nonequilibrium nature of the glassy state leads to the time-dependent embrittlement of glassy polymers by a process known as physical aging. Similarly, exposure of glassy polymers to an aggressive chemical environment may also lead to a time-dependent embrittlement process known as environmental stress cracking. In the latter process, the time required for brittle failure to occur as a result of chemical exposure depends on the nature of the polymer, its morphology, and the chemical environment. Since both physical aging and environmental stress cracking lead to essentially the same result – brittle failure – the question arises as to whether or not the physical aging process might affect the environmental stress cracking process. In this work, we demonstrate the accelerated effects of physical aging on craze initiation in a commercially important engineering thermoplastic, poly(ethylene terephthalate), both with and without exposure to aggressive chemical environments.

It has been well established that physical aging strongly affects the physical, mechanical, and transport properties of glassy polymers (*1,2*). In general, as a polymer is cooled at a finite rate through the glass transition temperature (T_g), the molecules are not able to respond within the time scale of the cooling process and are essentially trapped into a nonequilibrium state. Though the polymer molecules have greatly reduced mobility below the glass transition, their mobility is still finite and the polymer system continues to move toward equilibrium, though at an ever-decreasing rate, via conformational rearrangements as the material approaches the density corresponding to normal liquidlike packing. Both specific volume and molecular free volume have been observed to decrease during the physical aging process (*1-3*). The decreasing molecular free volume results in a reduction in molecular mobility, leading to a transition from ductile to brittle behavior. Yielding typically precedes ductile failure, while crazing precedes brittle failure (*4*). Several recent studies have addressed the effects of physical aging on craze initiation (*5-7*).

Polymeric materials exposed to aggressive or corrosive chemical environments are often observed to fail at much shorter times or at much lower loads than when not exposed to such environments. This phenomenon, typically referred to as environmental stress cracking (ESC), has been a persistent problem in the plastics industry and is a limiting factor in determining the applications for which a particular plastic might be used. Poly(ethylene terephthalate), a widely used engineering thermoplastic in either the amorphous or semicrystalline forms, is susceptible to ESC in certain chemical environments. In engineering applications this problem can become of critical importance since it threatens the ultimate performance of the material.

It has been noted that ESC failures of glassy polymers are higher in frequency during summer months or in warmer locations. This suggests that physical aging may be playing a role in the ESC process since it is well known that the physical aging process occurs more rapidly at temperatures nearer the glass transition temperature. Although, as mentioned, thermal and mechanical properties are well known to change as a result of physical aging, there has been very little study of how physical aging affects environmental stress crack behavior. Arnold and Eccott (8) recently investigated the effects of physical aging on craze initiation in polycarbonate exposed to ethanol and found that aging had relatively little effect, perhaps because of the relatively high levels of strain encountered in their tests.

In the work reported here, changes in molecular free volume and tensile properties of PET have been followed as a function of physical aging. Positron annihilation lifetime spectroscopy (PALS) was used to measure free volume. The effect of physical aging on the competition between yielding and crazing without exposure to corrosive chemical environments was also investigated. Finally, the effect of physical aging on craze initiation at low strains due to exposure to a variety of solvents was determined.

Experimental

EASTMAN 9921W poly(ethylene terephthalate) was used for the study. The polymer, in pellet form, was dried in dehumidified air at 60°C for 16 hours prior to injection molding. Tensile and flexural impact samples were injection molded into a cold mold (23°C) using a Toyo 90 injection molding machine. The cold mold resulted in amorphous specimens that were then aged at 40, 50, and 60°C in dry ovens and at 60°C and 95% relative humidity in an environmental chamber. The time of molding was taken as time zero for the aging experiments. For crazing tests, single-edge notch bend specimens were used.

The positron annihilation lifetime spectroscopy (PALS) apparatus used to measure free volume in this work consisted of an automated EG&G Ortec fast-fast coincidence system. The 1.3 MBq ^{22}NaCl source was a 2 mm spot source sandwiched between two Ti foils (2.54 μm foils). The source gave a two-component best fit to 99.99% pure, annealed, chemically polished aluminum ($\tau = 169 \pm 2$ ps, $I_1 = 99.2 \pm 0.4$ %, $\tau_2 = 850 \pm 250$ ps, $I_2 = 0.8 \pm 0.4$%). No source correction was used in the analysis of the data with the PFPOSFIT program (9). Measurements were made in air at 50% relative humidity with temperature control of $\pm 0.7°C$.

Tensile yield data were obtained following the procedures of ASTM D638 after aging for specified times at the various sub-T_g annealing temperatures. Tensile yield stresses at impact speed were estimated by multiplying by the empirical factor of 1.2. This was determined by measuring the yield stress at the highest strain rate attainable and dividing by the value obtained using ASTM D638. Obviously, impact properties estimated in this way are probably somewhat on the low side. The dry crazing tests for 9921W were conducted by using single-edge notch bend specimens machined from 3.2-mm thick injection-molded bars. Specimen width was 12.6 mm. The specimens were notched with a single-point fly cutter having a tip radius of 0.500 mm. The ratio of initial notch length to specimen width was 0.2. Izod tests were performed according to ASTM D256 at a rate of 3.46 m/s. The distance from the notch root to the point of craze initiation was identified from the specimen fracture surface by using an optical microscope. A minimum of five specimens was used to determine an average craze stress for each aging condition.

Measurements of critical strain for crazing during exposure to various organic solvents were made by strapping flexural bars to a Bergen elliptical strain rig (10) for which the strain at the surface of the specimen is known as a function of location. A strip of filter paper was placed on the surface of the specimen and solvent was applied with a dropper, taking care to not expose the edges of the specimen to the solvent. The filter paper was kept soaked with the solvent for 10 minutes, after which the filter paper and the residual solvent on the surface were removed. The location of the craze at the lowest strain was identified and recorded as the critical strain for crazing.

Results and Discussion

Positron Annihilation Lifetime Spectroscopy. The details of PALS will not be discussed here since several other chapters in this book are specifically concerned with this experimental technique and provide more than adequate description. Suffice it to say here that *ortho*Positronium (oPs), which is formed when a positron binds to an electron of parallel spin, tends to locate in free volume sites within a polymer, and the time required for it to annihilate via pickoff with an electron of anti-parallel spin residing in the surrounding material is related to the size of the free volume site. There are two parameters that are sensitive to polymer free volume: the oPs pickoff component lifetime, τ_3, which is related to the mean radius of the free volume cavities, and the oPs pickoff component intensity, I_3, which is related to the concentration or number of free volume cavities.

Because the physical aging experiments are time dependent experiments, it is important to establish that the oPs parameters τ_3 and I_3 are not dependent on contact time with the ^{22}Na source. Figure 1 displays the oPs parameters for PET as a function of source contact time at room temperature. There are no changes in τ_3 and I_3 parameters due to source contact. Figure 2 shows the temperature-dependent PALS data for PET between 23 and 87°C. The τ_3 vs. temperature plot shown in Figure 2a has two linear regions with a change in slope at T_g. These data were obtained by heating from room temperature and holding at each measurement temperature for 3.5 hours while the measurement was made. This plot is somewhat reminiscent of the specific volume - temperature plot that is often used to describe the thermodynamic aspects of glass formation. Figure 2b shows I_3 as a function of temperature in this same region. Again, a slope change is observed at T_g, and this slope change occurs several degrees below where it occurs for the τ_3 vs. temperature plot.

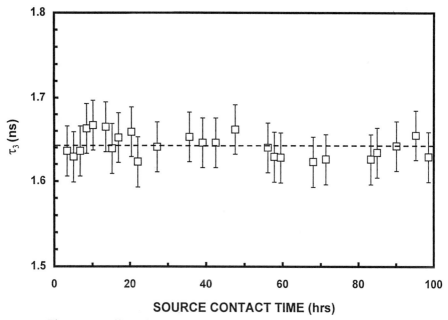

Figure 1a. Effect of source contact time on the PALS oPs parameter τ_3.

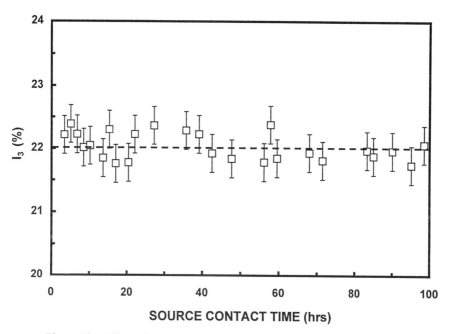

Figure 1b. Effect of source contact time on the PALS oPs parameter I_3.

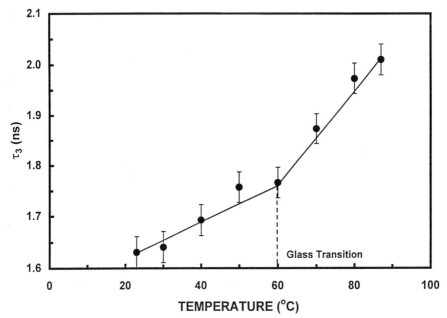

Figure 2a. Effect of temperature on τ_3, related to the mean free volume cavity radius.

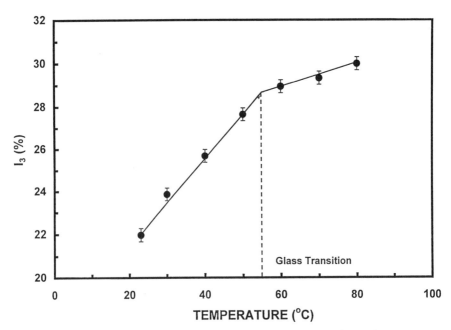

Figure 2b. Effect of temperature on I_3, related to the mean free volume concentration.

It should be noted that in plots of τ_3 and I_3 the change in slope with respect to temperature is observed in the range of 55-60°C. This is well below the glass transition temperature of this PET (~80°C) observed by DSC. The T_g determined by PALS has also been found to be lower than that observed by other techniques for other polymers as well. For example, Kasbekar (11) found that the change in the τ_3-temperature slope for atactic polystyrene occurs at 90°C while the change in the I_3-temperature slope occurs at about 60°C. It was suggested that the oPs free volume probe detects the changes in electron density due to local molecular motions that occur at temperatures lower than the cooperative motion of chains measured by other techniques.

To investigate the behavior of free volume during physical aging of the glass, an amorphous PET sample was heated from room temperature to 84°C where it was held for 10 minutes to erase previous physical aging. The sample was then cooled to 60°C or 40°C and held for approximately 100 hours in order to follow the time-dependent relaxation. The results for both τ_3 and I_3 are shown in Figures 3a and b, respectively. It is clear that τ_3, while a function of temperature, is not a function of time at either 60°C or 40°C. On the other hand, I_3 is approximately constant at 60°C but relaxes logarithmically with increasing aging time at 40°C. The fact that neither τ_3 nor I_3 changes during annealing at 60°C raises some important questions. Certainly it is well established that (1) PET undergoes volume relaxation at 60°C (12) and (2) the mechanical properties of PET measured at room temperature are strongly affected by how long the material has been aged at 60°C. It has recently been established that the viscoelastic properties of PET measured at 60°C indeed do change with time at that temperature (Scanlan, J. C., Eastman Chemical Company, personal communication, 1997). It seems possible that PALS is not sensitive to the changes in free volume with time at 60°C which affect the viscoelastic properties measured at this temperature. The PALS probe is sensitive to free volume cavities in a particular size range: large enough to accommodate oPs (~ 0.38 nm diameter) but not large enough to appear as internal surface (~>5 nm diameter) (13). Typical intra- and inter-molecular distances in polymers range from 0.2 to 0.7 nm making it most likely that only the smaller sites in the polymeric free volume distribution are excluded from the PALS measure of free volume. The oPs probe provides information on the dynamic free volume due to the time dependent nature of the probe. Typical oPs lifetimes vary from 1 to 3 ns such that the molecular motions of frequency less than 10^9 Hz sweep through dynamic free volume that is available for oPs localization. Faster motions make the free volume utilized in the motion appear "full" to the oPs. The sensitivity of oPs to the free volume relaxation at 40°C as compared to the constant parameters at 60°C suggests that free volume changes due to physical aging at 60°C occur via free volume sites not probed by oPs. These sites (or free volume elements) could be either small cavities or cavities previously occupied by molecular motions of frequency on the order 10^9 Hz. The latter explanation is most likely since the PALS insensitivity occurs at higher temperatures near the glass transition. Indeed the I_3 vs. temperature slope change occurs at 55°C for this PET, indicative of the increased mobility associated with the glass to rubber transition. The greater dependence of I_3 on temperature below T_g (as shown in Figure 2b) suggests that dynamic free volume available for oPs localization in the glass begins to appear "occupied" by molecular motion above T_g. It is thus

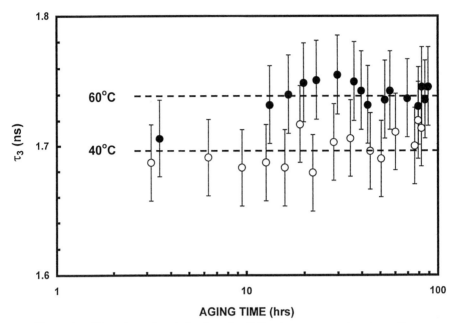

Figure 3a. Time-dependent behavior of τ_3 following a temperature jump from 84°C to 60°C or 40°C.

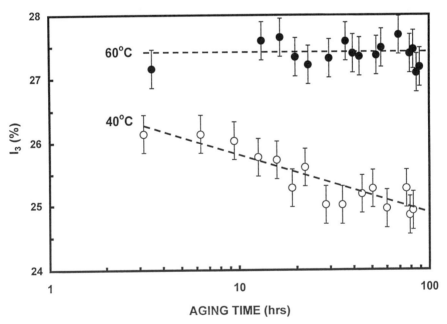

Figure 3b. Time-dependent behavior of I_3 following a temperature jump from 84°C to 60°C or 40°C.

proposed that the free volume relaxation at 60°C is dominated by a reduction in dynamic free volume inaccessible to oPs.

Hill et al. (3) found similar results for physical aging studies of polycarbonate ($T_g \sim$ 148°C), namely that the oPs pickoff parameter I_3 did not change with aging time at 120°C but did vary on cooling to lower temperatures. It was suggested that the constant free volume concentration indicated by the PALS I_3 parameter during aging at 120°C was due to the degree of molecular mobility near T_g as compared to the oPs timescale. The change in τ_3 vs. temperature slope occurs at 135°C and the change in I_3 vs. temperature slope occurs at 120°C for this polycarbonate. Thus it appears that the change in free volume elements during aging near T_g, which results in changes to viscoelastic properties at the aging temperature, occurs via a reduction in dynamic free volume sites (sites frequently occupied by molecular motions) inaccessible to oPs due to the frequency of the molecular motion. As these free volume sites decrease, the molecular motion is restricted and viscoelastic relaxation times increase. Similar changes occur at lower temperatures; however, because the frequency of molecular motion decreases with temperature, these dynamic free volume elements are accessible to oPs and the free volume relaxation during aging can be followed by PALS as shown in the present work for the aging temperature of 40°C.

Critical Stress for Crazing without Solvent Exposure. Hull and Owen (14) showed that crazing is the precursor to brittle fracture for polycarbonate, a typical ductile glassy polymer. The yield stress for polycarbonate and other glassy polymers is well known to increase during the physical aging process (1,2). On the other hand, Pitman et al. (15) have shown that physical aging does not significantly influence the craze stress of polycarbonate. Figures 4a and 4b show the effect of physical aging on both the craze stress and the yield stress of PET at 40° and 60°C, respectively. For the yield stress, both the measured values at low deformation rate and the estimated values at impact rate are shown. The impact values were estimated by multiplying the low rate data by a factor of 1.2, which was empirically determined. Error bars are not shown for the yield stress data because the standard deviation is within the size of the data points themselves. Clearly, the craze stress is essentially unaffected by physical aging at both aging temperatures. The craze stress was calculated from experimental data using Hill's slip line theory (16), given by the equation

$$\sigma_m = \frac{\sigma_y}{\sqrt{3}}\left[1 + 2\ln\left(1 + \frac{\overline{X}}{R_0}\right)\right]$$

where σ_m is the craze stress, σ_y is the tensile yield stress, R_o is the notch tip radius, and $\overline{X}$ is the distance from the notch tip to the craze origin. Kambour and Farraye (17) applied this approach to a number of glassy polymers for which the craze initiation resistance is greater than the shear flow resistance at the notch surface under the conditions of test. While the results shown in Figure 4 illustrate that the craze stress itself does not change, Figures 5a and 5b show that $\overline{X}$ decreases with aging. So while it might be assumed from the craze stress data shown in Figure 4 that there are no effects of aging on the crazing process, significant changes are indeed occurring. The decreasing distance from notch tip to craze origin and the increasing yield stress simply result in the craze stress itself remaining constant. Figure 4 indicates that at all

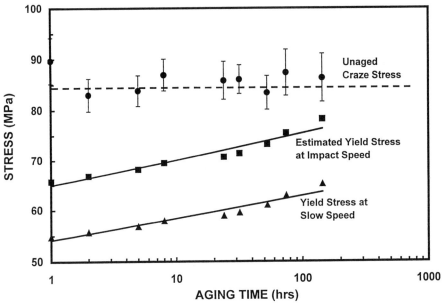

Figure 4a. Effect of physical aging at 40°C on yield and craze stresses. Filled circles indicate the craze stress as a function of aging time. The horizontal dashed line indicates the unaged craze stress.

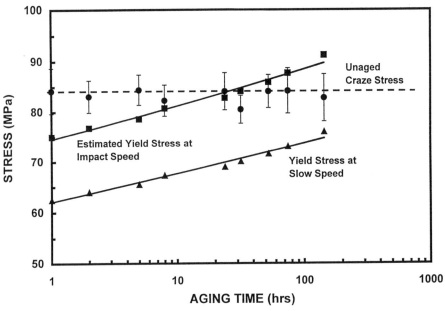

Figure 4b. Effect of physical aging at 60°C on yield and craze stresses. Filled circles indicate the craze stress as a function of aging time. The horizontal dashed line indicates the unaged craze stress.

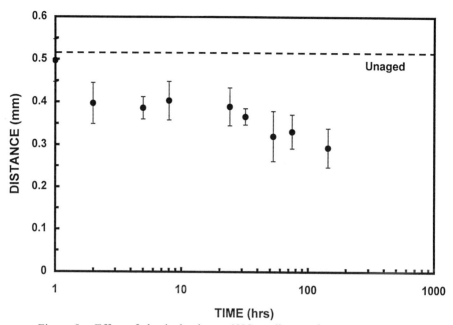

Figure 5a. Effect of physical aging at 40°C on distance from notch root to craze.

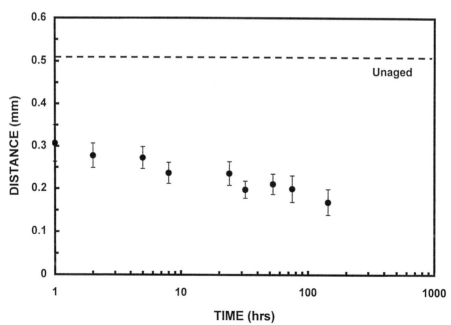

Figure 5b. Effect of physical aging at 60°C on distance from notch root to craze.

temperatures the yield stress of PET, at both slow deformation and impact rates, is initially well below the craze stress so that yielding is the favored failure mechanism and the material thus fails in a ductile manner. As physical aging progresses, the yield stress approaches the craze stress and crazing, leading to brittle failure, becomes the favored mechanism. This occurs at shorter times for impact rates and for higher aging temperatures.

Gusler and McKenna (5,7) found that physical aging tends to *decrease* the critical strain for crazing for both polystyrene and poly(styrene-*co*-acrylonitrile). The observation in this work that $\overline{X}$ for PET decreases with physical aging agrees with the result of Gusler and McKenna since $\overline{X}$ is related to critical strain. Kambour and Farraye (18) have suggested that it is unclear what is the true criterion for crazing. In the present work on PET, it has been found that the critical stress for crazing is independent of physical aging but that the strain at which crazing is observed is a function of physical aging.

Figure 6a shows the effect of aging at 60°C and 90% relative humidity on the craze stress and yield stress of PET. Initially there is an increase in yield stress at both low and high deformation rates. After about 10-20 hours of aging, both the craze stress and the low and high rate yield stresses begin to decrease. This results from the diffusion of water into the polymer and the resulting plasticization of the polymer by the absorbed water. Due to the similar values of the craze stress and the estimated yield stress at high deformation rate, the polymer might well be expected to exhibit brittle behavior after about 20 hours of aging under these conditions. It is interesting to note in Figure 6b that, although there is a decrease in the distance from the notch tip to the craze origin at intermediate times, the last data point, obtained at about 130 hours, suggests that this distance may increase at longer times.

Critical Strain for Crazing with Solvent Exposure. Figure 7 shows the effect of physical aging on the critical strain for crazing of PET in a wide variety of solvents. These solvents are listed in Table I along with their molar volume and solubility parameter, δ_t. In general, the critical strain is observed to decrease with aging, indicating an increased propensity for crazing and therefore likely an enhanced susceptibility to environmental stress cracking. This effect of physical aging on the critical strain for crazing probably results from the decrease in molecular mobility that occurs during the aging process. A craze itself is made up of highly oriented polymer, and previous work has shown that physical aging of PET does promote orientation in the form of strain-induced crystallization (19). Thus it is suggested that the increased molecular packing resulting from physical aging leads to higher local orientation at lower strains due to the more intimate contact between chains. This mechanism is intimately linked to free volume. The decrease in free volume measured by PALS during physical aging is due to a reduction in the concentration of free volume sites (I_3) which may be attributed to better local chain packing which retards mobility.

It should be emphasized that two fundamentally different types of craze tests were performed in this work. The test described initially, in which the craze stress below a notch was calculated from the slip line plasticity theory, without exposure to solvent, is a test in which the strain is changed as a function of time. The craze stress itself is calculated assuming that both slip line plasticity theory and the simple von Mises yield criterion are both applicable. The second test, used to determine the effect of solvent on crazing, is a surface crazing test under simple tension in which the strain

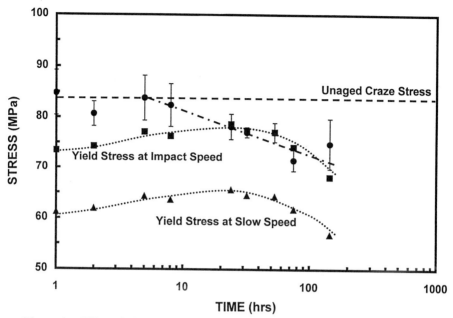

Figure 6a. Effect of physical aging at 60°C and 95% relative humidity on yield and craze stresses. Filled circles indicate the craze stress as a function of aging time. The horizontal dashed line indicates the unaged craze stress.

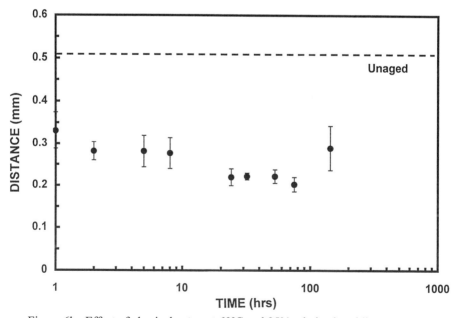

Figure 6b. Effect of physical aging at 60°C and 95% relative humidity on distance from notch root to craze.

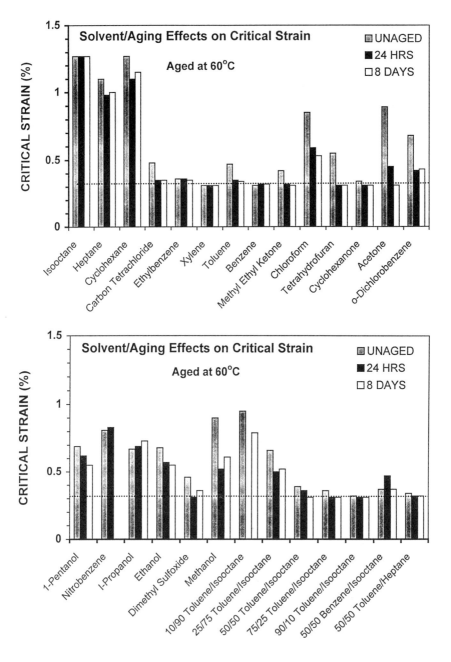

Figure 7. Effect of physical aging at 60°C on the critical strain for crazing.

Table I. Liquids Used in Critical Strain Tests

Solvent	Molar Volume (cm^3/mol)	δ_t (MPa$^{1/2}$)
Isooctane	166.1	14.3
Heptane	147.4	15.3
Cyclohexane	108.7	16.8
Carbon Tetrachloride	97.1	17.8
Ethylbenzene	123.1	17.8
Xylene	121.2	18.0
Toluene	106.8	18.2
Benzene	89.4	18.6
Methyl Ethyl Ketone	90.1	19.0
Chloroform	80.7	19.0
Tetrahydrofuran	81.7	19.4
Cyclohexanone	104.0	19.6
Acetone	74.0	20.0
o-Dichlorobenzene	112.8	20.5
1-Heptanol	141.9	21.5
1-Pentanol	109.0	21.7
Nitrobenzene	102.7	22.2
Aniline	91.5	22.6
m-Cresol	104.7	22.7
N-Methyl Pyrrolidone	96.5	22.9
n-Butanol	91.5	23.1
i-Propanol	76.8	23.5
Acrylonitrile	67.1	24.8
Dimethyl Formamide	77.0	24.8
Nitromethane	54.3	25.1
Butyrolactone	76.8	26.3
Ethanol	58.5	26.5
Dimethyl Sulfoxide	71.3	26.7
Methanol	40.7	29.6
Ethylene Glycol	55.8	32.9
Glycerol	73.3	36.1
Water	18.0	47.8
10% Toluene/90% Isooctane	160.0	14.7
25% Toluene/75% Isooctane	150.0	15.3
50% Toluene/50% Isooctane	136.0	16.3
75% Toluene/25% Isooctane	121.0	17.2
90% Toluene/10% Isooctane	113.0	17.6
50% Benzene/50% Isooctane	127.8	16.5
50% Benzene/50% Heptane	127.1	16.8

at each point remains constant while the stress is changing due to relaxation. Kambour and Farraye (*17*) and Kambour (*19*) did show, however, that properties measured from both tests correlate linearly with the product $\Delta T \cdot (CED)$, where $\Delta T = T_g - T_{test}$ and CED is the cohesive energy density of the solvent.

Conclusions

The temperature-dependent free volume of poly(ethylene terephthalate) has been measured by positron annihilation lifetime spectroscopy below and above the glass transition. A change in slope is observed for both the lifetime component, τ_3, and the intensity component, I_3. This change in slope occurs in the range of 55-60°C which is 15-20°C below the glass transition observed by standard techniques such as differential scanning calorimetry and dynamic mechanical analysis. Free volume relaxation due to physical aging at 60°C is not detectable by PALS, but free volume relaxation occurring at 40°C is detectable. Changes in the mechanical properties due to aging at both temperatures are observed in room temperature tests. It is suggested that the frequency of molecular motion renders the oPs probe insensitive to the relaxation of the dynamic free volume at 60°C. At lower temperatures the free volume relaxation is followed by the I_3 parameter which decreases logarithmically with aging time at 40°C similar to the logarithmic increase in yield stress for aging at 40°C. The reduction in free volume due to physical aging results in an increase in the stress needed to cause the chains to slip (yield stress). It was shown that physical aging can increase the yield stress to a value that exceeds the craze stress resulting in brittle failure.

The embrittlement of PET which results from physical aging has been shown to be due to the increase in yield stress during aging and the relative constancy of the craze stress during aging. The yield stress is initially less than the craze stress at both low and high strain rates but, with physical aging, increases to the level of the craze stress and beyond thus making crazing, rather than yielding, the more likely failure mechanism to be observed. This occurs at shorter aging times for specimens aged at higher sub-T_g temperatures as well as those tested at higher rates, and accounts for the transition from ductile to brittle failure which is well known to result from physical aging.

Physical aging of PET at 60°C prior to critical strain testing at 23°C generally causes a decrease in the critical strain for crazing for PET exposed to various organic solvents. This suggests that environmental stress cracking (ESC) may be more likely to occur after shorter exposures to stress and chemical environment for polymer that has been physically aged to a greater extent.

Literature Cited

1. Struik, L. C. E. *Physical Aging of Amorphous Polymers and Other Materials*, Elsevier: Amsterdam-New York, 1978.
2. Tant, M. R.; Wilkes, G. L. *Polym. Eng. Sci.* **1981**, *21*, 823.
3. Hill, A. J.; Heater, K. J.; Agrawal, C. M. *J. Polym. Sci. Polym. Phys. Ed.* **1990**, *28*, 387.

4. Kinloch, A. J.; Young, R. J. *Fracture Behaviour of Polymers*, Applied Science: London, New York, 1983.

5. Gusler, G. M.; McKenna, G. B. *Polym. Prepr.* **1995**, *36*(2), 63.

6. Arnold, J. C. *Polym. Eng. Sci.* **1995**, *35*, 165.

7. Gusler, G. M.; McKenna, G. B. *Annu. Tech. Conf. – Soc. Plast. Eng.* **1996**, *54th* (Vol. 2), 1537.

8. Arnold, J. C.; Eccott, A. R. In *High Performance Polymers and Polymer Matrix Composites*; Eby, R. K.; Evers, R. C.; Meador, M. A.; Wilson, D., Eds.; Mat. Res. Soc. Symp. Ser. Vol. 305; Materials Research Society: Pittsburgh, PA, 1993; pp 211-216.

9. Puff, W. *Comput. Phys. Commun.* **1983**, *30*, 359.

10. Bergen, R. L., Jr. *Soc. Plast. Eng. J.* **1962**, *18*, 667.

11. Kasbekar, A. D. M.Sc. Thesis, Duke University, Durham, NC, 1987.

12. Struik, L. C. E. *Internal Stresses, Dimensional Instabilities, and Molecular Orientations in Plastics*, John Wiley & Sons-New York, 1990; p. 15.

13. Venkateswaren, K.; Cheng, K. L.; Jean, Y. C. *J. Chem. Phys.* **1984**, *88*, 2465.

14. Hull, D.; Owen, T. W. *J. Polym. Sci. Polym. Phys. Ed.* **1973**, *11*, 2039.

15. Pitman, G. L.; Ward, I. M.; Duckett, R. A. *J. Mater. Sci.* **1975**, *10*, 1582.

16. Hill, R. *Q. J. Mech. Appl. Maths.* **1949**, *2*, 40.

17. Kambour, R. P.; Farraye, E. A. *Polym. Commun.* **1984**, *25*, 357.

18. Tant, M. R.; Wilkes, G. L. *J. Appl. Polym. Sci.* **1981**, *26*, 2813.

19. Kambour, R. P. *Polym. Commun.* **1983**, *24*, 292.

Chapter 18

Transitions, Properties, and Molecular Mobility During Cure of Thermosetting Resins

Jakob Lange, Roman Ekelöf, Nigel A. St. John, and Graeme A. George[1]

Centre for Instrumental and Developmental Chemistry, Queensland University of Technology, GPO Box 2434, Brisbane 4001, Australia

This chapter discusses the interrelation between mechanical properties, molecular mobility and chemical reactivity of curing epoxy-amine thermosets, illustrated by examples of how the charge recombination luminescence (CRL), heat-capacity and rate constants of chemical reactions are influenced by gelation and vitrification during isothermal cure. A comparison of dynamic mechanical, CRL and modulated temperature DSC data shows that vitrification is accompanied by an increase in CRL and a decrease in heat-capacity, and that the heat-capacity and CRL continue to change after the viscoelastic properties have levelled out. It is also shown how the rate constant of an intermolecular secondary amine reaction, measured by near infrared spectroscopy, is sensitive to gelation, whereas the intramolecular rate constant instead is sensitive to vitrification.

The solidification on cure plays a key role in determining the properties of a thermosetting polymer. Understanding the events and changes in the system during cure is essential if its full potential is to be employed. The cure process transforms the resin from a liquid to a glassy solid by the development of a three-dimensional covalently bound network. The two main events that may occur during isothermal cure are gelation and vitrification. Gelation, i.e. the liquid-to-rubber transition, corresponds to the formation of an infinite network, whereas vitrification, i.e. liquid or rubber-to-glass transition, occurs when the glass transition temperature (T_g) of the reacting system reaches the cure temperature. The sequence of events during isothermal cure at different temperatures for thermosetting systems has been extensively studied, e.g. by Gillham and Enns (*1*). Three main regimes of cure temperatures can be identified. On curing above the ultimate glass transition temperature ($T_{g\infty}$) of the polymer only gelation will occur. The system will then vitrify upon cooling. If the cure temperature is below $T_{g\infty}$ but above the temperature where gelation and vitrification occur simultaneously ($_{gel}T_g$) the system will first gel and then vitrify, whereas if the cure temperature is below $_{gel}T_g$ the polymer will not gel but only vitrify. Gelation and vitrification are associated with changes in the properties of the reacting material. On gelation the system acquires an

---

[1]Corresponding author.

equilibrium elastic modulus, and flow ceases to be possible. However, the relaxation time remains short and the molecular mobility high enough to permit chemical reactions not involving rearrangements of the network to continue. On vitrification the elastic modulus and relaxation time increase dramatically, and the molecular mobility strongly decreases. The rates of chemical reactions fall drastically as they become limited by the diffusion of reactive species.

When in the cured state below T_g, epoxy resins will form highly polar glasses. The actual chemical structure of the glass will depend on the T_g of the system, i.e. the overall conversion, since the network will consist of the products of three major chemical reactions; primary and secondary amine-epoxy addition and etherification, as depicted in Schemes 1-3. Near infrared spectroscopy has enabled the products of these reactions to be measured in real time during network formation (2) so providing both the absolute conversion and the rate of the reactions which result in gelation and vitrification. The chemical complexity of these reactions depends on the structure and purity of the resin and hardener. For example, one of the common high performance resins is based on tetraglycidyl diaminodiphenyl methane and thus has a theoretical functionality of four. In that case, in addition to the intermolecular reaction of secondary amine with epoxy on different molecules, which extends the network in three dimensions, leading to vitrification, an intramolecular reaction between the secondary amine and the second epoxide on the nitrogen atom of the same molecule may occur, leading to cyclisation (this is discussed in detail below). A similar process of intramolecular etherification may also occur. This results in a network with a lower cross-link density, higher free volume and higher polarity per cross-link. Efficient competition of etherification with amine reactions will often result in a high residual secondary amine content in the cured glassy resin. The high polarity from the -OH and -NH groups results in the sensitivity of T_g of cured epoxy glasses to the sorption of water. However, it also provides a useful probe of the structure of the glassy state of the resin by producing electron trapping sites following photo-ionisation of the resin. The subsequent detrapping and recombination with the positive centre on the nitrogen atom results in charge-recombination luminescence (3, 4).

The most commonly used technique to monitor the changes in physical properties during cure is dynamic mechanical analysis (1, 5). However, previous work has shown how a variety of physical phenomena are influenced by gelation and vitrification, and how the measurement of these phenomena can be used to follow the progress of cure. Thus changes in ultrasonic absorption and velocity as detected by ultrasonic measurements or by Brillouin scattering have been correlated with both gelation and vitrification (6). Changes in conductivity and dipolar relaxation obtained through dc conductivity measurements and dielectric spectroscopy have been correlated with viscosity, vitrification and relaxation time (7, 8), and spectral shifts in fluorescence have been linked to changes in microviscosity (9). The heat capacity measured by modulated differential scanning calorimetry has been shown to change on vitrification (10), and the line width of NMR signals has proven to be sensitive to the gel state (11). Apart from the general desire to increase the body of knowledge, the interest in exploring these indirect ways of following property change during cure has largely been driven by two current problems in cure monitoring. One is the lack of suitable techniques for following both the chemical and physical changes in the system during cure. Since no two

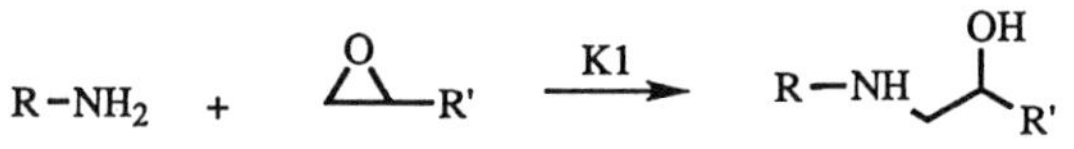

Scheme 1. Primary amine-epoxy addition.

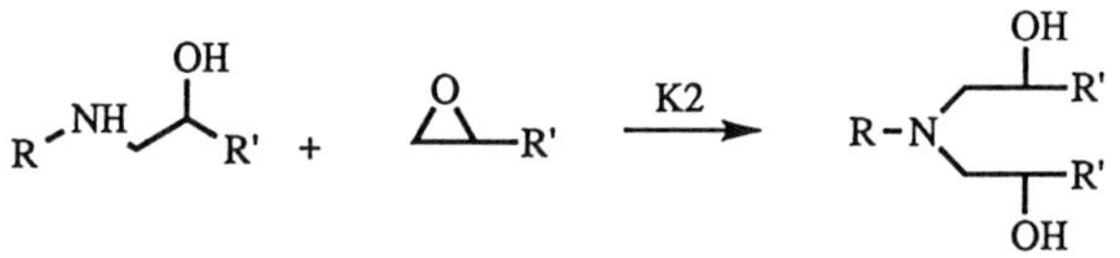

Scheme 2. Secondary amine-epoxy addition.

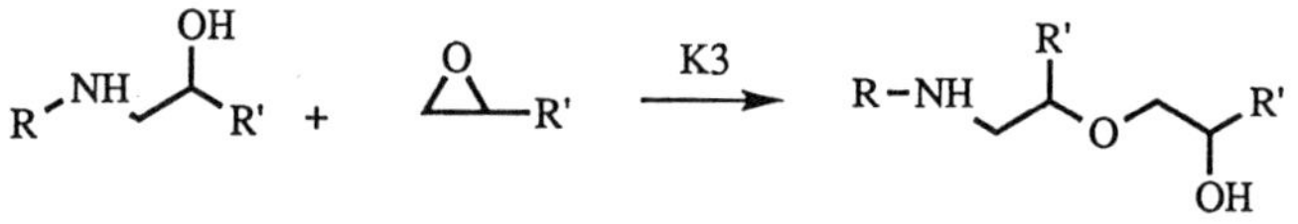

Scheme 3. Hydroxyl-epoxy addition (etherification)

techniques have exactly the same sample geometry and the same heating set-up, there will almost invariably be differences in reaction rate associated with exothermy and differences in heating rate. Techniques capable of following both the progress of the chemical reaction and the changes in physical properties in one single sample thus offer some major benefits. One example of this is the measurement of the absolute reaction rate with FTIR, simultaneously monitoring the heat flow by DSC. These experiments have, indirectly, detected the change in heat capacity during cure, and also exposed the difficulties in using DSC for evaluating cure kinetics (*12*). The other issue is the suitability for in-situ study, i.e. the monitoring of cure in real thermoset processing applications. Only a few techniques, particularly for the measurement of physical properties, are suited for in-situ measurements, in spite of the fact that the physical changes are of great importance to process control.

In this chapter the interrelation between mechanical properties, molecular mobility and chemical reactivity is discussed. Examples of how the changes in charge recombination luminescence, heat capacity and rate constants of chemical reactions can be related to the evolution of viscoelastic properties and the transitions encountered during isothermal cure of thermosetting materials are given. The possible application of the experimental techniques involved to in-situ cure process monitoring is also reviewed.

Experimental

Materials. The diglycidyl ether of bisphenol F (DGEBF), and N,N,N',N'-tetraglycidyl-4,4'-diaminodiphenylmethane (TGDDM), were received from Ciba Geigy. 4,4'-diaminodiphenylmethane (DDM), was obtained from Aldrich. 4,4'-diaminodiphenylsulphone (DDS), was provided from Sigma Chemical. All chemicals were used without further purification. The monomers are presented in Figure 1.

Methods. Dynamic mechanical analysis was performed in a Rheometrics RDS 2, using parallel plates of 8 mm diameter and samples of approximately 4 mm diameter and 1.5 mm thickness. The samples, stoichiometric mixtures of DGEBF and DDM, were inserted at room temperature and the rheometer then heated to the cure temperature at 15°C/min. The complex modulus was measured at regular intervals using a strain between 3 and 0.05% and a frequency of 1 Hz. After cure was completed, temperature scans at a cooling rate of about 5°C/min were run. The sample dimensions were checked after cure, and the moduli data re-calculated accordingly. Some runs were performed on TGDDM with 27% by weight of DDS added, using 17.5 mm diameter plates. A Perkin Elmer DMA 7 was used to measure property development after gelation on samples of TGDDM and 27% DDS, using 5 mm diameter pans and a 3 mm diameter probe, at a strain of 0.3%.

Charge recombination luminescence was measured in a set-up described in detail elsewhere (*13*). Stoichiometric mixtures of DGEBF and DDM in aluminium pans were taken to the cure temperature at 15°C/min and cured isothermally under nitrogen in a chamber covered by a quartz window. The sample was intermittently irradiated with a Kulzer Duralex UV-300 fibre optic wand for 60 s. After each irradiation the shutter of the photomultiplier was opened with a delay of 5 s, and the initial intensity of emitted light, I_0, was recorded.

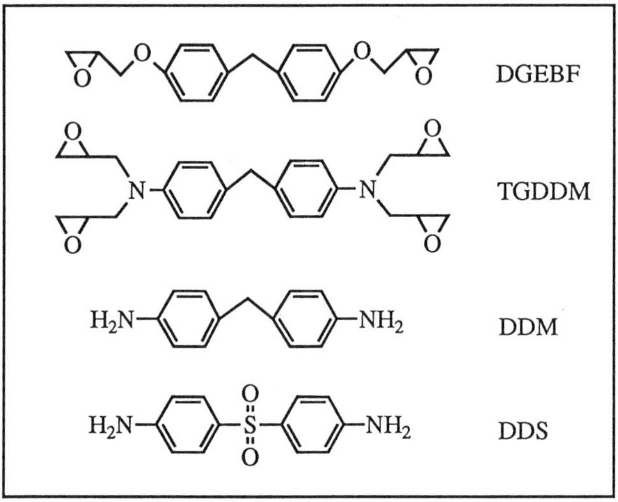

Figure 1. Epoxy and amine monomers.

Modulated temperature DSC scans were run on a Perkin Elmer Pyris DSC, on samples of DGEBF and DDM. After heating up at 15°C/min, isothermal runs at different temperatures with an imposed ±1°C variation at a scanning rate of 20°C/min were performed, which permitted the instantaneous heat-capacity to be calculated.

Near infrared spectra were collected using a Mattson Sirius 100 Fourier transform spectrometer with a heating block in the optical path (2). Samples of TGDDM with 27% by weight of DDS added, contained in 3 mm quartz ESR tubes, were cured in the spectrometer and spectra collected at regular intervals.

Results and Discussion

Dynamic Mechanical Analysis (DMA). DMA has been used extensively to study cure of thermosets in the laboratory and is something of a reference technique for the changes in physical properties. In this method a dynamic strain is imposed on the sample and the resulting stress and phase shift are measured, which permits the storage (G') and loss (G'') moduli to be calculated. Dividing the loss modulus by the storage modulus yields tan δ, which is proportional to the damping in the material. A recently developed extension of the technique employs a parallel plate instrument with controlled normal force to measure the sample shrinkage during cure, thus also giving information on the progress of the cure reaction (5).

With DMA, gelation during cure can be detected as a crossover between G' and G'', i.e. where tan δ equals 1, in stoichiometrically balanced systems. A more rigorous gelation criterion, valid for a wider range of systems, is the point at which tan δ becomes independent of measurement frequency (14, 15). Vitrification is in general detected as a peak in G'' or tan δ (16). Due to the time-dependent nature of this transition the definition of a vitrification point will always be somewhat arbitrary, and depend on the measurement frequency. It should be noted that it is difficult with dynamic mechanical analysis to measure accurately properties through the whole cure process (liquid-gel-glass) using one single geometry, since the stiffness typically changes seven orders of magnitude. If the geometry is chosen to give reliable data also in the rubbery-glassy domain, the readings before and during gelation will show significant scatter. The changes in G' and tan δ during isothermal cure at different temperatures of a stoichiometric mixture of DGEBF and DDM are shown in Figure 2. Early in the reaction the system is liquid, and the elastic modulus is low. As the system then gels and vitrifies, the modulus rises dramatically. Comparing the three graphs in Figure 2 illustrates the influence of cure temperature on the transitions encountered during cure. The $T_{g\infty}$ of the epoxy system is 150°C and the rubbery and glassy moduli are 20 and 1800 MPa, respectively. On cure at 170°C, i.e. above $T_{g\infty}$, only gelation is encountered. The modulus rises to the rubbery level, and after gelation tan δ remains low. When curing at 140°C, just below $T_{g\infty}$, after gelation the modulus first rises rapidly to the rubbery value and then increases more gradually to a level between the rubbery and glassy values. Tan δ is low after gelation but then exhibits a broad peak and is still high at the end of cure, indicating that the system remains in the transition region. As the cure temperature is lowered to 100°C, vitrification sets in soon after gelation: the peak in tan δ narrows and is shifted closer to the initial drop, and the final level decreases. The

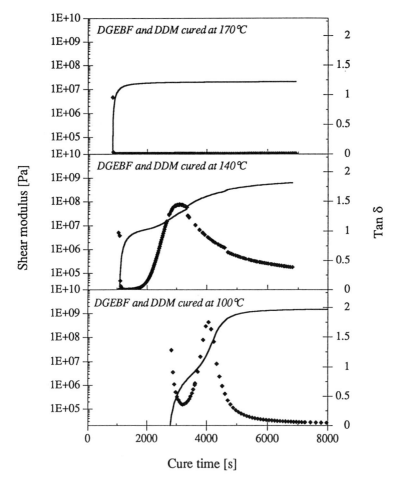

Figure 2. Modulus and damping (tan δ) during isothermal cure of DGEBF/DDM at three different temperatures.

modulus exhibits a clear two-step evolution and reaches the glassy value, showing that the system has passed the transition and entered the glassy state.

Charge-Recombination Luminescence (CRL). UV-irradiation of organic glasses causes photo-excitation of the chromophores present. These are typically aromatic ethers and amines from the resin and hardener. Depending on the energy of the radiation, as well as the polarity of the medium, ionisation may also occur. This process was first recognised for organic amines in alcohol glasses (*17*) and it was found that the resulting photo-electrons were trapped at either physical defects or chemical sites with a high electron affinity (such as a cage formed from hydroxyl groups). Studies of the photo-ionisation of amine-cured epoxy glasses (*4*) showed that similar processes could occur and the subsequent recombination of charges would under certain circumstances lead to emission of photons. The recombination may occur by tunnelling of the electron or by the disruption of the cage and is of much longer duration than photoluminescent processes such as phosphorescence. The intensity of the emitted light was found to depend on the mobility in the system (*4*). It was also shown that the CRL could be distinguished from chemiluminescence as it was observed in an inert atmosphere and was insignificant before vitrification (*4*). This CRL phenomenon can therefore be employed to monitor the later stages of formation of the network during cure of thermosetting resins. CRL does not provide any direct information on the chemical changes during cure, except insofar as the polarity of the network changes the nature and efficiency of traps, but does have the potential to be used in-situ with the aid of fibre-optics.

The change in emitted light after UV-irradiation, collected at regular intervals during cure of DGEBF/DDM mixtures at different temperatures, is shown in Figure 3. At all cure temperatures the level of emitted light is high at first. On heating up the CRL intensity decreases as the system becomes more mobile. As the temperature then stabilises, the curves differ between the cure temperatures. At the highest cure temperature, 170°C, the intensity drops to a low level and remains constant throughout the cure. Curing at lower temperatures yields an increase in the intensity in the later parts of cure. At 140°C there is a small and gradual increase, whereas cure at 100°C exhibits a steep rise and a levelling out at a high intensity. Comparing the CRL curves with the rheological data in Figure 2 shows the influence of the transitions encountered during cure on the amount of emitted light. Cure at 170°C, i.e. above $T_{g\infty}$, yields only gelation. As can be seen there is no change in the CRL intensity as the resin passes gelation. However, at lower temperatures, where vitrification occurs, there is a change in intensity. A detailed comparison of the CRL and rheology graphs reveals that the rise in intensity correlates well with the onset of vitrification. At 140°C vitrification starts after about 2000 s cure time, as indicated by the rise in tan δ. At the same time the CRL intensity begins to rise and continues to increase all through the remainder of the cure process. This can be related to the resin entering into the transition region, but not quite passing on into the glassy state. The possible significance of the continuing increase in the CRL signal will be discussed below. Cure at 100°C instead shows a sharp increase in CRL intensity at 4000 s, which also is where there is a rapid increase in tan δ, followed by a levelling out after about 6000 s, which correlates with the sharp decrease in tan δ. The constant level of the CRL towards the end of cure can be related to the high

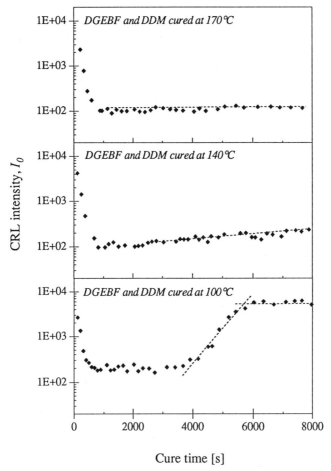

Figure 3. Charge-recombination luminescence intensity during isothermal cure of DGEBF/DDM at three different temperatures.

elastic modulus and low value of tan δ, which together indicate that the material has entered the glassy state. The final CRL intensities at 100 and 140°C are quite different. This is due partly to the incomplete vitrification of the system at 140°C, but also to the influence of temperature on the CRL from the glassy polymer, which in separate studies has been found to increase significantly as the temperature decreases.

These results clearly show that the presence of CRL is linked to the occurrence of a glassy state, and that the increase in CRL signal correlates closely with the onset of vitrification during cure. The findings thus confirm that this technique has the potential of being used as an in-situ sensor for vitrification during cure.

Modulated Temperature Differential Scanning Calorimetry (MT-DSC). Standard DSC monitors the heat-flow from the cure reaction, which yields information on the kinetics and progress of the chemical reactions. MT-DSC involves imposing a small temperature oscillation on the normal temperature program, which makes it possible to also measure the instantaneous heat capacity (C_p) of the sample (18). It is well-known that the C_p of polymeric materials changes at T_g, and MT-DSC has been used to detect isothermal vitrification during cure (10). There are two main types of MT-DSC instruments, based on heat flow and power compensation, respectively. Power compensated MT-DSC's have only recently become available, and the data in the literature on this method is thus still limited. However, in spite of different approaches to data collection and evaluation, the two methods are expected to produce similar results (19). Figure 4 shows the evolution of the C_p during cure of DGEBF/DDM mixtures at different temperatures measured using a power compensated MT-DSC. On cure at 170°C there is no vitrification, and consequently very little change in C_p is observed. At the lower cure temperatures a decrease in C_p during cure is detected. At 140°C the decrease sets in after about 3000 s cure time, which can be compared with the onset of vitrification at 2000 s as measured by tan δ (Figure 2). The decrease in C_p continues throughout the rest of cure. At 100°C the decrease in C_p starts after about 3800 s and ends at around 6000 s cure time, to be compared with the onset and end of vitrification at 3000 and 5000 s, respectively (Figure 2). Vitrification is thus detected somewhat later in terms of change in C_p than in terms of tan δ. The other vitrification criterion used in the literature, peak in G'', occurs after the peak in tan δ (16). It appears that the calorimetric determination of vitrification might correlate better with G''.

Further comparisons can be made between the change in C_p and the changes in CRL intensity during cure. There are clear similarities between the curves, in that at the higher temperature no change in signal on cure is detected, at the intermediate temperature a change occurs, but there is no levelling out of the signal, and, finally, at the lowest temperature a distinct change between two levels can be observed. Particularly the behaviour at the intermediate temperature is interesting. Although the material remains in the transition region at the end of cure, the DMA data (Figure 2) indicate that the properties are stabilising, as shown by a clear tendency towards levelling out in modulus and damping. The fact that both the C_p and the CRL continue to change suggests that these parameters are more sensitive to changes deep in the glassy state than are viscoelastic properties as measured by DMA.

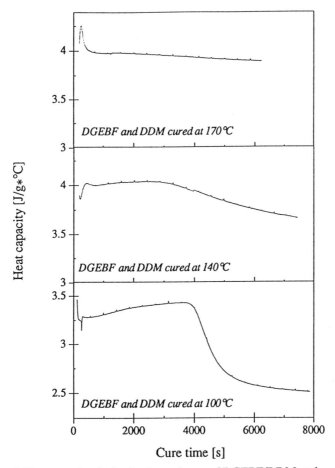

Figure 4. Heat-capacity during isothermal cure of DGEBF/DDM at three different temperatures.

Fourier Transform Near-Infrared Spectroscopy (FTNIR). FTNIR spectroscopy has been used for the quantitative determination of epoxy content of resins for many years (*20*), but only recently has it been used for real-time monitoring of the reaction kinetics of network formation (*2, 21*). Because it is possible to determine absolute concentration of functional groups and thus absolute reaction rates, subtle kinetic effects of gelation and vitrification on cure chemistry may be observed. Kozielski et al (*22*) in a study by FTNIR of reaction rates of commercial polyfunctional epoxy resins found that there was a drop in the secondary amine reaction rate at the onset of vitrification which could then be fitted with the inclusion of a diffusion-controlled term. They were unable to observe any effect of gelation. From molecular modelling, this was attributed to the possibility of efficient intramolecular reaction, but no further kinetic analysis was performed. The kinetic effect of both gelation and vitrification on the secondary amine reaction rate has been observed by us for the cure of TGDDM by the aromatic amine DDS.

Schemes 4 and 5 show the two possible reactions of the secondary amine group of DDS with either another epoxide on a different molecule or on the same molecule. By using a statistical analysis of the epoxy reaction probability, it is possible to separate the two processes from the kinetic FTNIR data. Figure 5 shows the changes in the rate coefficients as a function of cure time. For this system, gelation and onset and end of vitrification were detected at 4700, 5500 and 7500 s cure time, respectively, using two different DMA techniques. It is apparent that the intermolecular reaction rate coefficient, K4, decreases at gelation while the intramolecular, K5, term is unaffected until the onset of vitrification and then decreases smoothly. Inspection of the reaction in Scheme 4 shows that the intermolecular reaction requires significant movement of reactive species, and it is therefore not surprising that the rate of this reaction is affected by the increase in viscosity on gelation. However, Scheme 5 shows that only a molecular rotation is required to achieve the intramolecular reaction, and it appears that this can continue into the glassy state although the rate coefficient drops by a factor of 10.

The changes in reaction rates of different species measured by FTNIR can thus be utilised as "molecular probes" of gelation and vitrification, providing simultaneous information on the progress of the chemical reaction and the transitions encountered during cure. However, the technique requires significant data analysis and is therefore not readily adapted to on-line monitoring.

Conclusions

Comparison of dynamic mechanical, charge-recombination luminescence and heat-capacity data obtained during isothermal cure of a DGEBF/DDM epoxy resin shows that an increase in CRL and a decrease in heat-capacity occur as the material vitrifies, and that the changes in heat-capacity and CRL correlate well with the onset of vitrification as indicated by the increase in damping. A detailed real time FTNIR investigation of rate constants for a TGDDM/DDS epoxy system shows an influence of gelation and vitrification on the secondary amine reaction rate, where the rate of the intermolecular reaction decreases on gelation, and the rate of intramolecular reaction falls as vitrification is encountered.

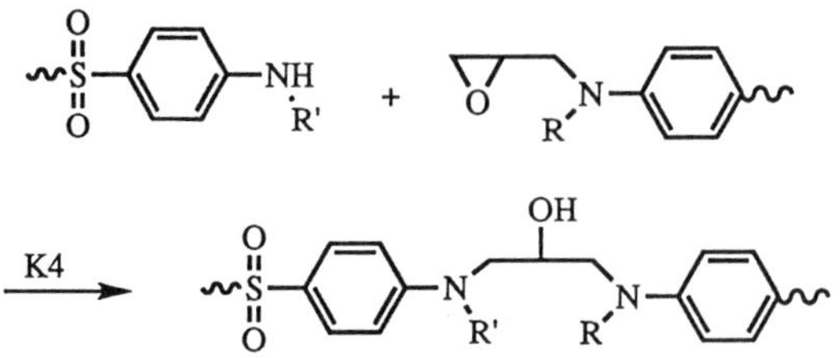

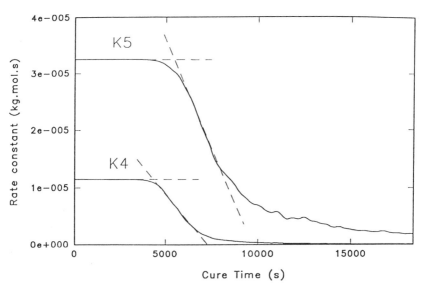

Scheme 4. Intermolecular secondary amine-epoxy addition.

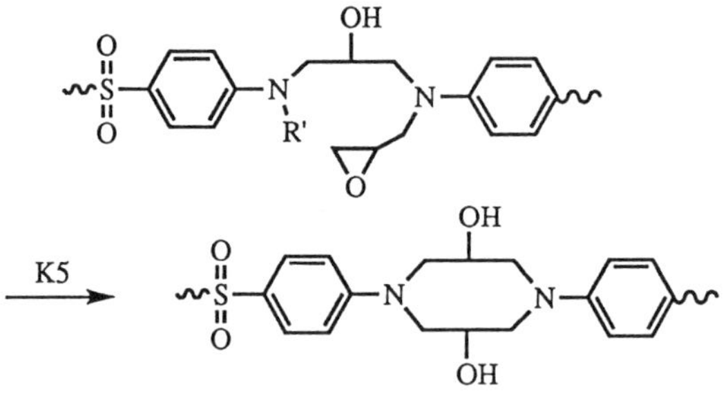

Scheme 5. Intramolecular secondary amine-epoxy addition.

Figure 5. Effective values of rates for intermolecular (K4) and intramolecular (K5) secondary amine reaction versus cure time for TGDDM/DDS.

Acknowledgment

Financial support from the Wennergren Foundation, the Swedish Institute and the Swedish Fund for Occupational Health is gratefully acknowledged.

Literature Cited

1. Gillham, J. K. and Enns, J. B., *Trends Polym. Sci.* **1994**, *2*, 406.
2. St John, N. A. and George, G. A., *Polymer* **1992**, *33*, 2679.
3. Albrecht, A. C., *Acc. Chem. Res.* **1970**, *3*, 238.
4. Billingham, N., C., Burdon, J. W., Kozielski, K. A. and George, G. A., *Makromol. Chem.* **1989**, *190*, 3285.
5. Lange, J., Toll, S., Månson, J. -A. E., and Hult, A., *Polymer* **1995**, *36*, 3135.
6. Alig, I., Lellinger, D., Nancke, K., Rizos, A. and Fytas, G., *J. Appl. Polym. Sci.* **1992**, *44*, 829.
7. Simpson, J. O. and Bidstrup, S. A., *J. Polym. Sci.: Part B: Polym. Phys.* **1995**, *33*, 55.
8. Mangion, M. B. M. and Johari, G. P., *J. Polym. Sci.: Part B: Polym. Phys.* **1991**, *29*, 1127.
9. Song, J. C. and Neckers, D. C., *Polym. Eng. Sci.* **1996**, *36*, 394.
10. Van Mele, B., Van Assche, G., Van Hemelrijk, A. and Rahier, H., *Thermochim. Acta* **1995**, *268*, 121.
11. Stöver, H. D. H. and Fréchet, J. M. J., *Macromolecules* **1991**, *24*, 883.
12. De Bakker, C. J., St John, N. A. and George, G. A., *Polymer* **1993**, *34*, 716.
13. St John, N. A., *PhD Thesis*, Univ. of Queensland 1993.
14. Winter, H. H. and Chambon, F. *J. Rheol.* **1986**, *30*, 367.
15. Muller, R.; Gérard, E.; Dugand, P.; Rempp, P. and Gnanou, Y. *Macromolecules* **1991**, *24*, 1321.
16. Serrano, D. and Harran, D., *Eur. Polym. J.* **1988**, *24*, 675.
17. Lewis, G. N. and Lipkin, D., *J. Am. Chem. Soc.* **1942**, *64*, 2801.
18. Boller, A., Jin, Y. and Wunderlich, B., *J. Therm. Anal.* **1994**, *42*, 307.
19. Schawe, J. E. K., *Thermochimica Acta* **1995**, *260*, 1.
20. Dannenberg, H, *SPE Trans.* January **1963**, 78.
21. George, G. A., Cole-Clarke, P. A., St John, N. A. and Friend, G., *J. Appl. Polym. Sci.* **1991**, *42*, 643.
22. Kozielski, K. A., George, G. A., St John, N. A. and Billingham, N. C., *High Perform. Polym.* **1994**, *6*, 263.

Chapter 19

A Dynamic Probe of Tribological Processes at Metal–Polymer Interfaces: Transient Current Generation

J. T. Dickinson, L. Scudiero, and S. C. Langford

Physics Department, Washington State University, Pullman, WA 99164–2814

We present measurements of transient electrical currents delivered to a stainless steel stylus during the abrasion of a high density polyethylene (HDPE) and two commercial forms of polymethylmethacrylate (PMMA). At low normal forces, abrasion proceeds continuously and the magnitude of these currents at a given stylus velocity is independent of normal load. At higher normal loads, stick-slip motion is observed, where the "slip" portion of the cycle is accompanied by dramatic increases in current. In HDPE, stick-slip motion is due to localized fibril formation and rupture. In PMMA, stick-slip motion is principally due to localized fracture. The sign and magnitude of these transient currents are sensitive to polymer chemistry. Careful analysis of both magnitude and fluctuations in the current thus provides a real-time probe of the micromechanics of asperity/surface interactions.

Wear of polymers during sliding contact is generally attributed to adhesive transfer between the two sliding surfaces, abrasive cutting, and fatigue (*1*). An important component of friction and wear during rubbing of two surfaces arises in the rapid, transient making and breaking of adhesive bonds between asperities. In the case of polymers rubbing against metals and ceramics, adhesive forces contribute dramatically to the wear of the softer polymer. Our interests are in probing both spatially and temporally the various damage mechanisms during tribological loading (sliding contact) (*2*). When a grounded conductor is drawn across a polymer, charge transfer during contact and detachment can deliver significant currents to the conductor. Although the charge exchange mechanisms are not well understood (*3*), we are finding interesting correlations between the micromechanics of abrasion and the measured signals. Our goals are to show how these currents and emissions are related to the extent of substrate damage, to examine the contribution of adhesion to the frictional force, and to elucidate the physics of contact charging and other sources of charge (*4*).

This work describes a new probe of dynamic processes accompanying the tribological loading of metal/polymer interfaces. We instrument the polymer substrate and a conducting stylus to measure the transient electrical currents generated as the stylus is moved across the substrate under normal load. Simultaneous measurements of the lateral and normal forces on the stylus are also performed. Both sets of measurements are readily made on ms to µs time scales. To date, we have measured currents accompanying the abrasion of several insulators, including single crystal inorganics (including MgO and NaCl), polymers (including polymethylmethacrylate (PMMA), polycarbonate, polystyrene, and polyethylene), and metallic surfaces separated by organic fluids, e.g., perfluoropolyether lubricants (5). We have also made a number of related studies of the charged particles emitted during sliding contact (an extension of our work on fracto-emission—particle emission generated during fracture) (6). In the case of high density polyethylene (HDPE), we have shown previously that electron emission during abrasion is many orders of magnitude more intense than emission during fracture (7). This is consistent with the strong electrical effects often observed when insulating materials are separated from other insulators or conductors.

The current measurements described here are also related to previous observations of transient currents generated by cracks propagating along metal-polymer interfaces (5). We have shown, for example, that significant currents can be generated during interfacial crack propagation along a stainless steel surface embedded in an epoxy matrix (8). Similar currents are generated when a pressure sensitive adhesive is peeled from a copper substrate (9). The magnitude of these currents indicate that electrostatic forces can significantly enhance deformation in the adhesive along the peel front, thereby significantly enhancing the peel energy (10). In addition, the current fluctuations during peeling have been shown to be chaotic in the deterministic sense (11), corroborating that the mechanics of peel are also chaotic. The transient currents generated during abrasive, sliding contact appear to be similarly "rich" in structure and information content.

Experiment

Two types of PMMA sheet were employed in this work: Rohm and Haas Plexiglas G, which is a cell cast, high molecular weight polymer (> 1,000,000 weight average), and Plexiglas MC, which is a melt calandered (extruded) material with a much lower molecular weight (usually less than 200,000 weight average) (12). Plexiglas MC also contains a co-monomer (ethyl acrylate) to improve flow properties and a small amount of lubricant to facilitate extrusion. Both materials are predominantly syndiotactic, but Plexiglas MC has a measurably higher isotactic content. The HDPE employed in this work was 3 mm thick sheet (BASF 5205) with a density of 0.95 g/cm^2 and a crystallinity of 85%.

The experimental apparatus for current, charge, and force measurements is shown in Figure 1. Abrasion was performed by translating the polymer substrate past a stationary, conducting stylus (copper or stainless steel). Current delivered to the conducting stylus was detected with a Keithley 602 electrometer. The conducting stylus was electrically isolated from the supporting cantilever and the lead from the stylus to the electrometer was shielded. Two 1-cm long portions of the cantilever were machined to provide thin, rectangular sections to localize the bending moments applied by the cantilever. Strain gauges attached to the machined portions were calibrated to indicate the lateral and normal forces applied to the tip of the conducting stylus. Strain gauge outputs were digitized with a LeCroy 6810 digitizer at 100 μs intervals.

Surface charge measurements were made with a conducting probe translated along the wear track, with a gap of about 1 mm between the probe and the polymer surface. Charge remaining on the polymer surface induces a corresponding charge on the probe as it passes over the charged region. This charge was detected with a Keithley 602 electrometer, and the electrometer output digitized with a LeCroy 6810 waveform digitizer as above. The experiment was mounted in a vacuum system, which provided good electrical shielding and permitted controlled atmospheres (e.g., Ar, N_2) as well as vacuum or air. The stylus tip radii ranged from 10-500 μm. Some measurements were made in an indentation mode, making and breaking contact with vertical (normal) motion only.

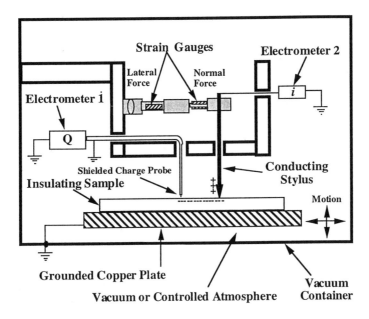

Fig. 1.   Schematic of experimental arrangement for measuring transient currents.

Transient Current Measurements: HDPE

Contact charging between HDPE and a wide variety of conducting materials (e.g., copper, stainless steel, tungsten carbide) leaves the HDPE negatively charged. This is consistent with the position of HDPE near the "negative end" of the triboelectric series [13]. While contact is maintained, charge transfer is localized to the interfacial region. This charge transfer (across an intact interface) does not produce a measurable current signal. However, when contact is broken, residual charges on the insulating surface "induce" readily detected currents in the conducting phase.

Figure 2 shows a typical load (normal force) and transient current signals when a stainless steel stylus (450 μm in diameter) is brought directly down onto a HDPE sheet, held, then raised. In this case, a small positive current is observed from the stainless steel soon after contact. With small fluctuations, this current is maintained throughout contact. As contact itself cannot generate such currents, we expect that this current is produced when stress relaxation in the HDPE around the stylus draws small patches of material away from the stylus. Stress relaxation is apparent in the slowly decaying normal force after loading and before unloading, when the downward stylus displacement has ceased. Similar relaxation while the tip is still moving down is also expected (during loading). When the stylus is subsequently raised, a large positive current is observed. This current continues until the upward motion of the stylus is arrested (the magnitude of the induced charge ceases to change), or when the distance between the stylus and the substrate becomes much larger than the stylus diameter (induced charge becomes vanishingly small).

When the stylus is translated across the HDPE in a friction or wear geometry, contact and detachment occur in a more or less continuous fashion. Then significant currents can be generated and sustained. Figure 3 shows typical transient current signals generated when a small stainless steel stylus (60 μm in diameter) is translated across a fresh HDPE surface. Transient current signals for three normal loads (0.2, 0.5, and 1.0 N) are shown, where the data are taken at the beginning of new wear tracks. In each case, a relatively steady current of 20-30 pA is established, independent of load. At the highest load, however, the current shows large fluctuations, primarily in the form of current spikes 40-50 pA in magnitude. For very long wear tracks, the current (and the lateral force) eventually decrease somewhat as the stylus becomes coated with polymer.

At still higher loads, these fluctuations become larger. Simultaneous lateral force and current data taken at a normal force of 5 N are shown in Figure 4. The "steady state" current in this case is comparable to that at lower loads (about 20 pA), but positive current spikes at least 500 pA high are observed. (The peak currents are actually off scale in this measurement.) Each current spike is accompanied by a sudden, transient drop in the lateral force, consistent with a stick-slip process. This sudden drop in the lateral force allows the stylus to slide rapidly forward. During this "slip" portion of the stick-slip cycle, the instantaneous tip velocity (and thus "rate of detachment") can be very high, resulting in very high transient currents.

The abruptness of the transition to slip, as well as the strong correlation between the peak current and the drop and lateral force, are especially clear when a small portion of the current signal in Figure 4 is displayed on an expanded time scale [Figure 5]. The duration of the peak current is less than 10 ms, and corresponds well with the duration of the sudden drop in lateral force that accompanies it. Note also that both the lateral force and transient current signals require some hundreds of milliseconds to recover their normal, steady state values. Further, the lateral force begins to drop (and the current begins to rise) 100-300 ms before the onset of the slip event.

SEM micrographs of the wear track clarify the sequence of events leading up to the slip event. Micrographs of typical portions of wear tracks formed at loads of 2 N and 5 N, respectively, are shown in Figure 6. Both wear tracks show plowing and periodic fibril formation along either side of the wear track, where broken fibrils

276

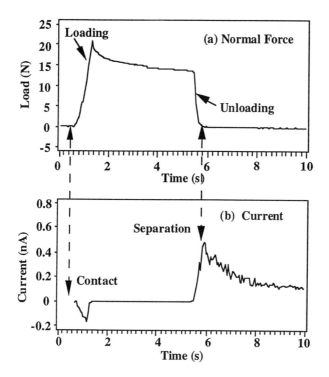

Fig. 2   (a) Load (normal force) and (b) transient current signals during indentation and detachment of a stainless steel stylus from a HDPE surface.

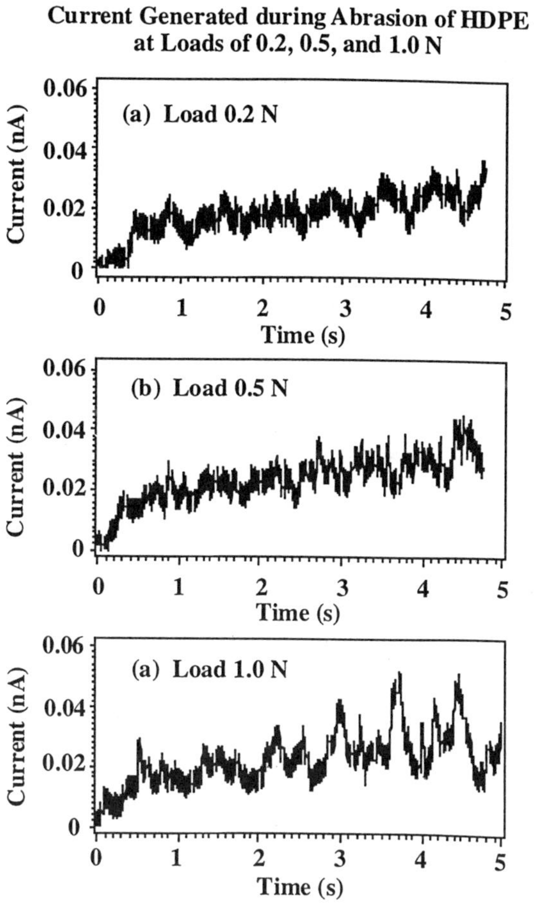

**Current Generated during Abrasion of HDPE
at Loads of 0.2, 0.5, and 1.0 N**

(a) Load 0.2 N

(b) Load 0.5 N

(a) Load 1.0 N

Fig. 3. Current signals generated during abrasion of HDPE with stainless steel at normal loads of (a) 0.2 N, (b) 0.5 N, and (c) 1.0 N. The stylus velocity is 1.6 mm/s.

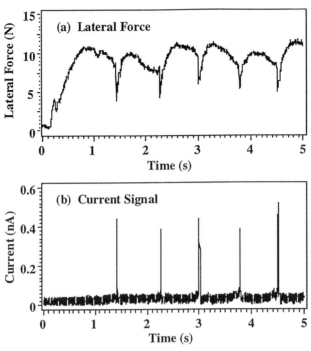

Fig. 4. (a) Lateral force and (b) current signal during abrasion of HDPE with stainless steel at a normal load of 5 N. The stylus velocity is 1.6 mm/s.

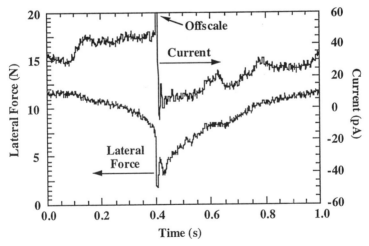

Fig. 5. Plots of transient current and lateral force on a faster time scale showing in detail one stick-slip event. The normal force was 5 N.

SEM Micrographs of Wear Tracks in HDPE Made with a Stainless Steel Stylus

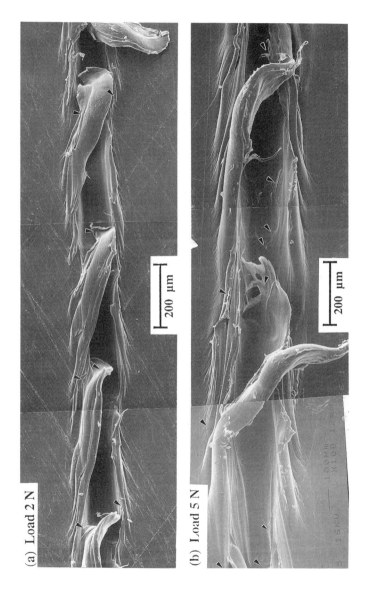

(a) Load 2 N

200 μm

(b) Load 5 N

200 μm

Fig. 6.   Scanning electron micrographs of two single-pass wear tracks formed at normal loads of (a) 2 N and (b) 5 N.   Stylus motion was from left to right.

appear at roughly periodic intervals. The spacing of the broken fibrils is consistent with the distance between slip events inferred from transient current and lateral force measurements and the average stylus velocity (1.6 mm/s). For instance, the spikes in

Figure 4 are about 750 ms apart, corresponding to a spacing of 1.2 mm. This agrees well with the distance between fibrils in Figure 6b. At lower normal forces, the slip events are spaced more closely. The average distance between slip events inferred from the current signals at a normal load of 2 N is 0.8 mm, in good agreement with the fibril spacing observed by SEM (about 0.7 mm).

The fibril orientation in Figure 6 indicates that they were formed from material piled up in front of stylus due to plowing. Since this material is anchored to the sides of the wear track, it becomes stretched and oriented into long fibrils. These fibrils make a large contribution to the lateral force exerted on the stylus. The high strength of oriented polyethylene ensures that failure initiates in the relatively unoriented material where the fibril is anchored to the side of the wear track. One end of the fibril is essentially torn from the matrix. Catastrophic fibril failure produces a large, rapid drop in lateral force and allows the stylus to jump forward. This results in the sharp current spike coincident with the load drop.

Going back to Fig. 5, we note that prior to the current spike (the slip event) there is in fact a *precursor*, namely a slowly increasing current and a simultaneous reduction in lateral force prior, i.e., the stylus is accelerating prior to the catastrophic event. Since polyethylene does not normally strain soften, we speculate that there is significant deformation at the fibril root or anchor prior to failure. This would allow the stylus to displace slowly and generate the increasing current along with a corresponding decrease in lateral force. It would be very interesting to determine if this yielding can be linked to the locus of failure.

Transient Current Measurements: PMMA

The currents accompanying sliding contact depend strongly on the chemical composition as well as the physical properties of the materials involved. Two common types of commercial PMMA, for instance, yield opposite currents when abraded with a metal stylus. (PMMA is much closer to the "middle" of the triboelectric series than HDPE, and may accept either positive or negative charge from typical metals, depending on the stylus material.) Typical charge signals accompanying the translational motion of a diamond stylus over Plexiglas G (cell cast) and Plexiglas MC (extruded) samples appear in Figure 7. Moving the charge probe across freshly abraded portions of the wear track yields a positive signal for Plexiglas G (representing positive charge on the PMMA) and a negative signal for Plexiglas MC (representing negative charge on the PMMA). When the diamond stylus is brought directly down onto the surface and lifted straight up, with no translational motion, Plexiglas G becomes positively charged and Plexiglas MC becomes negatively charged, consistent with charge transfer during abrasion.

Several factors may contribute to the different charging behavior of these materials. For instance, the Plexiglas G is a considerably more pure version of PMMA than Plexiglas MC. Extruded material (Plexiglas MC) also tends to be more oriented, has a lower glass transition temperature (T_g) and (in this case) is slightly more isotactic. Sharma and Pethrick have shown that tacticity has a strong influence on the triboelectric charging of PMMA during contact with ferrite (Fe_2O_4) beads (*14*). In Table I we compare our results with those of Sharma and Pethrick:

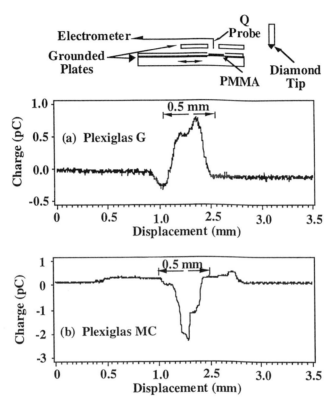

Fig. 7.   Surface charge measurements vs. displacement on two different types of PMMA generated by several cycles of lateral sliding of a diamond stylus over the polymer surfaces. In vacuum, these traces are stable for minutes.

Table I

Type of PMMA/stylus	T_g	Sign of Charge on PMMA
"G"(cast)/diamond	95 °C	+
"MC" (extruded)/diamond	85 °C	−
syndiotactic/ferrite(*14*)	120 °C	+
isotactic/ferrite(*14*)	50 °C	−

We note that the T_g for the two types of PMMA rank in the same order as values reported for syndiotactic (T_g = 115 C) and isotactic (T_g = 45 C) material (*15*). We propose that structural differences, perhaps related to tacticity, in these polymer surfaces are responsible for the observed differences in charge transfer with the conducting stylus. Obviously, studies employing better characterized polymer and stylus surfaces are required.

Using conducting stainless steel as the stylus yields the same signs on the two types of PMMA as using a diamond stylus. Measuring currents dynamically during lateral motion of the stylus again yields fluctuating current signals, similar to the HDPE results. The effect of varying the normal force on the current delivered to a stainless steel stylus (radius 60 µm) during a single pass across a fresh Plexiglas MC surface is shown in Figure 8 for an average lateral speed of 1.8 mm/s. At all three forces, the current from the tip is predominately positive—consistent with the removal of positive charge from the surface and thus the net negative surface charge observed during abrasion with diamond in Figure 7b. At the two lower normal forces, the current fluctuates around an average of about 500 pA. However, at the highest normal forces, the current also displays large positive spikes, 1-5 nA in amplitude on a background of several hundred pA. Again, simultaneous lateral force measurements show that the current spikes coincide with dramatic, rapid drops in the lateral force consistent with stick-slip motion. The temporal period of these large fluctuations is ~160 ms which corresponds to an average spatial separation of 0.3 mm.

SEM micrographs of PMMA single pass wear tracks generated by a stainless steel stylus moving at 1.6 mm/s are shown in Fig. 9 for three normal forces. As expected, the width of the wear track increases with increasing load. Whereas the transient current traces in Fig. 8 show interesting fluctuations at all normal loads, little corresponding periodicity is seen by SEM for 1 N (9a) and 2 N (9b). Nevertheless, these wear tracks display considerable microstructure, including strong evidence of plastic flow due to plowing. In contrast, at 5 N (9c) very strong periodic features are observed with a spatial period of 0.29 mm, in good agreement with that predicted from the transient current spikes. Here the wear track appears to alternate between plowing and fracture, which together generate the 0.3 mm periodicity. (Fracture debris (ejecta—(*16*)) along the sides of the wear track were removed prior to SEM imaging.) We therefore believe that the rapid stress release accompanying these fracture events results in slip. The central portion of the track shows a corresponding fluctuation in the depth suggesting that the stylus is deflected downwards (digs in) and upon fracture the stylus jumps upward and forward (the slip event) to begin a new cycle. The "stick" part of the cycle is when the stylus has reached its deepest point. At this point we are not sure of how stationary the stylus is at this instant; the non-zero current between spikes in Figure 8c indicates that stylus motion does not stop between spikes.

Conclusion

The transient current signals generated during sliding contact between a stainless steel stylus and a polymer substrate are strongly affected by the micromechanics of the sliding interface (fibril formation, fracture) and the surface chemistry along the interface. These currents are especially sensitive to surface

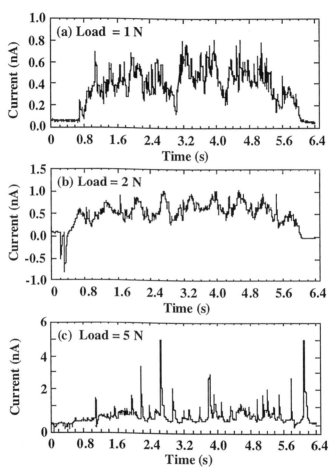

Fig. 8.   Current signals produced by abrasion of Plexiglas MC with stainless
steel at normal loads of (a) 1 N, (b) 2 N, and (c) 5 N.

SEM Micrographs of Wear Tracks in PMMA
Made with a Stainless Steel Stylus

(a) Load 1 N

(b) Load 2 N

(c) Load 5 N

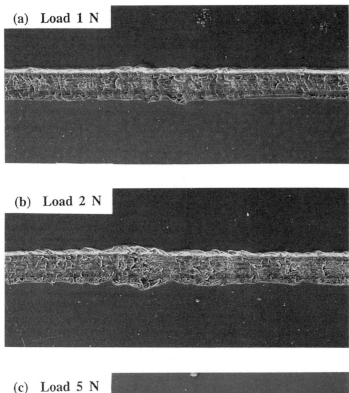

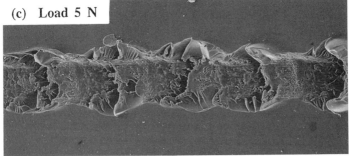

├─── 300 μm ───┤

Fig. 9.   Scanning Electron Micrographs of single pass wear tracks at normal load of (a) 1N, (b) 2 N, and (c) 5 N. Stylus motion was from left to right at a speed of 1.8 mm/s.

chemistry and the rate and nature of "debonding" processes associated with adhesive detachment during wear. Consequently, stick-slip motion, where detachment is catastrophic in nature, has a very strong effect on the resulting current. Careful analysis of both magnitude and fluctuations in the current thus provides a real-time probe of the micromechanics of asperity/surface interactions. Ongoing studies are extending the size of the stylus in two directions: towards (i) smaller, (sub-micron) dimensions and (ii) larger dimensions (~1 cm), representing the extremes of a single asperity vs. a collection of many asperities. In addition, correlating transient current measurements with fracto-emission signals (e.g., emission of electrons, positive ions, neutral atoms and molecules) is providing additional information on the dynamics and wear mechanisms of metal/polymer tribological interactions.

Acknowledgments

This work was supported by the National Science Foundation Surface Engineering and Tribology Program under Grant CMS-9414405. We would like to thank Mary Dawes, Washington State University, for her assistance in the laboratory.

Literature Cited

1. Ludema, K. C.; in *Friction, Lubrication, and Wear Technology*, ; Blau, P. J., Ed.; ASM International: Materials Park, Ohio, USA, 1992; Vol. 18, pp 236-241.
2. Dickinson, J. T.; Langford, S. C.; Nakahara, S.; Scudiero, L.; Kim, M.-W.; Park, N.-S.; in *Fractography of Glasses and Ceramics III*, ; Varner, J. R., Fréchette, V. D. and Quinn, G. D., Ed.; American Ceramic Society: Westerville, Ohio, 1996; Vol. 64, pp 193-256.
3. Lee, L.-H.; in *Adhesive Bonding*, ; Lee, L.-H., Ed.; Plenum: New York, 1991, pp 1-30.
4. Dickinson, J. T.; Langford, S. C.; Waultersack; Yoshizaki, H. submitted to *Wear* **1997**.
5. Dickinson, J. T.; Scudiero, L.; Yasuda, K.; Kim, M.-W.; Langford, S. C. *Tribology Lett.* **1997**, *3*, 53-67.
6. Dickinson, J. T.; in *Non-Destructive Testing of Fibre-Reinforced Plastic Composites*, ; Summerscales, J., Ed.; Elsevier Applied Science: London, 1990, pp 429-482.
7. Dickinson, J. T.; Jensen, L. C.; Dion, R. P. *Appl. Phys.* **1993**, *73*, 3047.
8. Zimmerman, K. A.; Langford, S. C.; Dickinson, J. T. *J. Appl. Phys.* **1991**, *70*, 4808.
9. Dickinson, J. T.; Jensen, L. C.; Lee, S.; Scudiero, L.; Langford, S. C. *J. Adhes. Sci. Technol.* **1994**, *8*, 1285-1309.
10. Scudiero, L.; Langford, S. C.; Dickinson, J. T. In *Proc. ICAST, Amsterdam, October 1995*, ; VSP: Zeist, The Netherlands, 1997, in press.
11. Scudiero, L.; Langford, S. C.; Dickinson, J. T.; In *Proc. ICAST, Amsterdam, October 1995*, ; VSP: Zeist, The Netherlands, 1997, in press.
12. Cholod, M. S., personal communication, **1997**.
13. Montgomery, D. J. *Solid State Phys.* **1959**, *9*, 139.
14. Sharma, V. K.; Pethrick, R. A. *J. Electrostatics* **1990**, *25*, 309.
15. Mark, J. E.; Eisenberg, A.; Graessely, W. W.; Manderlkern, L.; Koenig, J. L. *Physical Properties of Polymers*; American Chemical Society: Washington, D.C., 1984.
16. Donaldson, E. E.; Dickinson, J. T.; Bhattacharya, S. K. *J. Adhesion* **1988**, *25*, 281-302.

Chapter 20

The Effect of Orientation by Solid State Processes on the Amorphous Regions in Poly(vinyl alcohol) and High Density Poly(ethylene)

Anita J. Hill[1], T. J. Bastow[1], and R. M. Hodge[2]

[1]CSIRO, Division of Manufacturing Science and Technology, Private Bag 33, South Clayton MDC, Clayton, Victoria 3169, Australia
[2]Materials Engineering and Technical Support Services, 720G Lakeview Plaza Boulevard, Columbus, OH 43085

The effect of solid state drawing processes on the amorphous regions of two semicrystalline vinyl polymers has been investigated. The amorphous region of poly(vinyl alcohol) PVOH is a glass at room temperature as compared to the rubber amorphous region of high density poly(ethylene) HDPE. The state of the amorphous region is a function of the draw ratio with a maximum relative free volume occurring at approximately 53% of the maximum achievable draw ratio λ_{max} for both polymers, or at a relative draw ratio λ_R of 30% as defined by $\lambda_R = 100[\lambda_{exp}-1]/[\lambda_{max} -1]$. Maximum constraint on the amorphous regions due to drawing occurs at 53% of λ_{max} for HDPE (rubber) and 87% of λ_{max} for PVOH (glass). It is postulated that the reorientation of crystals due to drawing results in the initial dilation of the amorphous region free volume followed by local orientation in the glass and relaxation in the rubber.

Production of high modulus films and fibres from vinyl polymers has been achieved by various methods (1). In most cases relating to solid state production, drawing is carried out at temperatures in between the glass transition temperature and the melting temperature of the polymer in order to enhance molecular mobility of the amorphous regions during the drawing process. Increasing molecular mobility during drawing can facilitate molecular orientation in the drawn polymer, since the chains are better able to untangle and change conformation in response to the applied drawing stress. Samuels (2) has demonstrated the importance of amorphous orientation to tenacity of vinyl polymers fibres, whilst Ward (3) has shown that Young's modulus (in the draw direction) depends on molecular conformations of the chains irrespective of whether they are in crystalline or amorphous regions. The presence of crystals, crosslinks, hydrogen bonds, or entanglements is usually necessary to maintain the conformational changes and orientation of the amorphous regions once the applied stress is removed.

Characterization of polymer orientation is most often accomplished via X-ray techniques which are suited to crystalline and paracrystalline regions (3-6). However, semicrystalline polymers present a complex system of crystalline, amorphous, and intermediate phases (4-6) and complete characterization of semicrystalline polymers can only be achieved by application of a variety of techniques sensitive to particular aspects of orientation. As discussed by Desper (4), one must determine the degree of orientation of the individual phases in semicrystalline polymers in order to develop an understanding of structure-property relationships. Although the amorphous regions of oriented and unoriented semicrystalline polymers are primarily responsible for the environmental stress cracking behaviour and transport properties of the polymers, few techniques are available to examine the state of the amorphous material at the submicroscopic level.

Sonic modulus, birefringence, linear dichroism, release stress, solid state nuclear magnetic resonance (NMR) and positron annihilation lifetime spectroscopy (PALS) methods have been used to characterize orientation of amorphous regions in glassy and semicrystalline polymers (3,7-12). In a glassy single phase polymer, orientation of amorphous material is inferred from the increase in sonic modulus in the draw direction, increase in release stress, decrease in free volume, decrease in molecular mobility as measured by NMR, increase in glass transition temperature, and decrease in gas permeability (7,8), whilst the orientation function can be directly determined via birefringence (13). In semicrystalline polymers the characterization of orientation in the different phases is dependent on the use of techniques which can accurately separate the contributions of the amorphous and crystalline regions. In the present work, a range of characterization techniques are employed: wide angle and Laue X-ray diffraction which probe the crystalline phase, PALS which probes the amorphous phase and crystal/amorphous interface, solid state NMR which probes the amorphous and crystalline phases (and in some cases the contributions can be separated), differential scanning calorimetry using T_g to characterize the amorphous regions and the enthalpy of melting to quantify the crystalline fraction, and density measurements. The samples studied in the present work compose two systematic series of glassy and rubbery semicrystalline vinyl polymers drawn in the solid state over a range of draw ratios which vary orientation angle from 180 to 5 degrees without dramatically varying the degree of crystallinity. This procedure highlights morphological changes associated with the amorphous regions and crystal/amorphous interface due to the crystal reorientation and microfibrillation caused by drawing.

An understanding of the morphological changes associated with the crystalline material in a polymer undergoing drawing is well established (1). Semicrystalline polymers initially undergo a process of lamellae rotation towards the direction of applied stress during tensile deformation. This is followed in the latter stages of deformation by the crystal lamellae unravelling and extending in the direction of the applied stress to form extended, highly aligned chain molecules. The orientation process in the crystals does not occur in isolation, however, as the amorphous regions interspersed between crystals are intimately associated with the crystalline regions due to the presence of interlamellar tie molecules within the amorphous regions. Crystalline rotation must proceed in a cooperative fashion with deformation and conformational changes of the amorphous molecules. It may be expected, therefore, that the state of the amorphous material, whether glassy or rubbery, will exert considerable influence over the drawing process and will also determine to a large degree, the extent and nature of any morphological changes occurring in the amorphous material itself.

The drawing behaviour of PVOH is compared to that of HDPE, since both polymers have similar crystal structures and have almost identical extended chain stiffness (*14*). The structural similarities between PVOH and poly(ethylene) (HDPE) offer the opportunity to study the role played by the amorphous region during drawing of two structurally similar polymers, one with the amorphous region in a glassy state (PVOH) at room temperature, and the other (HDPE) in a rubbery state. The nature of the response of the amorphous region to the applied stress depends on the rate of strain with respect to the relaxation times of the possible modes of molecular mobility (*13*). The maximum drawability in PVOH is limited with respect to that in HDPE due to increased effective entanglement density and limited molecular mobility in the amorphous regions promoted by inter- and intra-chain hydrogen bonding between hydroxyl side groups (*15-18*). Despite the strong structural similarity between the two polymers, the dry T_g of PVOH is significantly higher than that of HDPE because of hydrogen bonding. The dry T_g of PVOH is ~ 85°C as detected by PALS (*19*) whilst the T_g of HDPE is ~ -17°C as detected by PALS (*20*), although assignment of the glass transition in polyethylenes is a matter of debate (*21*). The amorphous phase is of particular importance in PVOH and HDPE films and fibres as it has considerable influence over transport properties and chemical resistance (*1*).

Experimental

Materials

The poly(vinyl alcohol) (PVOH) used in this study is Elvanol grade HV from DuPont chemicals. The material is derived from poly(vinyl acetate) (PVAc) by hydrolysis and is in powder form as received. The degree of hydrolysis for Elvanol HV is 99.8% (0.2% residual acetate groups). The molecular weight of the material used in this investigation is Mw =105,600 - 110,000. PVOH films were produced as described elsewhere (*14*). Samples were drawn in tension at a strain rate of 1.4×10^{-3} sec^{-1} to a maximum draw ratio λ_{max} = 3.08 at 22°C. As-cast material has an equilibrium water content of 5.3 wt% at 22°C and 50% relative humidity (RH) giving a T_g value of 31°C. The HDPE is Hoechst unfractionated Hostalen GF 7660, M_w = 1.06×10^5, M_n = 1.01×10^4. Hydrostatic extrusion, with extrusion temperatures from 60-100°C and strain rates from 2.5×10^{-3} sec^{-1} to 1.4×10^{-2} sec^{-1}, was used to vary draw ratio (die size was varied) with a maximum draw ratio λ_{max} = 16. The maximum draw ratios were limited by fracture.

Methods

The degree of orientation in drawn samples was determined from Laue diffraction patterns obtained with a Phillips PW1030 flat plate camera using unfiltered Cu-Kα radiation in transmission. The azimuthal intensity profiles from the diffraction patterns were obtained by measuring light transmission through the developed flat plate film, in an azimuthal trace around the circumference matching that of the {110} reflection using a Joyce Loebl double beam recording microdensitometer Mk IICS equipped with a rotating stage. The wide angle X-ray diffraction (WAXD) spectra were collected for 2θ from $5°$ to $85°$ using a Siemens D500 diffractometer using a step size of $0.04°$ and a speed of $1.0°$ min^{-1}. The ^{13}C CPMAS spectra were collected using a Bruker MSL400 at an operating frequency of 101.6 MHz in 4mm PSZ rotors at a spinning frequency of 6 kHz. For HDPE the ^{1}H/^{13}C contact time was 1msec and the pulse sequence repetition time was 5 sec to ensure full relaxation. Lineshape simulations were made with the Bruker LINESIM program. PALS spectra were collected at 22°C and 50% RH using an automated EG&G Ortec fast-fast coincidence system. Two identical samples sandwiched a Ti foil ^{22}Na source. The source was moved between runs for the HDPE due to irradiation induced positronium inhibition. The error bars are population standard deviations for 3-5 runs. The PALS spectra for both polymers gave a best fit using a three component analysis (22). The lifetime of the shortest component was fixed at 0.125 ns. The differential scanning calorimetry (DSC) data were collected using a DuPont 2200 DSC from -20°C to 200°C using a heating rate of 20°C min^{-1}. Density measurements were made at 22°C in nitrogen using an AccuPyc 1330 pycnometer by Micromeritics.

Results and Discussion

Figure 1 shows the variation in tensile modulus with orientation angle for the PVOH and HDPE samples. The degree of orientation in a material may be defined in terms of an orientation angle (OA), which is estimated from the width at half height of peaks contained in the intensity profile taken from an azimuthal trace of the Laue reflections (23). Smaller angles indicate samples with a high degree of orientation. In this investigation, the OA was calculated from the strong {110} reflections from oriented crystals. Figure 1 indicates that tensile modulus increases monotonically as the crystallites become increasingly oriented parallel to the tensile direction (characterized by a smaller orientation angle) and progressively more tensile load is borne directly along the extended chain axis. The form of the curve in Figure 1 is similar to that reported elsewhere (16).

In Figure 2, the relationship between orientation angle and draw ratio for PVOH and HDPE is illustrated. The maximum draw ratio achieved in the PVOH was $\lambda_{max} = 3.08$, compared with that in the HDPE of $\lambda_{max} = 16.00$, reflecting the increased molecular mobility and absence of entangled amorphous regions in the HDPE samples with respect to the PVOH samples. Draw ratio is a process parameter and does not convey the structural information inherent in the measurement of orientation angle. Indeed, Samuels (2) has illustrated that draw ratio may indicate very little about degree of orientation in semicrystalline polymers; however, in this work we are interested in the use of solid state drawing processes to vary materials properties and have shown in Figure 2 that for these two polymers and these solid state drawing processes, the relationship between orientation angle and draw ratio is well established. Thus, in order to directly compare morphological changes in both polymers at similar stages in the drawing process, the data in Figure 2 will be normalized to express the relative

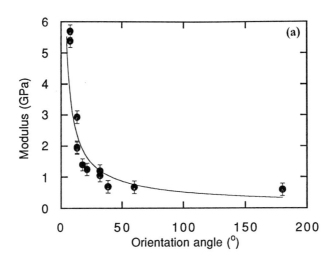

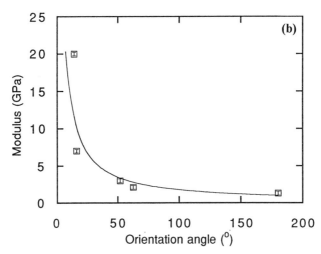

Figure 1. Young's modulus in tension as a function of orientation angle for (a) PVOH and (b) HDPE. Curve drawn to indicate trend.

degree of molecular drawing λ_R; defined as the percent ratio of the amount of drawing measured, λ_{exp}, to the maximum achievable draw ratio, λ_{max}, in the following manner:

$$\lambda_R = 100[\lambda_{exp}-1]/[\lambda_{max}-1] \tag{1}$$

Thus, for undrawn samples $\lambda_{exp} = 1$ and $\lambda_R = 0$ % and for fully drawn samples $\lambda_{exp} = \lambda_{max}$ and $\lambda_R = 100$ %.

The data in Figure 3 indicate that tensile drawing of both polymers causes the crystallites to become oriented towards the direction of strain and shows that maximum molecular orientation in both polymers is essentially achieved at comparable stages in both polymers. The major part of the orientation angle decrease occurs by $\lambda_R = 30$%. Beyond $\lambda_R = 30$%, little further orientation is achieved indicating that drawing beyond this point is accomplished by chain slippage in the crystalline regions, or deformation/conformation changes in the amorphous regions in the polymer, or a combination of both mechanisms. In order to determine the nature of the changes occurring in the amorphous regions of the polymers during drawing, it is necessary to measure the relative amounts of amorphous and crystalline material present at various stages of the drawing process.

WAXD and DSC were used to measure the amount of crystalline and amorphous phase in each of the samples. Figure 4 shows the crystallinity of the samples as a function of draw ratio. The values of crystallinity from DSC and WAXD are not in exact agreement; however, the trends show no significant change in crystallinity due to solid state drawing. Similar results have been reported for semicrystalline polymers with degree of crystallinity measured by DSC, WAXD, NMR, and density (5,6,24), and the discrepancy between crystallinity values has been discussed in terms of either dilation of amorphous free volume by crystal constraint (6,24) or presence of a paracrystalline intermediate phase (5,6).

The amorphous regions of the polymers can be probed by the PALS technique, which measures two free volume related parameters (25,26). The orthoPositronium (oPs) lifetime, τ_3, is related to the mean free volume cavity radius (assuming a model of spherical free volume cavities), and the parameter I_3 gives information relating to the relative number of free volume cavities in the polymer. In the present work, the relative total free volume is given by the parameter $\tau_3{}^3 I_3$ which gives a number proportional to the relative mean volume of the cavities multiplied by their relative concentration (27). Table I contains the PALS oPs parameters for PVOH and HDPE samples measured after solid state drawing. Suzuki et al. (28) have reported a four component analysis for HDPE whilst other researchers (20,29) have reported a three component analysis. In both types of analysis, the longest lifetime component is attributed to oPs annihilations taking place in the amorphous regions of semicrystalline HDPE.

Figure 5 indicates that the free volume probed by oPs increases over the same stages of drawing where the orientation angle undergoes the major part of the change ($\lambda_R \leq 30$ %). For the glassy amorphous regions of the PVOH samples, the free volume decreases after $\lambda_R = 30$ % before appearing to reach a constant value - noticeably lower than that in the undrawn polymer - when the draw ratio approaches λ_{max}. It is proposed that the rotation of crystal lamellae in the direction of the applied stress during the initial stages of drawing is responsible for the initial increase in free volume,

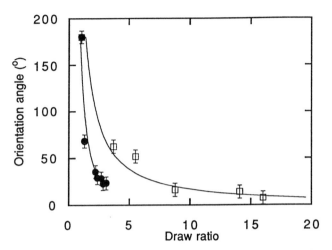

Figure 2. Orientation angle as a function of draw ratio for PVOH (●) and HDPE
(□). Curve drawn to indicate trend.

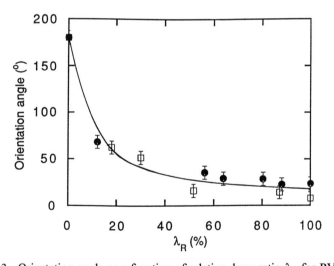

Figure 3. Orientation angle as a function of relative draw ratio λ_R for PVOH (●)
and HDPE (□). Curve drawn to indicate trend.

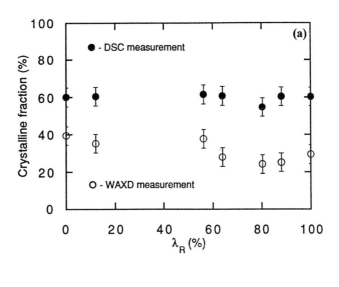

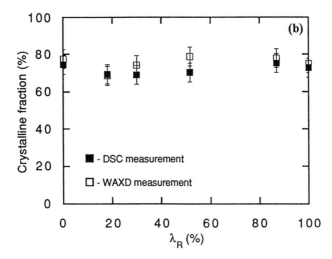

Figure 4. Degree of crystallinity measured by DSC and WAXD as a function of relative draw ratio for (a) PVOH and (b) HDPE.

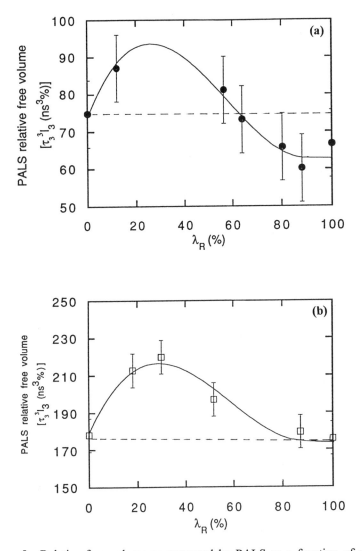

Figure 5. Relative free volume as measured by PALS as a function of relative draw ratio for (a) PVOH and (b) HDPE. Curve drawn to indicate trend.

through a mechanism of dilation of the amorphous regions in the immediate vicinity of the crystals. In the latter stages, the effect of drawing on the amorphous free volume is one of constraint, as the extended crystal chains cause the glassy amorphous material to align locally in the strain direction. Since the amorphous regions of the PVOH are in a glassy state, they have insufficient mobility to revert to a completely random state during elastic recovery. This free volume behaviour can be compared to that of HDPE for which the amorphous regions are rubbery. Crystal reorientation has a similar effect of dilation of the amorphous free volume with the latter stages of drawing resulting in local orientation of the amorphous regions. The result, that the relative amorphous free volume is similar in the fully drawn state to that found prior to dilation, suggests that the rubbery amorphous regions have sufficient mobility to reach a relaxed equilibrium state once crystal rotation is complete.

The relative mobility of the rubbery amorphous phase in the HDPE samples can be inferred from the linewidth of the ^{13}C chemical shift for the amorphous phase. Figure 6 displays the ^{13}C spectrum for $\lambda_{exp} = 1$, ($\lambda_R = 0\%$). The ^{13}C NMR spectra for all samples can be fitted with three peaks characteristic of the carbons in monoclinic (M) crystals, orthorhombic (O) crystals, and amorphous (A) material. The crystalline regions (M and O) have similar linewidths that do not vary systematically as a function of draw ratio ($\Delta\nu = 40 \pm 9$ Hz; 100.61 Hz/ppm). The amorphous region linewidth is plotted as a function of draw ratio in Figure 7. The increase in linewidth with initial drawing indicates decreased mobility of the amorphous phase. It is interesting to note that the molecular mobility of the amorphous regions is most reduced at $\lambda_R = 30\%$. For this reason, it is postulated that the increase in amorphous free volume measured by PALS (displayed in Figure 5) is due to the action of crystalline regions on the adjacent amorphous regions as the crystals reorient. This action places a constraint on the amorphous phase, dilating free volume and reducing mobility. The ^{13}C NMR spectra for PVOH do not display separate peaks characteristic of the crystalline and amorphous regions, making it difficult to measure changes in mobility associated solely with the amorphous regions.

Table I shows that the dilation of the PALS free volume in the initial stages of drawing is due to an increase in the relative mean size of the free volume cavities, τ_3, as opposed to their relative number I_3. Zipper et al. (30) have shown that constraint of the amorphous phase by crystals in a polyester results in an increase in the τ_3 parameter as well as an increase in T_g. The T_g for the PVOH samples was measured using DSC and can be used to give an indication of the mobility of the amorphous regions. Figure 8 shows the T_g of the PVOH samples as a function of draw ratio. The major part of the increase in T_g occurs after crystal reorientation is complete and as the PALS free volume decreases below the value characteristic of the undrawn material. The maximum T_g at $\lambda_R = 80\%$ corresponds to the minimum τ_3 and I_3 (see Table I). Thus it is postulated that the increase in T_g in the latter stages of drawing is caused by local orientation of amorphous regions. Ito and Hatakayama (31) have discussed the maximum in T_g which occurs as a function of draw ratio (at approximately $\lambda_R = 50\%$) in glassy polycarbonate in terms of a decrease in intramolecular entropy caused by the increase in molecular alignment along the chain axis. These authors also point out that a change in free volume can reflect the configurational structure of the glass, and one would expect an increase (maximum) in sub-T_g relaxation time and decrease

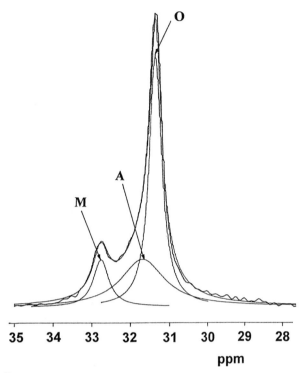

Figure 6. ^{13}C NMR spectrum for undrawn HDPE showing the three components obtained from spectrum simulation. M = monoclinic crystalline, O = orthorhombic crystalline and A = amorphous.

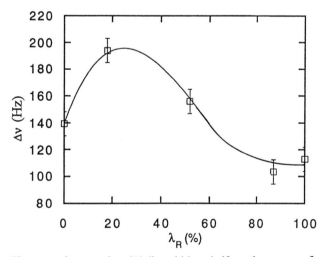

Figure 7. The amorphous region (A) linewidth at half maximum as a function of relative draw ratio for HDPE. Curve drawn to indicate trend.

Table I PALS oPs parameters as functions of experimental draw ratio λ_{exp}

PVOH			HDPE		
Draw Ratio	τ_3 (ns) ±0.03	I_3 (%) ±0.3	Draw Ratio	τ_3 (ns) ±0.03	I_3 (%) ±0.3
1.00	1.496	22.350	1.00	2.092	19.450
1.25	1.568	22.588	3.69	2.148	21.470
2.17	1.537	22.355	5.49	2.142	22.380
2.33	1.487	22.278	8.76	2.160	18.280
2.67	1.411	21.418	14.00	2.045	21.000
2.83	1.436	22.250	16.00	2.036	20.860
3.08	1.455	21.658			

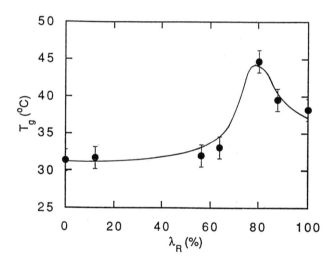

Figure 8. The glass transition temperature as a function of relative draw ratio for PVOH. Curve drawn to indicate trend.

(minimum) in free volume at the T_g maximum (31). There are some similarities between the present work and that of Ito and Hatakayama (31) in that a maximum in T_g is observed as a function of draw ratio and in the present work this maximum coincides with a minimum in the amorphous free volume probed by PALS. Comparison of the present work with that of Zipper et al. (30) highlights the difference in constraint of the amorphous region by crystals which seems to dilate the free volume elements (increase in τ_3, increase in T_g) and orientation of the amorphous regions in semicrystalline polymers by macroscopic strain (decrease in τ_3, increase in T_g). The data present a consistent picture of the stages of solid state drawing in terms of the effect on the amorphous properties in these vinyl polymers: initial dilation due to constraint imposed by crystal reorientation, followed by local orientation of the amorphous regions.

It should be noted that Jean et al. (10,32,33) have shown via angular correlation of annihilation radiation (ACAR) that the approximately spherical free volume elements in amorphous polycarbonate, polymethylmethacrylate, and polyaryletheretherketone become elliptical due to orientation of the glassy state. Shelby (7) discusses the possible effects of free volume element (or "hole") anisotropy on the PALS free volume parameter $\tau_3^3 I_3$. He suggests the existence of a distribution of hole eccentricities due to only partial alignment of chain segments in the stretch direction and further suggests a higher probability of oPs annihilation in the larger and more spherical free volume holes. Shelby (7) concludes that although the effect of free volume hole anisotropy on the PALS parameters (oPs formation probability and pickoff lifetime) is not completely understood at this time, the isotropic PALS measure of relative free volume in amorphous regions, $\tau_3^3 I_3$, is reasonably accurate when applied to anisotropic material.

Density measurements showed no systematic variation as a function of draw ratio (Figure 9) which indicates that the amorphous and crystalline densities are either both approximately constant or both changing in an inversely coupled way. Possible effects of microfibrillation and/or crazing in the samples in the latter stages of drawing also make the density data difficult to interpret. In light of the PALS results, the data may indicate that these density measurements are not sensitive to the changes in amorphous region free volume. The PALS oPs probe can have greater sensitivity than density measurements to free volume effects on the molecular level (27). The free volume sites available for oPs localization typically range in size from 0.2 to 0.7 nm and are located inter- and intra-chain. In semicrystalline polymers the amorphous regions and interfacial regions offer sites of lowest electron density which are most favorable for the oPs localization. Hence, as demonstrated by Zipper et al. (30) and in the present work, the oPs probe can be used to evaluate the constraint on the amorphous regions imposed by crystals, with a sensitivity much greater than that found with density measurements.

In summary, it is postulated that the initial increase in free volume is due to the constraint of the amorphous phase by the crystals as they rotate in order to accommodate the strain. The macroscopic strain used in the present study appears to constrain and then locally orient the amorphous regions giving a window of maximum free volume and reduced molecular mobility. Because the free volume goes through a maximum, it is possible that drawing can be used to increase the free volume of the

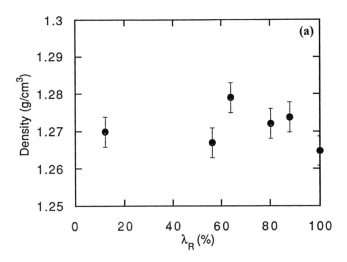

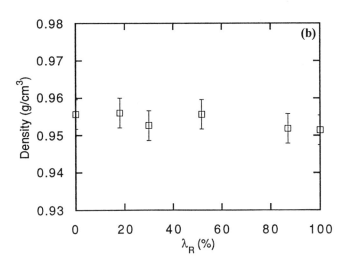

Figure 9. Density as a function of relative draw ratio for (a) PVOH and (b) HDPE.

amorphous regions and hence tailor transport properties in semicrystalline polymers. Wang and Porter (34) have reported a maximum in gas permeability at $\lambda_R = 30\%$ for HDPE samples subjected to varying draw ratios via solid state co-extrusion using a strain rate of approximately 0.5 sec^{-1}, much faster that that used in the present work. As mentioned in the introduction, the response of the amorphous regions to the applied stress will depend on the molecular relaxation times as compared to the strain rate. At the temperatures used for extrusion in this work and that of Wang and Porter (60-100°C) the samples are well above T_g and the frequency of motion of the amorphous material is orders of magnitude higher than the strain rate employed. The variation of gas permeability with draw ratio in the HDPE samples of Wang and Porter was attributed to varying degree of crystallinity. The degree of crystallinity of the samples was measured via DSC, and these results are compared to those of the present work in Figure 10. The solid state extrusion processes result in similar degrees of crystallinity for samples extruded at very different strain rates to the same relative draw ratios. The gas permeability results of Wang and Porter (34) are reproduced in Figure 11. Comparison of the gas permeability (Figure 11) and the PALS free volume (Figure 5b) behaviors for these HDPE samples prepared by solid state extrusion suggests that whilst degree of crystallinity (or amorphous fraction) is important to transport properties, the effect of drawing on the amorphous free volume also plays an important role in gas transport. Our future work will examine transport properties of these vinyl polymers and attempt to determine the relationship between PALS free volume and gas transport given the complex morphologies of drawn semicrystalline polymers.

Conclusions

Tensile drawing and hydrostatic extrusion were used to vary the draw ratio, orientation angle, tensile modulus, and amorphous properties of PVOH and HDPE. The amorphous free volume probed by PALS was maximized with initial drawing and subsequently reduced by further drawing of the polymer. The mobility of the glassy amorphous regions in PVOH (as indicated by T_g) was slightly reduced due to the constraint imposed by crystal reorientation and most reduced by local orientation of the glass. In contrast the mobility of the rubbery amorphous regions in HDPE (as indicated by ^{13}C NMR linewidth) was most reduced due to the constraint caused by crystal reorientation. The PALS free volume in PVOH was reduced in the latter stages of drawing below that found in the undrawn material, whilst the rubbery amorphous regions of HDPE had similar PALS free volume in the undrawn and fully drawn samples. It was postulated that the orientation of the glassy amorphous regions (PVOH) resulted in the reduced PALS free volume whilst the rubbery amorphous regions (HDPE) were able to relax. The implications of these findings for mechanical (environmental stress cracking) and transport properties will be further investigated.

Acknowledgments. The authors would like to thank Natasha Rockelman, David Hay and Kieren Rogan (CSIRO), Marty Tant, Peter Shang and Judy Dean (Eastman Chemical Company), Phillip Wilksch (Royal Melbourne Institute of Technology), and Kang Zhi Lin and Graham Edward (Monash University) for technical support and helpful discussions.

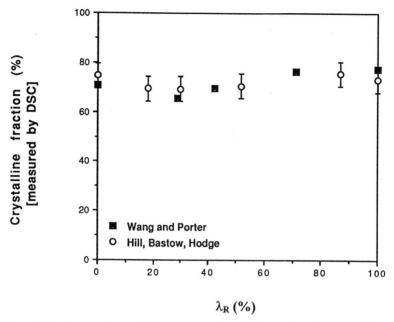

Figure 10. Degree of crystallinity measured by DSC as a function of relative draw ratio for HDPE including the data of Wang and Porter (*34*).

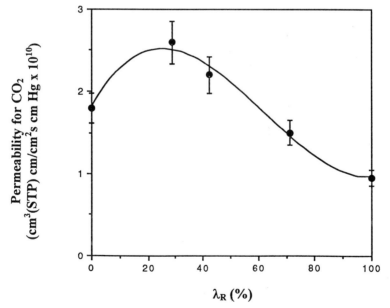

Figure 11. Permeability for CO_2 as a function of relative draw ratio for HDPE samples. Adapted from ref. 34. Curve drawn to indicate trend.

302

Literature Cited
1. Ward, I. M. *Advances in Polymer Science* **1985**, *70*, 1.
2. Samuels, R. J. *J. Macromol. Sci. Phys.* **1970**, *B4*, 701.
3. Ward, I. M. *Advances in Polymer Science* **1985**, *66*, 82.
4. Desper, C. R. *J. Macromol. Sci. Phys.* **1973**, *B7*, 105.
5. Cheng, J., Fone, M., Reddy, V. N., Schartz, K. B., Fisher, H. P., and Wunderlich, B. *J. Polym. Sci. B* **1994**, *32*, 2683.
6. Fischer, E. W. and Fakirov, S. *J. Mater. Sci.* **1976**, *11*, 1041.
7. Shelby, M. D. *PhD Dissertation* **1996**, Virginia Polytechnic Institute and State University.
8. Shelby, M. D., Wilkes, G. L., Tant, M. R., Zawada, J., Bastow, T. J., and Hill, A. *J. Polym. Mater. Sci and Engr.* **1997**, *76*, 485.
9. Wilkes, G. L. *Advances in Polymer Science* **1971**, *8*, 91.
10. Jean, Y. C, Rhee, Y., Lou, Y., Shelby, M. D., Wilkes, G. L. *J. Polym. Sci. B* **1996** *34*, 2979.
11. Aleksanyan, G. G., Berlin, A. A., Gol'danskii, A. V., Grineva, N. S., Onishchuk, V. A., Shantarovich, V. P., and Safonov, G. P. *Sov. J. Chem. Phys. (UK)* **1990**, *5*, 2196.
12. Botev, M, Judeinstein, P., Neffati, R., and Rault, J. *Macromolecules* **1996**, *29*, 8538.
13. Wimberger-Friedl, R. *Prog. Polym. Sci.*, **1995**, *20*, 369.
14. Hodge, R. M., Edward, G. H., and Simon, G. P., *Plast. Rubb. Comp. Proc. App.* **1996**, *25*, 459.
15. Conibeer, C. J., Lin, K. Z. and Edward, G. H. *Mater. Forum* **1990**, *14*, 224.
16. Cebe, P and Grubb, D. T. *J. Mater. Sci.* **1985**, *20*, 4465.
17. Garrett, P. D. and Grubb, D. T. *J. Mater. Res.* **1986**, *1*, 561
18. Garrett, P. D. and Grubb, D. T. *J. Polym. Sci. B. Polym. Phys.* **1988**, *26*, 2509.
19. Suda, Y., Uedono, A., and Ujihira, Y., *Mater. Sci. Forum* **1992**, *105-110*, 1721.
20. Lin, D. And Wang, S. J., *J. Phys. Condens. Matter* **1992**, *4*, 3331.
21. McCrum, N. G., Read, B. E., and Williams, G., *Anelastic and Dielectric Effects in Polymeric Solids*, Wiley: New York, **1967**.
22. Puff, W., *Comput. Phys. Commun.* **1983**, *30*, 359.
23. Alexander, L. E. *X-ray Diffraction Methods in Polymer Science,* Wiley: New York, **1969**.
24. Marand, H., Velikov, V., Verna, R. K., Cham, P. M., Prabhu, V., and Dillard, D. *Polym. Prepr.* **1995**, *36*, 263.
25. Jean, Y. C. *Microchem. J.* **1990**, *42*, 72.
26. Zipper, M. D. and Hill, A. J. *Mater. Forum* **1994**, *18*, 215.
27. Hill, A. J., Weinhold, S. , Stack, G. M., and Tant, M. R. *Euro. Polym. J.* **1996**, *32*, 843.
28. Suzuki, T., Oki, Y., Numajiri, M, Miura, T., Kondo, K., and Ito, Y., *J. Polym. Sci. B* **1992**, *30*, 517.
29. Stevens, J. R. And Edwards, M. J., *J. Polym. Sci. C* **1970**, *30*, 297.

30. Zipper, M. D., Simon, G. P., Cherry, P., and Hill, A. J. *J. Polym. Sci. B* **1994**, *32*, 1237.
31. Ito, E. and Hatakeyama, T. *J. Polym. Sci. B* **1975**, *13*, 2313.
32. Jean, Y. C, Nakanishi, H., Hao, L. Y., and Sandreczki, T. C, *Phys. Rev. B* **1990**, *42*, 9705.
33. Jean, Y. C. and Shi, H., *J. Non-Crystalline Solids* **1994**, *172-174*, 806.
34. Wang, L. H. and Porter, R. S. *J. Polym. Sci. B* **1984**, *22*, 1645.

TRANSPORT PROPERTIES

Chapter 21

Free Volume and Transport Properties of Barrier and Membrane Polymers

B. D. Freeman[1] and Anita J. Hill[2]

[1]Department of Chemical Engineering, North Carolina State University, Raleigh, NC 27695–7905
[2]CSIRO Manufacturing Science and Technology, Private Bag 33, South Clayton MDC, Clayton, Victoria 3169, Australia

Ultrahigh permeability polymers are stiff chain glassy polymers that pack very poorly in the solid state, having high values of fractional free volume and a high degree of connectivity of free volume elements. High barrier materials are prepared from stiff chain glassy polymers that pack efficiently in the solid state, with low fractional free volume values. Manipulation of solid state chain packing changes permeability coefficients over many orders of magnitude. Control of chain chemistry and packing structure of stiff chain glassy polymers permits rational tailoring of permeation properties between those of high barriers and those of extremely permeable membranes via systematic manipulation of free volume and free volume distribution.

The transport of small molecules in polymers plays a key role in the use of membranes for liquid, gas, and vapor separations; barrier plastics for packaging; controlled drug delivery devices; monomer and solvent removal from formed polymers; and in the study of physical aging of glassy polymeric materials. In applications ranging from gas separation to barrier packaging, glassy polymers have permeation and separation characteristics superior to those of rubbery polymers and, as a result, glassy polymers are used commercially in these applications.

Very high permeability membranes and low permeability barrier films derive from stiff chain, glassy polymers. High permeability membranes result from rigid disordered materials, and low permeability barrier films are fabricated from rigid locally ordered materials. For example, poly(1-(trimethylsilyl)-1-propyne) [PTMSP], a stiff chain polymer with a glass transition temperature in excess of 300°C, is the most permeable polymer known (1). Its oxygen permeability coefficient at ambient conditions is approximately 1×10^{-6} cm³(STP) cm/(cm² s cm Hg), which is more than *ten* times the oxygen permeability of highly flexible, rubbery poly(dimethylsiloxane) [PDMS], the most permeable rubbery polymer (2-5). Bulky substituents and double bonds along the PTMSP backbone lead to rigid, twisted polymer chains which pack very poorly, resulting in extraordinarily high solubilities, diffusivities, and, in turn, high permeabilities. In complementary contrast, glassy liquid crystalline polymers [LCPs] such as poly(p-phenyleneterephthalamide) [PPTA] and poly(p-hydroxybenzoic acid-co-6-hydroxy-2-naphthoic acid) [HBA/HNA] pack very efficiently in the solid state and exhibit extremely high barrier properties. For example, at 35°C the oxygen permeability of HBA/HNA73/27, a random copolymer containing 73 mole percent HBA, and PPTA are 47×10^{-15} and 80×10^{-15} cm^3(STP) cm/(cm^2 s cm Hg), respectively (6-11). These values are *eight orders of magnitude* lower than the oxygen permeability coefficient of PTMSP, demonstrating the extraordinary range of permeation properties between those of low free volume glassy polymers, which enjoy efficient chain packing in the solid state, and high free volume glassy polymers, where chain packing in the solid state is strongly frustrated.

Free volume is a convenient concept to characterize the amount of space in a polymer matrix that is not occupied by the constituent atoms of the polymer and is available to assist in the molecular transport of small penetrant molecules. As noted by Adam and Gibbs (12), the concept of free volume embodies inter- and intramolecular interaction as well as the topology of molecular packing in the amorphous phase. In this regard, transport properties of polymers have been described by Peterlin (13) as having static and dynamic free volume contributions, thereby acknowledging the importance of packing-related free volume as well as cooperative segmental chain dynamics to penetrant transport. In developing new polymers for membrane and barrier applications and in the systematic processing-induced manipulation of permeation properties of polymers, free volume is often used to rationalize experimentally observed structure/property relations. Therefore, the relationship among polymer backbone structure as well as higher order structure, such as nematic order in LCPs, and free volume and free volume distribution in the solid state is important.

In this chapter, we examine the effect of chain packing on permeation properties for three families of polymers that span the range from extremely low permeability, liquid crystalline barrier materials to intermediate permeability, amorphous gas separation membrane materials to ultrahigh permeability polymers. All of the polymers discussed are glasses at the ambient measurement conditions. Chain packing in these systems is characterized by density-based estimates of free volume and by positron annihilation lifetime spectroscopy [PALS], which permits an estimate of both the size and concentration of free volume elements in a polymer matrix. The effect of free volume on permeability of a range of penetrants of different sizes is presented. For the highest permeability glassy polymers, subtle variations in free volume distribution result in extraordinary changes in permeation properties.

Background

The permeability of a polymer film of thickness ℓ to a penetrant, P, is (14):

$$P = \frac{N \, \ell}{p_2 - p_1}, \tag{1}$$

where N is the steady state gas flux through the film, and p_2 and p_1 are the upstream and downstream penetrant partial pressures, respectively. When the downstream pressure is much less than the upstream pressure, permeability is often written as (14):

$$P = S \times D, \tag{2}$$

where S, the apparent solubility coefficient, is the ratio of the dissolved penetrant concentration in the upstream face of the polymer to the upstream penetrant partial pressure in the contiguous gas or vapor phase, and D is the concentration averaged penetrant diffusion coefficient (14).

Permeability properties are very sensitive to chain packing in the solid state. Chain packing is often characterized in terms of free volume. However, a simple, direct, unambiguous measurement of free volume in polymers is not available. The most common characterization of solid state chain packing is fractional free volume, FFV, which is defined as follows (*14*):

$$FFV = \frac{V - V_o}{V}.$$ (3)

In this expression, V is the polymer specific volume, and V_o is the so-called occupied volume of the polymer, which is commonly estimated as 1.3 times the van der Waals volume of the constituent monomers (*15*). The van der Waals volumes of monomer units are usually estimated using Bondi's group contribution method (*16*).

As permeability depends on both solubility and diffusivity, one may consider the effect of free volume on solubility and diffusion coefficients individually when assessing the impact of free volume on permeation properties. As discussed in more detail below, the effect of chain packing on gas diffusion coefficients in amorphous glassy polymers is stronger than the effect of chain packing on gas solubility coefficients. Thus, correlations of permeability with free volume often closely resemble correlations of diffusion coefficients with free volume. In polymers with highly ordered regions, such as semicrystalline or liquid crystalline polymers, the effect of free volume on solubility is stronger than in amorphous materials. In ordered or partially ordered polymers, solubility is sensitive to the amount of ordered material and the efficiency of packing in the ordered regions (*6, 10, 17*).

The effect of free volume on penetrant diffusion coefficients in polymers is often described using concepts from the Cohen and Turnbull model (*18*). This statistical mechanics model provides a simplistic description of diffusion in a liquid of hard spheres. A hard sphere penetrant is considered to be trapped in a virtual cage created by its neighbors. Free volume is defined as the volume of the cage less the volume of the penetrant. Free volume fluctuations, which occur randomly due to thermally-stimulated Brownian motion of neighboring hard spheres, provide opportunities for the penetrant to execute a diffusion step if the gap (*i.e.* free volume fluctuation) occurs sufficiently close to the penetrant to be accessible and is of sufficient size to accommodate it. The diffusion coefficient of a penetrant is given by:

$$D = \int_{v^*}^{\infty} F(v)p(v)dv$$ (4)

where v is the size of a free volume element, v^* is the minimum free volume element size which can accommodate the penetrant, $F(v)$ is the contribution of free volume elements of size v to the total diffusion coefficient, and $p(v)$ is the probability of finding a free volume element of size between v and v+dv. The distribution of free volume is obtained, using standard techniques of statistical mechanics, by maximizing, at fixed number of molecules and fixed total free volume, the excess entropy resulting from the distribution of free volume. The result is:

$$p(v) = \frac{\gamma}{\langle v_f \rangle} \exp\left[-\frac{\gamma}{\langle v_f \rangle} v \right]$$ (5)

where γ is an overlap parameter introduced to avoid double counting of free volume elements shared by more than one hard sphere, and $\langle v_f \rangle$ is the average free volume.

As average free volume in a polymer matrix cannot be measured directly, $<v_f>$ in Equation 5 is usually replaced by fractional free volume, which is calculated based on polymer density and an estimate of the occupied volume in the polymer. Free volume distributions calculated using Equation 5 are presented in Figure 1 for several values of average free volume. From Figure 1, this theory predicts that smaller free volume elements are more numerous than larger ones. Qualitatively, this prediction is consistent with molecular simulations of free volume distributions in glassy polymers such as atactic poly(propylene)(*19*) and poly(vinylchloride) (*20*). As indicated in Figure 1, the theory also predicts that the probability of finding a free volume element of a particular size increases as average free volume increases.

Using these results, the Cohen and Turnbull model provides the following expression for the penetrant diffusion coefficient (*18*):

$$D = A \exp\left(-\frac{\gamma}{\langle v_f \rangle} v^*\right) \tag{6}$$

where A is a pre-exponential factor which depends weakly on temperature. Penetrant size is strongly correlated with v^*. From Equation 6, penetrant diffusion coefficients decrease exponentially with increasing penetrant size (v^*), which is qualitatively consistent with experimental observations for infinite dilution penetrant diffusion coefficients in glassy polymers (*14*). This model, based on diffusion of spheres, cannot provide insight into the effect of penetrant shape on diffusion coefficients. Equation 6 also predicts that, in two polymer matrices with very different average free volume, the effect of penetrant size on diffusion coefficients is weaker in the higher free volume polymer. While this model is obviously a crude approximation of the complex cooperative segmental chain dynamics which govern penetrant transport in polymers, it provides an intuitively useful qualitative rationale for the effect of free volume and penetrant size on transport properties of polymers.

Low Permeability Liquid Crystalline Polymers

Liquid crystalline polymers such as PPTA and HBA/HNA copolymers have remarkably high barrier properties (*6, 10, 11, 21-23*). Values of oxygen permeability for several glassy LCPs and for conventional, non-liquid crystalline high barrier glassy polymers are presented in Table 1. Oxygen permeability is at least an order of magnitude lower in LCPs than in common glassy polymers such as PET and PVC and can be comparable to oxygen permeability in PAN, a noted barrier polymer.

The influence of chain packing (*i.e.* free volume) on solubility, diffusivity and permeability in liquid crystalline polymers can be studied by comparing properties of LCPs in the disordered, isotropic state with those in the ordered, liquid crystalline state. HIQ-40 is a random, glassy, thermotropic, nematogenic terpolymer synthesized from 40 mole percent *p*-hydroxybenzoic acid and 30 mole percent each of isophthalic acid and hydroquinone. The chemical structures of the constituent monomers for HIQ-40 are:

| *p*-hydroxybenzoic acid | isophthalic acid | hydroquinone |

HIQ-40 may be dissolved in volatile solvents and cast into thin, metastable, glassy, optically isotropic, transparent films at ambient conditions. The rapid evaporation of volatile solvent molecules, in comparison with the characteristic rate of mesogen ordering, results in the preparation of solvent-free, isotropic, and disordered amorphous films which pack as illustrated in the cartoon of Figure 2a. Subsequent heating in the range of the glass transition temperature, T_g, confers sufficient chain mobility to permit the development of axial (*i.e.* nematic liquid crystalline) order and small amounts (<10%) of three-dimensional crystalline order (*cf.* Figure 2b), which persist upon cooling the sample to ambient conditions (*6, 21, 23, 24*).

Table I. Oxygen permeability in glassy barrier polymers and liquid crystalline polymers at 35°C

Polymer	Permeability $\times 10^{15}$ [cm^3(STP) cm/(cm^2 s cm Hg)]
Conventional, Non-Liquid Crystalline Glassy Polymers:	
Amorphous Poly(ethyleneterephthalate) [PET] (*25*)	11,000
Poly(vinylchloride) [PVC] (*26*)	9,400
Poly(acrylonitrile) [PAN] (*27*)	54
Liquid Crystalline Polymers:	
HBA/PET 80/20[a] (*25*)	500
PPTA	80
HBA/HNA73/27	47

[a] HBA/PET 80/20: a liquid crystalline copolyester composed of 80 mole percent *p*-hydroxy-benzoic acid [HBA] and 20 mole percent PET.

In a nematic liquid crystalline polymer such as HIQ-40, rod-like mesogenic units of the polymer chains are locally aligned, somewhat imperfectly, in a common direction, described by a vector called the director, **n**, as shown in Figure 2b. Nematic domains formed by regions where the director value is constant or changes smoothly and continuously with position are separated by defects or domain boundaries, where the director and, therefore, molecular orientation change abruptly, as shown in Figure 2b. The characteristic width of these domain boundaries is believed to be several hundred Angströms in nematic thermotropic polymers (*28*). The nature of the molecular structure in boundary regions is not well understood. However, several reports suggest that boundary regions may contain relatively high concentrations of chain ends (*28, 29*).

By comparing the sorption and transport behavior of small molecules in an as-cast, disordered, isotropic sample with those of an annealed, ordered, frozen liquid crystalline sample, the effect of axial ordering on sorption and transport properties may be determined unambiguously. Moreover, the influence of axial ordering on other properties (*e.g.* density, fractional free volume, glass transition temperature, and free volume accessible to orthoPositronium) may be determined.

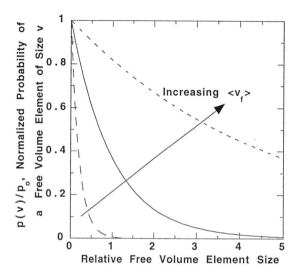

Figure 1. Free volume distribution from the Cohen-Turnbull model. p_o is the probability of finding an infinitely small free volume element, $\gamma/<v_f>$. The relative free volume element size is $\gamma v/<v_f>$.

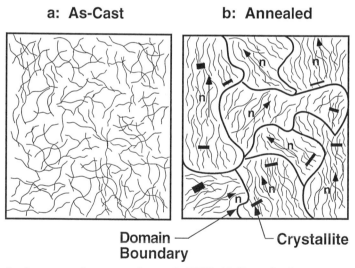

Figure 2. Structure of as-cast and annealed HIQ-40 films. In the as-cast sample, the mesogenic units of the polymer chains are kinetically trapped in a disordered, isotropic arrangement. In the annealed sample, arrows in the domains represent the director, **n**, and point in the direction of orientation of the domains.

Acetone Sorption and Diffusion in HIQ-40. Kinetic gravimetric sorption experiments provide a convenient and sensitive experimental probe of both solubility and diffusivity in thin polymer films of uniform thickness (30). Uniform HIQ-40 films were cast from solution in an organic solvent mixture to provide a set of as-cast, isotropic samples (6, 21). These as-cast films were then annealed for approximately one hour at temperatures ranging from 100°C to 330°C to induce ordering. The films were quenched to ambient conditions, and kinetic gravimetric sorption experiments were performed using acetone as a probe penetrant. The acetone solubility and diffusivity determined from these measurements are shown in Figure 3. Both solubility and diffusivity are highest in the as-cast film. Solubility and diffusivity decrease monotonically as the annealing temperature is increased to 200°C. There is little change in solubility and diffusivity as the annealing temperature is raised to 300°C from 200°C. Annealed samples exhibit solubilities and diffusivities that are up to approximately an order of magnitude lower than in the unannealed, as-cast sample.

Most of the change in acetone solubility and diffusivity occurs in the temperature range of the glass-rubber transition. In the as-cast sample, the glass transition is observed at 42°C (31). In the sample annealed at 200°C, the glass transition of the nematic regions is observed at 139°C. A cold crystallization exotherm centered near 150°C is observed by differential scanning calorimetry. This exotherm is associated with the development of low levels of three-dimensional crystallinity. Samples annealed between 170°C and 300°C exhibit crystallinity detectable by wide angle X-ray diffraction (31). A broad melting endotherm is centered at about 310°C; this endotherm marks the melting of crystallites and the transformation to a nematic fluid. As-cast film samples are optically isotropic at 25°C, those annealed at 200°C exhibit marked birefringence, suggesting the development of axial order, and those annealed at 300°C and 330°C exhibit textures indicative of a nematic liquid crystalline phase (31).

The ordering of the as-cast films due to thermal annealing is accompanied by an increase in density for annealing temperatures up to 300°C and a slight decrease in density in the sample annealed at 330°C, which is above the crystalline melting point in HIQ-40. Figure 4 presents the correlation of acetone solubility and diffusivity with fractional free volume in the as-cast and annealed samples. There is a strong dependence of both solubility and diffusivity on chain packing.

Based on these results, mesogenic order efficiently frustrates penetrant solubility and diffusivity in glassy HIQ-40. Myers et al.(17) observed that even the smallest penetrant is not soluble in the crystalline regions of common glassy and rubbery polymers. Experimental determinations of gas and vapor sorption levels in most semi-crystalline polymers are well-described by an analytical model which assumes that the polymer may be divided into a non-crystalline phase which accommodates penetrant sorption and a crystalline phase which does not. This two-phase model has been extended to liquid crystalline polyesters with good success (8, 32). Some of the HIQ-40 samples in this study contain low levels of conventional three dimensional crystalline regions (<10%) and highly organized liquid crystalline regions which could preclude dissolution of penetrant molecules. Based on the two-phase model, the fraction of conventional crystallinity would have to be approximately 90% to explain the observed sorption behavior of acetone in HIQ-40 samples that had been annealed at temperatures between 200°C and 300°C (6). However, DSC and X-ray measurements of crystallinity in HIQ-40 suggest that only a very small fraction of the material is crystalline, clearly too low to account for the very high fraction of inaccessible regions in HIQ-40 suggested by the ten-fold decrease in acetone solubility (6, 21).

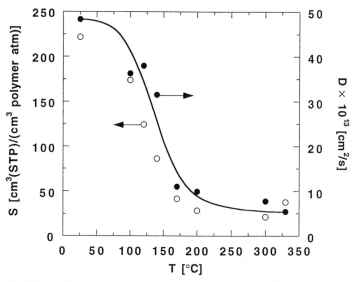

Figure 3. Effect of annealing temperature on acetone solubility (open circles) and diffusivity (filled circles) in HIQ-40 samples at 35°C and an acetone relative pressure of 0.15.

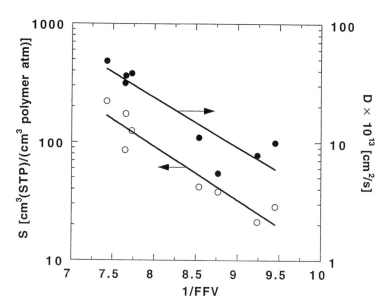

Figure 4. Correlation of acetone solubility and diffusivity in HIQ-40 with fractional free volume (6, 23). The solubility and diffusivity were determined at 35°C and an acetone relative pressure of 0.15.

If the ordering of the liquid crystalline phase precludes solubility within the HIQ-40 domains just as conventional crystallinity typically precludes sorption and transport in the crystallites, then penetrant molecules would be largely restricted to boundary regions between domains. Since the domain boundaries account for a small fraction of the sample mass (28), penetrant solubilities based upon the overall sample weight would appear to be quite low relative to solubilities in conventional amorphous polymers. Moreover, the domains could act as impenetrable barriers to diffusion of small molecules, providing a tortuous path for the penetrant molecules. In HBA/HNA73/27, gas and vapor solubility and transport data are consistent with models suggesting that only domain boundary regions separating nematic liquid crystalline domains are accessible for penetrant sorption and transport (8, 22).

Permeability of Gases and Acetone in HIQ-40. Figure 5 presents the relative permeability of HIQ-40 to a series of gases as a function of annealing temperature. The permeability of annealed samples are reported relative to the permeability of an unannealed (i.e. as-cast) sample. Acetone permeability is also presented in this figure and is calculated as the product of acetone solubility and diffusivity according to Equation 2. The numbers in parentheses are the kinetic diameters of the penetrant molecules. Kinetic diameter is a common measure of penetrant size. As shown in Figure 5, the permeability of all penetrants decreases with increasing annealing temperature up to 300°C. The larger penetrants are more strongly affected by annealing-induced ordering than the smaller penetrants. The results of the simple free volume theory of diffusion (Equation 6) suggest that since the larger penetrants require more free volume to execute a diffusion step, diffusion coefficients of larger penetrants should be more strongly influenced by a reduction in free volume than the diffusion coefficients of small penetrant molecules.

Positron Annihilation Lifetime Spectroscopy of HIQ-40 Films. Positron annihilation lifetime spectroscopy has emerged as a sensitive technique to probe free volume in polymers (33, 34). PALS uses orthoPositronium [oPs] as a probe of free volume in the polymer matrix. oPs resides in regions of reduced electron density, such as free volume elements between and along chains and at chain ends (33). The lifetime of oPs in a polymer matrix, τ_3, reflects the mean size of free volume elements accessible to the oPs. The intensity of oPs annihilations in a polymer sample, I_3, reflects the concentration of free volume elements accessible to oPs. The oPs lifetime in a polymer sample is finite (on the order of several nanoseconds), so PALS probes the accessibility of free volume elements on nanosecond timescales (35).

Table II presents PALS results and other physical property data for an as-cast HIQ-40 sample and for a sample that was annealed for one hour at 200°C. The annealing protocol results in a 2.5% increase in density, which corresponds to a 17% decrease in fractional free volume. The acetone diffusion coefficient decreases almost five-fold and acetone solubility decreases by approximately 90% as a result of the ordering induced by the annealing protocol. The oPs lifetime decreases by 14%, suggesting that the average free volume cavity size decreases due to annealing. Based on the oPs lifetime, the mean free volume cavity diameter may be estimated (36); these values are reported in parentheses in Table II. The PALS I_3 parameter, which reflects the relative concentration of free volume elements in the polymer matrix, is approximately 22% lower in the annealed, liquid crystalline sample.

The free volume accessible to oPs is more than 50% lower in the annealed sample. This decrease is much larger than the decrease in free volume probed by density, and suggests that both the amount of free volume and the accessibility of free volume over the lifetime of the oPs probe decrease as the isotropic as-cast sample orders towards the liquid crystalline state. This finding is consistent with the much

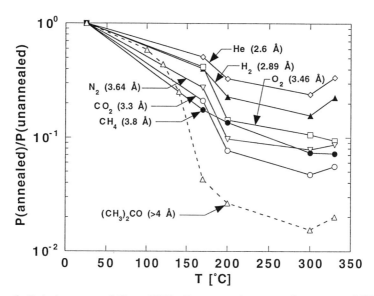

Figure 5. Relative permeability of HIQ-40 to several gases and acetone at 35°C (*24*). The numbers in parentheses are the penetrant kinetic diameters.

higher glass transition temperature, in the annealed, nematic, liquid crystalline state than in the as-cast, isotropic state. The large increase in T_g upon annealing suggests rather profound decreases in segmental mobility upon ordering the polymer from the isotropic to the liquid crystalline state.

These composite results indicate that liquid crystalline ordering in aromatic polyesters such as HIQ-40 can strongly decrease penetrant solubility and diffusivity while improving static packing efficiency, as characterized by density and FFV. Transport properties of larger penetrant molecules are more strongly affected by this ordering process than small penetrant molecules. The improvement in packing efficiency is accompanied by sharp decreases in segmental motions important in the glass-rubber transition and perhaps by decreases in polymer chain motion important for oPs accessibility in the polymer matrix.

Table II. Physical properties and PALS results for HIQ-40 films

Property	As-Cast	Annealed	% Change
Density [gm/cm^3] ±0.001 (6)	1.374	1.408	+2.5
FFV (6)	0.128	0.106	-17
T_g [K] (31)	315	412	+31
Relative Free Volume Cavity Size, τ_3 [ns] ± 0.03(37)	1.80 (5.3Å)[a]	1.54 (4.7Å)[a]	-14
Relative Free Volume Concentration, I_3 [%] ± 0.3(37)	15.2	11.8	-22
Relative Free Volume Accessible to Positrons, $\tau_3^3 I_3^{b}$ [ns^3 %] (37)	89.4	42.9	-52
Acetone Solubility [cm^3(STP)/(cm^3 atm)] (6)	220	20	-87
Acetone Diffusivity [cm^2/s] (6)	48×10^{-13}	10×10^{-13}	-79

[a] Mean free volume cavity diameter calculated according to the free volume model of Jean (36, 38).
[b] Based on a model of spherical free volume cavities, the free volume accessible to oPs may be estimated from the product of the cube of the cavity size, τ_3, and the concentration of free volume cavities, I_3 (33).

Intermediate Permeability Amorphous Polymers

In contrast to the LCP results just presented, in glassy polymers used as gas separation membranes, free volume influences diffusion coefficients much more than solubility coefficients. Figure 6 provides an example of this effect. In this figure, the solubility, diffusivity, and permeability of methane in a series of glassy, aromatic, amorphous poly(isophthalamides) [PIPAs] are presented as a function of the fractional free volume in the polymer matrix. (More complete descriptions of the transport properties of this family of materials are available elsewhere (39, 40)). The fractional free volume is manipulated systematically in this family of glassy polymers by synthesizing polymers with different substituent and backbone elements as shown in

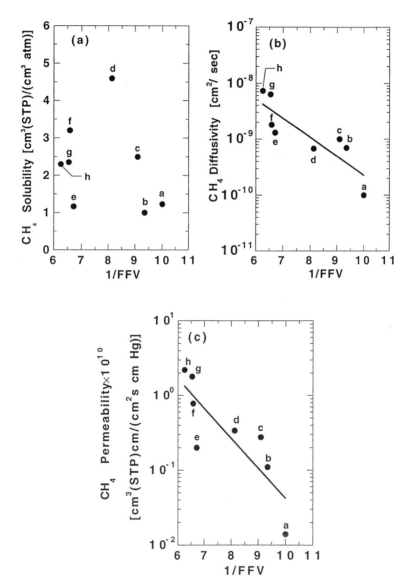

Figure 6. Methane solubility, diffusivity, and permeability of a series of low free volume polyisophthalamides. The sorption and transport data were determined at 35°C and an upstream pressure of 3 atmospheres.

Table III. From Figure 6a, there is no systematic trend of gas solubility with fractional free volume in these materials. Other studies have found a small decrease in gas solubility with decreasing free volume (14). The interchain gaps which are the locus of sorption in polymers are probably more accessible to penetrants in amorphous polymers than in crystalline or liquid crystalline regions, where these gaps may be absent or so small as to be inaccessible or the gaps may be locked into rigid, ordered regions where the chain packing is too energetically difficult to disturb to permit dissolution of penetrant molecules. The tendency of a penetrant molecule to dissolve in an amorphous polymer matrix depends strongly on penetrant condensability and the strength of specific polymer-penetrant interactions and weakly on the free volume of the polymer matrix (14).

Table III. Chemical structure, physical properties, and PALS results for poly(isophthalamides) presented in Figure 6.

Symbol	X	R	Tg [°C]	FFV	τ_3 [ns]	I_3 [%]	τ_3^3 [ns^3]
a	H	SO$_2$	323	0.100	1.60	18.3	4.07
b	C$_6$H$_5$	SO$_2$	328	0.107	1.69	17.4	4.84
c	C(CH$_3$)$_3$	SO$_2$	337	0.110	1.81	19.6	5.90
d	Si(CH$_3$)$_3$	SO$_2$	273	0.123			
e	H	C(CF$_3$)$_2$	297	0.149	2.11	20.7	9.35
f	C$_6$H$_5$	C(CF$_3$)$_2$	311	0.152	2.24	20.2	11.3
g	C(CH$_3$)$_3$	C(CF$_3$)$_2$	315	0.153	2.55	20.3	16.6
h	Si(CH$_3$)$_3$	C(CF$_3$)$_2$	272	0.156	2.64	20.5	18.5

Figures 6b and 6c present the effect of free volume on methane diffusivity and permeability, respectively. Over the range of free volume explored ($0.1 \leq FFV \leq 0.16$), diffusivity decreases by roughly two orders of magnitude as free volume decreases. The change in permeability with increasing free volume essentially mirrors that of diffusivity, indicating that the most important effect of free volume on transport properties in this family of materials is the impact of free volume on diffusion coefficients.

PALS results allow a comparison of the effect of polymer substituent and backbone chemistry on the relative size and concentration of free volume elements. The methane solubility is not strongly correlated with the PALS free volume parameters (similar to the result shown for fractional free volume in Figure 6a). The methane diffusivity and permeability of these polyisophthalamides are strongly correlated with the relative free volume element size probed by oPs as shown in Figures 7a and 7b. The PALS parameter $\tau_3{}^3$ represents the relative volume of an equivalent equiaxed free volume element. Including the I_3 values (representative of the relative concentration of free volume sites) does not improve the correlation. The PALS results suggest that increases in fractional free volume in this family of polymers are dominated by increases in mean free volume element size rather than concentration.

In marked contrast to the results presented in Figure 4 for mesogenic HIQ-40, there is a negligible dependence of solubility on chain packing in the amorphous PIPAs, as characterized by fractional free volume, even though the maximum change in free volume in HIQ-40 due to mesogenic ordering (17%) is much less than that observed in the PIPAs (37%). This result highlights the efficacy of liquid crystalline ordering to reduce penetrant solubility and supports the notion that liquid crystalline ordering precludes or strongly decreases penetrant sorption in nematic domains.

High Permeability Polymers

High permeability membranes result from stiff-chain, glassy, disordered materials. As mentioned previously, poly(1-(trimethylsilyl)-1-propyne) [PTMSP], a stiff chain polymer with a glass transition temperature in excess of 300°C, is the most permeable polymer known (2-5). This polymer has the lowest density of organic polymers, approximately 0.75 gm/cm^3 (1). This extremely low density leads to extraordinarily high fractional free volume, 0.29. Other high permeability glassy polymers may be prepared from amorphous, random copolymers based on tetrafluoroethylene [TFE] and 2,2-bistrifluoromethyl-4,5-difluoro-1,3-dioxole [PDD]. Poly(tetrafluoroethylene) [PTFE] is a low permeability, semicrystalline polymer (41). Random copolymerization of TFE with sufficient amounts of PDD results in the formation of wholly amorphous, high glass transition copolymers which are among the most permeable polymers known (42, 43).

The chemical structure of TFE/PDD copolymers and PTMSP are:

TFE/PDD   **PTMSP**

Properties for several TFE/PDD copolymers and PTMSP are compared in Table IV. Density and glass transition temperatures for the TFE/PDD copolymers were obtained from Buck and Resnick (44), and the density and glass transition temperature for PTMSP are from the study of Nakagawa et al.(1). Among the fluoropolymers in this table, PTFE homopolymer exhibits the lowest glass transition temperature, the lowest oxygen permeability coefficient, and the lowest fractional free volume. In the polymers in Table IV, the PALS results suggests a bimodal distribution of free volume elements consisting of smaller cavities characterized by τ_3 and I_3 and larger cavities characterized by τ_4 and I_4. This observation is consistent with that of Consolati et al.(45), Alentiev et al.(46), and Yampolskii et al.(47) for PTMSP

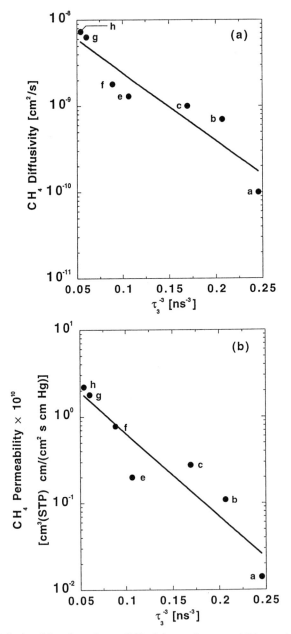

Figure 7. Relationship of methane diffusivity and permeability and PALS relative free volume element size. The transport data were determined at 35°C and an upstream pressure of 3 atmospheres, and the PALS data were collected at ambient conditions.

andTFE/PDD copolymers. τ_3 is markedly smaller in TFE/PDD65 than in TFE/PDD87 and PTMSP and is similar to that for PTFE. τ_3 for PTMSP and TFE/PDD87 are similar. This result suggests that addition of even 65% PDD does not

Table IV. Physical properties, permeability and PALS properties for PTFE, TFE/PDD copolymers, and PTMSP at 25°C

Property	PTFE	TFE/PDD65	TFE/PDD87	PTMSP
mole percent PDD	0	65	87	-
T_g [°C]	-73	160	240	>300
Density	2.18[b], 2.00[c]	1.81	1.74	0.75
FFV	0.23[c]	0.28	0.30	0.29
Oxygen Permeability $\times 10^{10}$ [cm^3(STP) cm/(cm^2 s cm Hg)]]	4.5[b], 10[e]	365	1380	9,860[d]
τ_3 [ns] ± 0.03	1.00[f]	1.03	1.32	1.31
I_3 [%] ± 0.3	15.8	5.97	4.87	9.64
$\tau_3{}^3 I_3$ [ns^3 %]	15.8	6.5	11.2	21.7
τ_4 [ns] ± 0.1	2.8	4.6	5.8	7.0
I_4 [%] ± 0.6	4.3	15.0	13.6	30.1
$\tau_4{}^3 I_4$ [ns^3 %]	94.4	1,460	2,650	10,320
$\tau_3{}^3 I_3 + \tau_4{}^3 I_4$	110	1,467	2,661	10,342

[a] The numerical suffix gives the mole percent PDD used to prepare these copolymers.
[b] Density and permeability are reported for a sample with a density of 2.18 gm/cm^3 (*41*). Based on a correlation between density and crystalline content, this density value corresponds to 34 volume percent crystallinity (*48*).
[c] Estimated values for a hypothetical 100% amorphous PTFE sample. Amorphous density is from van Krevelen (*15*).
[d] data from the study of Pinnau and Toy. T=23 °C(*49*)
[e] Amorphous permeability based on: P (amorphous) = P (semicrystalline)$/\phi_a^2$, where ϕ_a is the volume fraction of amorphous material (*10*).
[f] PALS parameters for PTFE were determined using a commercial PTFE film from DuPont. This film contains 85 weight percent crystallinity as estimated from wide angle X-ray diffraction. The PALS parameters (I_3 and I_4) are not corrected for crystallinity in this table.

markedly change the size of the small free volume elements, but that 87% PDD increases their size by approximately 30%. However, the smaller free volume elements contribute only 14% of the total free volume accessible to oPs,

$(\tau_3{}^3I_3+\tau_4{}^3I_4)$, in PTFE, only 0.4% in TFE/PDD 65 and TFE/PDD87 and only 0.2% in PTMSP. Thus, the vast majority of the free volume accessible to oPs is in the larger free volume elements.

For the TFE/PDD copolymers, τ_4 is about 20% larger in TFE/PDD87 than in TFE/PDD65 suggesting that the larger free volume elements accessible to oPs have an average volume $(\tau_4{}^3)$ approximately 100% larger in TFE/PDD87 than in TFE/PDD65. τ_4 is substantially higher in PTMSP than in TFE/PDD65, TFE/PDD87 or PTFE, suggesting that the largest free volume elements accessible to oPs in PTMSP are significantly larger than the largest free volume elements in the TFE/PDD copolymers or PTFE.

The relative concentration of the free volume elements, I_3 and I_4, is 20% and 10% lower, respectively, in TFE/PDD87 than in TFE/PDD65. Thus, while the size of these elements is higher in the copolymer with more PDD, the concentration is lower. The net relative free volume accessible to oPs $(\tau_3{}^3I_3+\tau_4{}^3I_4)$ is 80% larger in TFE/PDD87 than in TFE/PDD65.

TFE/PDD65 has been examined using PALS by Davies and Pethrick (5 Q. Whilst these authors found two oPs localization sites in the polymer, they did not comment on the possible free volume structure responsible for these two oPs components. The size of the larger cavities was postulated to be the controlling factor in the high gas diffusion rates of TFE/PDD65. The size and concentration of the large free volume elements, and consequently the overall free volume, available for penetrant transport are much larger in the TFE/PDD copolymers than in PTFE. This result is consistent with the notion that the addition of PDD to TFE frustrates chain packing, rendering the resulting copolymers totally amorphous and dramatically more permeable than PTFE.

PTMSP has the largest τ_4 and I_4 of all of the polymers considered and, as a result, has an enormously higher accessible free volume to oPs than the other materials. This result is consistent with the oxygen permeability data, where PTMSP is more than seven times as permeable to oxygen as TFE/PDD87, the most permeable TFE/PDD copolymer. These results suggest that the free volume accessible to oPs in the copolymers is much lower than that in PTMSP, which is intriguing since TFE/PDD87 and PTMSP have essentially the same fractional free volume (as characterized by density and group contribution estimates of occupied volume). Relative to the copolymers, PTMSP has a much higher concentration of both large and small free volume elements accessible to oPs over the nanosecond time scale of the experiment, and the largest free volume elements in PTMSP are substantially larger than free volume elements in the TFE/PDD copolymers. Most of the difference in relative free volume estimated by PALS is due to the contribution from the large free volume elements in PTMSP (4 Q. In this regard, Consolati *et al(4 5* and Yampol'skii *et al(4 7* have attributed the smaller cavities in PTMSP to a channel structure and the larger cavities to conventional inter-and intra-chain free volume.

These composite results suggest that the distribution and availability of free volume in PTMSP and the TFE/PDD copolymers are very different. Both PTMSP and the TFE/PDD copolymers are high T_g, stiff chain materials, so it is unlikely that the vast differences in accessible free volume and permeability coefficients is solely related to great differences in segmental dynamics between these materials which would render the free volume in PTMSP much more accessible on the time scales appropriate for PALS and permeation. Rather, it seems more likely that free volume elements in PTMSP are interconnected and span the sample, providing extremely efficient pathways for penetrant diffusion. In fact, the notion of interconnected free volume elements in PTMSP has been invoked to explain the unusual transport properties in this material (*49, 51, 52*). In many respects, gas and vapor transport in PTMSP is more similar to transport in microporous carbon than to transport in conventional glassy polymeric gas separation membranes. In the TFE/PDD copolymers, the free volume elements may be much more finely dispersed than in PTMSP and are not interconnected on the timescale of the PALS measurement. In fact, we have observed that gas sorption into the nonequilibrium excess volume in TFE/PDD copolymers is only 25% of the value expected based on macroscopic dilatometry measurements, suggesting that a substantial amount of free volume in the TFE/PDD copolymer is inaccessible to even the smallest penetrants.(*53*)

Conclusions

The most permeable polymers are amorphous glassy polymers in which chain packing is sufficiently poor to permit penetrant access. These polymers are distinguished by their high values of free volume. Moreover, the distribution of free volume may be important to the permeability in such polymers. At the same average free volume, polymers with interconnected free volume elements may have extraordinarily high transport properties relative to polymers in which the free volume distribution does not favor permanently connected free volume elements. In this regard, PALS provides a very useful probe of free volume distribution in such materials and provides results which are more consistent with gas permeability results than density-based fractional free volume. The least permeable polymers are glassy polymers in which the backbone structure permits very efficient chain packing in the solid state. Liquid crystalline polymers provide one example of such classes of materials. As liquid crystalline order is perfected in the glassy copolyester considered and free volume is reduced in the polymer matrix, both penetrant diffusivity and solubility decrease strongly, and the permeability properties of larger penetrants are influenced to a larger extent than those of smaller penetrants. Systematic manipulation of free volume and free volume distribution via backbone structure and processing results in many orders of magnitude change in permeability properties.

Acknowledgments

The authors would like to acknowledge partial support of this work by the National Science Foundation (Young Investigator Award CTS-9257911-BDF).

Literature Cited

1. Nakagawa, T.; Saito, T.; Asakawa, S.; Saito, Y. *Gas Separation and Purification*, **1988**, *2*, 3.
2. Ichiraku, Y.; Stern, S. A.; Nakagawa, T. *J. Membrane Sci.*, **1987**, *34*, 5.
3. Nagai, K.; Higuchi, A.; Nakagawa, T. *J. Polym. Sci.: Polym. Phys. Ed.*, **1995**, *33*, 289.
4. Morisato, A.; Shen, H. C.; Sankar, S. S.; Freeman, B. D.; Pinnau, I.; Casillas, C. G. *J. Polym. Sci.: Polym. Phys. Ed.*, **1996**, *34*, 2209.
5. Morisato, A.; Freeman, B. D.; Pinnau, I.; Casillas, C. G. *J. Polym. Sci.: Polym. Phys. Ed.*, **1996**, *34*, 1925.
6. Cantrell, G. R.; Freeman, B. D.; Hopfenberg, H. B.; Makhija, S.; Haider, I.; Jaffe, M. In *Liquid Crystalline Polymers*, C. Carfagna , Ed., Pergamon Press: Oxford, 1994; pp 233.
7. Weinkauf, D. H.; Paul, D. R. *J. Polym. Sci.: Polym. Phys. Ed.*, **1991**, *29*, 329.
8. Weinkauf, D. H.; Paul, D. R. *J. Polym. Sci.: Polym. Phys. Ed.*, **1992**, *30*, 817.

9.   Weinkauf, D. H.; Paul, D. R. *J. Polym. Sci.: Polym. Phys. Ed.*, **1992**, *30*, 837.
10.   Weinkauf, D. H.; Paul, D. R. In *Barrier Polymers and Barrier Structures*, W. J. Koros , Ed., American Chemical Society: Washington, D.C., 1990; pp 60.
11.   Weinkauf, D. H.; Kim, H. D.; Paul, D. R. *Macromolecules*, **1992**, *25*, 788.
12.   Adam, G.; Gibbs, J. H. *J. Chem. Phys.*, **1965**, *43*, 139.
13.   Peterlin, A. *J. Macromol. Sci. B*, **1975**, *11*, 57.
14.   Ghosal, K.; Freeman, B. D. *Polymers for Advanced Technologies*, **1994**, *5*, 673.
15.   VanKrevelen, D., *Properties of Polymers*; Elsevier: Amsterdam, 1990; pp 875.
16.   Bondi, A. *J. Phys. Chem.*, **1964**, *68*, 441.
17.   Myers, A. W.; Rogers, C. E.; Stannett, V.; Szwarc, M. *TAPPI*, **1958**, *41*, 716.
18.   Cohen, M. H.; Turnbull, D. *J. Chem. Phys.*, **1959**, *31*, 1164.
19.   Greenfield, M. L.; Theodorou, D. N. *Polymer Preprints*, **1992**, *33*, 689.
20.   Shah, V. M.; Stern, S. A.; Ludovice, P. J. *Macromolecules*, **1989**, *22*, 4660.
21.   Morisato, A.; Miranda, N. R.; Willits, J. T.; Cantrell, G. R.; Freeman, B. D.; Hopfenberg, H. B.; Makhija, S.; Haider, I.; Jaffe, M. In *Crystallization and Related Phenomena in Amorphous Materials: Ceramics, Metals, Polymers, and Semiconductors*, M. Libera, P. Cebe, T. Haynes and J. Dickinson , Eds., Materials Research Society: Washington, DC, 1994; pp 81.
22.   Miranda, N. R.; Willits, J. T.; Freeman, B. D.; Hopfenberg, H. B. *J. Membrane Sci.*, **1994**, *94*, 67.
23.   Cantrell, G. R.; Freeman, B. D.; Hopfenberg, H. B. *ACS Polymer Preprints*, **1993**, *34*, 894.
24.   Park, J. Y.; Paul, D. R.; Haider, I.; Jaffe, M. *J. Polym. Sci.: Polym. Phys. Ed.*, **1996**, *34*, 1741.
25.   Weinkauf, D. H.; Paul, D. R., ACS Symposium on Barrier Polymers, Washington, D.C., 3 (1989).
26.   Tikhomirov, B. P.; Hopfenberg, H. B.; Stannett, V. T.; Williams, J. L. *Die Makromol. Chemie*, **1968**, *118*, 177.
27.   Allen, S. M.; Fujii, M.; Stannett, V. T.; Hopfenberg, H. B.; Williams, J. L. *J. Membrane Sci.*, **1977**, *2*, 153.
28.   Donald, A. M.; Windle, A. H., *Liquid Crystalline Polymers*; Cambridge University Press: Cambridge, 1992; pp 310.
29.   Kléman, M. In *Liquid Crystalline Polymers*, A. Ciferri , Ed., VCH Publishers: New York, 1992; pp 365.
30.   Crank, J.; Park, G. S. In *Diffusion in Polymers*, J. Crank and G. S. Park , Eds., Academic Press: New York, 1968; pp 1.
31.   Cantrell, G. R.; McDowell, C. C.; Freeman, B. D.; Noël, C. *J. Polym. Sci.: Polym. Phys. Ed.*, **in press**.
32.   Miranda, N. R.; Morisato, A.; Freeman, B. D.; Hopfenberg, H. B.; Costa, G.; Russo, S. *ACS Polymer Preprints*, **1991**, *32*, 382.
33.   Hill, A. J.; Weinhold, S.; Stack, G. M.; Tant, M. R. *Eur. Polym. J.*, **1996**, *32*, 843.
34.   Kobayashi, Y.; Haraya, K.; Hattori, S.; Sasuga, T. *Polymer*, **1994**, *35*, 925.
35.   Hill, A. J. In *High Temperature Properties and Applications of Polymeric Materials*, M. R. Tant, J. W. Connell and H. L. N. McManus, Eds., ACS Books: Washington, DC, 1995; pp 65.
36.   Nakanishi, H.; Wang, S. J.; Jean, Y. C. In *International Symposium on Positron Annihilation Studies of Fluids*, S. C. Sharma , Ed., World Scientific: Singapore, 1987; pp 292.

37. Hill, A. J.; Freeman, B. D.; McDowell, C. C.; Jaffe, M., International Membrane Science and Technology Conference, Sydney, Australia, PC.4 (1996).
38. Jean, Y. C. *Microchem J.*, **1990**, *42*, 72.
39. Ghosal, K.; Freeman, B. D.; Chern, R. T.; Alvarez, J. C.; delaCampa, J. G.; Lozano, A. E.; deAbajo, J. *Polymer*, **1995**, *36*, 793.
40. Morisato, A.; Ghosal, K.; Freeman, B. D.; Chern, R. T.; Alvarez, J. C.; delaCampa, J. G.; Lozano, A. E.; deAbajo, J. *J. Membrane Sci.*, **1995**, *104*, 231.
41. Pasternak, R. A.; Christensen, M. V.; Heller, J. *Macromolecules*, **1970**, *3*, 366.
42. Nemser, S. M.; Roman, I. C., Perfluorinated Membranes, U.S. Patent 5,051,114 to Air Liquide.
43. Pinnau, I.; Toy, L. G. *J. Membrane Sci.*, **1996**, *109*, 125.
44. Buck, W. H.; Resnick, P. R., 183rd Meeting of the Electrochemical Society, Honolulu, HI, (May 17, 1993).
45. Consolati, G.; Genco, I.; Pegoraro, M.; Zanderighi, L. *J. Polym. Sci.: Polym. Phys. Ed.*, **1996**, *34*, 357.
46. Alentiev, A. Y.; Yampolskii, Y. P.; Shantarovich, V. P.; Nemser, S. M.; Plate, N. A. *J. Membrane Sci.*, **1997**, *126*, 123.
47. Yampol'skii, Y. P.; Shantorovich, V. P.; Chernyakovskii, F. P.; Zanderleigh, L. *Journal of Applied Polymer Science*, **1993**, *47*, 85.
48. Sperati, C. A. In *Polymer Handbood, Third Edition*, J. Brandrup and E. H. Immergut , Eds., John Wiley & Sons: New York, 1989; pp V/35.
49. Pinnau, I.; Toy, L. G. *J. Membrane Sci.*, **1996**, *116*, 199.
50. Davies, W. J.; Pethrick, R. A. *Eur. Polym. J.*, **1994**, *30*, 1289.
51. Pinnau, I.; Casillas, C. G.; Morisato, A.; Freeman, B. D. *J. Polym. Sci.: Polym. Phys. Ed.*, **1996**, *34*, 2613.
52. Srinivasan, R.; Auvil, S. R.; Burban, P. M. *J. Membrane Sci.*, **1994**, *86*, 67.
53. Bondar, V.; Singh, A.; Freeman, B. D. *AIChE Topical Conference on Separation Science and Technologies*, **1997**, *2*, 831.

Chapter 22

The Segmental Motion and Gas Permeability of Glassy Polymer Poly(1-trimethylsilyl-1-propyne) Membranes

T. Nakagawa, T. Watanabe, and K. Nagai

Department of Industrial Chemistry, Meiji University Higashi-mita, Tama-ku, Kawasaki 214, Japan

The permeability of poly[1-trimethylsilyl-1-propyne], PMSP, to light gases is higher than that of any other nonporous, synthetic polymer at ambient temperatures. PMSP is in the glassy polymer state at ambient temperatures. One problem with PMSP is a decrease of its gas permeability with age. During the aging process, C'_H decreased. The parameter C'_H represents the maximum concentration of penetrant gas in the unrelaxed (Langmuir) domains, that is microvoids, of glassy polymers. The spin-lattice relaxation time of ^{13}C (T_1) of the backbone chain carbons in the PMSP increased due to aging, i.e. molecular motion slowed, whereas molecular motion of the side-chain carbons did not change. A copolymer with 1-phenyl-1-propyne (PP) and a blend of PMSP and PPP showed smaller C'_H values for CO_2 and C_3H_8, and the decay of the permeability with aging was much improved for these polymers, especially for the blend PMSP/PPP 95/5 polymer membrane. T_1 of the backbone carbons for the blend membrane were higher in the unaged state than for PMSP homopolymer but increased with aging to a similar value. The results suggest that the initial nonequilibrium state of the glassy PMSP, copolymer, and blend membranes was different and each glass approached a more stable state with age.

Poly[1-(trimethylsilyl)-1-propyne] (PMSP) is a typical glassy polymer at room temperature that was first synthesized by Masuda and Higashimura in the 1980's (1). Recently, membranologists have studied their gas permeation properties. The PMSP membrane has the highest gas permeability of all polymeric membranes. Therefore, this polymer is expected to have potential utility in industrial applications such as the separation of oxygen and nitrogen from air.

An industrial gas separation system requires high gas permeability, a high separation factor, and durability. However, the biggest problem of PMSP polymer is a decrease in its gas permeability with age. Therefore, the authors have been studying the aging process and stabilization of the gas permeability of the PMSP membrane (2 - 7). It has been found that membrane contamination is a dominant factor causing the change in gas permeability and in the absence of contaminations, the aging is due solely to the relaxation of the unrelaxed volume (the unrelaxed free volume in the glassy state or excess free volume below the glass transition temperature). The volume relaxation can be attributed to physical aging, which is most likely related to a change in molecular motion. Normally the physical aging of glassy polymers depends on temperature, and the temperature effect on the physical aging of PMSP is remarkable (2).

The high permeability of PMSP is due to the high diffusivity, which depends on a large excess free volume compared with other polymers. However, as this large excess free volume decreases, the diffusivity is reduced by physical aging. Controlling the free volume relaxation allows stabilization of the PMSP membrane through copolymerization with 1-phenyl-1-propyne (PP) (6,8) and blending with poly(1-phenyl-1-propyne) (8).

The objective of this investigation is to describe the effect of physical aging on the molecular motion of the membranes of PMSP and poly(1-trimethyl-1-propyne-co-1-phenyl-1-propyne) [poly(TMSP-co-PP] and blend polymer of PMSP with poly(1-phenyl-1-propyne) (PPP).

The chemical structures of PMSP, poly(TMSP-co-PP), and PPP are shown in Figure 1.

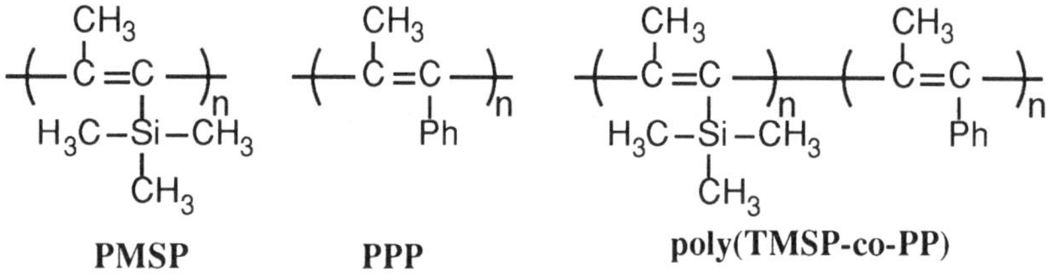

PMSP   **PPP**   **poly(TMSP-co-PP)**

Figure 1. Chemical structures of PMSP, PPP, and poly(TMSP-co-PP).

Experimental

Materials The polymers used were the same PMSP, poly(TMSP-coPP), and PPP previously synthesized (6,8), having an average molecular weight between 80 x 10^4 and 100 x 10^4. All membranes, including blends of PMSP and PPP, were cast on a horizontal glass plate from a solution of the polymer in toluene. They were immersed in methanol just before several measurements to prevent aging of the membranes.

The drying conditions influenced the gas permeation properties of the PMSP membranes; therefore, the membranes were dried under vacuum according to the same method as previously described (6).

Gas sorption and permeation measurements The gas sorption and permeation measurements were performed according to the same method described in previous studies (6,7). The gas sorption isotherms and the gas permeabilities were determined by the gravimetric method and the vacuum-pressure method, respectively. A silicone rubber packing that contained no blending agents was used for the gas permeation measurement.

Aging condition The PMSP membranes were aged at 30°C in an evacuated vessel by an oil vacuum pump and a cold trap with liquid nitrogen which was placed between the pump and the vessel. The vacuum pressure was controlled between 10^{-2} and 10^{-3} mmHg. The vacuum pump oil used was Sanvac oil No.160 (the Asahi Vacuum Chemical Industry Co., Ltd.), having a vapor pressure of 4.1 x 10^{-6} mmHg at 10°C.

NMR measurements The high resolution, solid state NMR measurements were performed at 30°C on a JEOL JNM 400 spectrometer. Sample membranes were cut into small pieces having an area of about 5 x 5 mm and packed into each sample tube. The spin lattice relaxation time (T_1) of carbons and silicon were obtained by the CPT_1 pulse sequence, operating at a spinning speed of 6 kHz.
The measurements were carried out for more than three different membranes prepared at the same time from the same cast solution.

Results and discussion

Polymer Characterization The DSC curves of the PMSP, poly(TMSP-co-PP) showed no thermal change up to 280°C (6). The blends of the PMSP and PPP also had no thermal change during the DSC measurements over the same temperature range. The results of dynamic viscoelastic measurement by Masuda also showed no glass transition between -150 and 250°C. It is suggested that the Tg of these polymers is in excess of 280°C such that these polymers were glassy under the conditions used in the gas permeation, gas sorption, and solid-state NMR measurements in this study.

Sorption isotherm for the aged PMSP Sorption isotherms for N_2, O_2, CH_4, and CO_2 in PMSP membrane at 35°C have been reported by one of the authors (9). Sorption isotherms for C_3H_8 in the initial and aged PMSP at 35°C are shown graphically in Figure 2 in the form of plots of the concentration of c of the penetrant dissolved in the polymer versus the penetrant pressure p. Such isotherms are accurately described by the dual-mode sorption model (10):

$$c = c_D + c_H = k_D p + C'_H bp/(1 + bp) \quad (1)$$

where c_D is the concentration of the penetrant population in the Henry's domain and c_H is the penetrant population dissolved in the unrelaxed domain or Langmuir's mode domain in the form of pre-existing microcavities; k_D is a solubility coefficient in the Henry's mode domain, and C'_H and b are a microcavities saturation constant and Langmuir affinity constant in the Langmuir's mode domain. C'_H can be taken as a measure of the excess free volume in this domain.

As can be seen, with increasing the aging, the sorption of propane in the aged PMSP decreased. These isotherms were analyzed using the dual-mode sorption

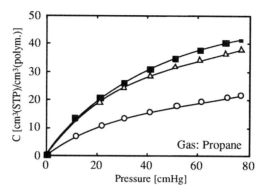

Figure 2. Sorption isotherms for propane in the initial and aged PMSP membranes at 35°C; initial (■), aged for 14 days (△), and for 30 days (○). Aging condition: stored in a vacuum vessel for 14 and 30 days.

model. The dual-mode sorption parameter C'_H as calculated using equation 1 and decreased with time: that is 54.4 in $cm^3(STP)/cm^3$(polymer) for one day, 40.0 for 14 days, and 22.3 for 30 days. A decrease in the C'_H means that the membranes undergo excess volume relaxation with aging time.

Effect of aging on molecular motion of PMSP Figure 3 shows the ^{13}C T_1 in the initial and aged (14 days) PMSP membranes. The error-bar shows the maximum and the minimum values through the repeated measurements. The side chain protonated carbons of C_a and C_b have smaller T_1 values compared to the backbone chain non-protonated carbons of C_c and C_d (11), which means the molecular motion of carbon C_a and C_b was much faster than that of C_c and C_d. During aging, the change in the T_1 values of the side chain carbons was smaller than those of the

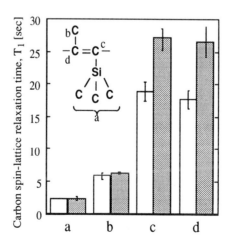

Figure 3. ^{13}C T_1 in the initial (open) and aged (filled) PMSP. Aging condition: stored in a vacuum vessel for 14 days.

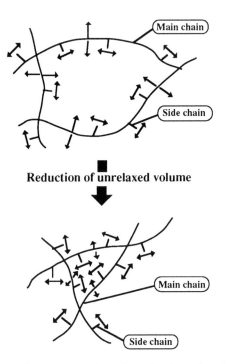

Figure 4. Schematic representation of mechanism of molecular chain relaxation.

backbone chain carbons. These results suggest that the molecular motions in methyl groups in side chains were not reduced by aging. Although relaxation of the unrelaxed volume occurred, the aged PMSP maintained a larger unrelaxed volume compared to other conventional glassy polymer. T_1 value was reversible when the aged sample was re-cast.

This NMR result may be due to a change in the degree of twisting of the backbone chains during aging. A schematic representation of the morphological change for molecular chain relaxation is shown in Figure 4.

Effect of aging on the permeability and molecular motion of the membranes of PMSP, poly(TMSP-co-PP) and blend of PMSP/PPP Glassy polymers, such as PMSP, are nonequilibrium materials and their permeation and sorption properties drift over time as thermally driven, small-scale polymer segmental motions cause a relaxation of nonequilibrium excess free volume. The microcavities of large size which are present in PMSP membrane have been considered to be responsible for the decay of C'_H and the gas permeability (4). Therefore, it is possible to stabilize the gas permeability by control the C'_H by copolymerization or blending with the other acetylene derivatives such as PP and PPP, respectively.

Fig. 5 shows the effect of aging time on the permeability coefficient for oxygen in PMSP, poly(TMSP-co-PP), and PMSP/PPP blend membranes at 30°C. This figure also shows the effect of PP content in the copolymer and PPP content

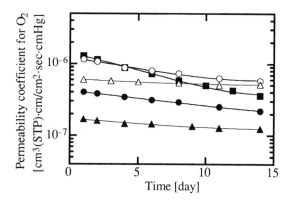

Figure 5. Effect of aging on the permeability coefficient for O_2 in PMSP, copolymer, and blend membranes. Aging condition: stored in a vacuum vessel at 30°C. PMSP (■), poly(TMSP-co-PP) 95/5 (○), 80/20 (●), blend PMSP/PPP 95/5 (△), and 80/20 (▲).

Table I Dual-mode sorption parameter, C'_H, for CO_2 and C_3H_8 in membranes homopolymer, copolymer, and blend polymer at 35°C

Polymer	C'_H [cm³(STP)/cm³(polymer)]	
	CO_2	C_3H_8
PMSP	151	50.1
PMSP[a]	51.0	40.0
Poly(TMSP-co-PP) 95/5	116	45.0
Blend PMSP/PPP 95/5	134	40.0
PPP	28.5	12.5

a: Aged under vacuum for 14 days.

in the blend polymer. Of course, the higher the PP or PPP content, the lower the permeability coefficient, because the gas permeability of homopolymer PPP is lower than that of PMSP (6,8).

The most stable membrane in this study for the oxygen permeability was the blend 95/5 (PMSP/PPP). The phase separation was observed in the blend PMSP/PPP which contained the higher content of PPP than 5 wt%.

The C'_H of these membranes for CO_2 and C_3H_8 are summarized in Table I. Interestingly, although the initial C'_H value of CO_2 in the blend polymer, which is the most stable polymer, is higher than that of copolymer, the opposite results appeared in the C_3H_8 case. However, C'_H value is always total volume for the related microvoids. The microvoids of glassy polymers have a size distribution. A CO_2 molecule is small relative to a C_3H_8. If we compared with the C'_H value for C_3H_8, C'_H in the blend is smaller than that of the copolymer. This result suggests that there are a small number of large size microvoids in the blend, because C_3H_8

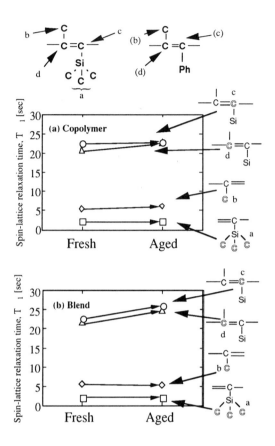

Figure 6. Effect of aging on ^{13}C T_1 in poly(TMSP-co-PP) 95/5 and blend PMSP/PPP 95/5. (a) poly(TMSP-co-PP), (b) blend PMSP/PPP.

molecules would not be able to sorb in the small size microvoids.

Figure 6 shows the comparison of the spin-lattice relaxation time, T_1, for the aged poly(TMSP-co-PP) and the aged blend PMSP/PPP with those of the corresponding fresh polymers. As can be seen, the T_1 of the C_c and C_d in both the copolymer and the blend was not changed,　similar to the result found for PMSP. The T_1 of the C_c and C_d in the copolymer was not changed either. The volume relaxation may be prohibited by the added phenyl group, suggesting a stacking effect of the phenyl groups. It is considered that the copolymer still has the enough space for the molecular motion of the backbone carbons after aging. However, a distinct increase in T_1 was observed at the C_c and C_d, backbone carbons, in the blend. Considering the results of C'_H, the very small morphological change due to aging in the blend, it is thought that the decrease of the microvoids of the larger size in the blend caused the slower molecular motion.

The gas permeability of these membranes with rich PMSP structure depended on the larger free volume　rather than on molecular motion in the T_1 measurement.

Conclusions

Poly(1-trimethylsilyl-1-peopyne), PMSP, contains an unusually large excess free volume, which may be also responsible for the high gas permeability and its decay with aging. Molecular motion, which is evaluated by the spin-lattice relaxation time, T_1, of the backbone carbons was increased with age, but molecular motion of side chain was very stable after aging. This means that the larger molecular scale gap between polymer segments still exists in the aged PMSP membrane.

Modification of PMSP for the stabilization of gas permeability by the copolymerization or blending with phenyl-containing acethylene derivative was successful. Especially, the blend PMSP/PPP 95/5 is very stable with high gas permeability. The change of molecular motion of the backbone carbons in the blend after aging is different than that of the copolymer. This suggests that the initial non-equilibrium state of the glassy PMSP, copolymer and blend membranes was different and approached a more stable state with age.

Literature Cited
1. Masuda, T; Isobe, E.; Higashimura, T.; Takada, K. *J. Amer. Chem. Soc.* **1983**, 105, 7473.
2. Nakagawa, T.; Saito, T.;Asakawa, S.; Saito, Y. *Gas Separation and Purification* **1988**, 2, 3.
3. Asakawa, S.; Saito, Y.; Waragai, K.; Nakagawa, T. *Gas Separation and Purification* **1989**, 3, 117.
4. Nakagawa, T.; Fujisaki, S.; Nakano, H.; Higuchi, A. *J. Memb. Sci.* **1994**, 94, 183.
5. Nagai, K.; Higuchi, A.; Nakagawa, T. *J. App. Polym. Sci.* **1994**, 54, 1353.
6. Nagai, K.; Higuchi, A.; Nakagawa, T. *J. Polym. Sci., Polym. Phys. Ed.* **1995**, 33, 289.
7. Nagai, K.; Nakagawa, T. *J. Memb. Sci.* **1995**, 105, 261.
8. Nagai, K.; Mori, M.; Watanabe, T.; Nakagawa, T. *J. Polym. Sci., Polym. Phys. Ed.* **1997**, 35, 119.
9. Ichiraku, Y.; Stern, S. A.; Nakagawa, T. *J. Memb. Sci.* **1987**, 34, 5.
10. Vieth, W. R.; Howell, J. M.; Hsieh, J. *J. Memb. Sci.* **1976**, 1, 177.
11. Costa, G.; Grosso, A.; Sachi, M. C.; Stein, P. C.; Zetta, L. *Macromolecules* **1991**, 24, 2858.

Chapter 23

Vacancy Spectroscopy of Polymers Using Positronium

Yasuo Ito

Research Center for Nuclear Science and Technology, The University of Tokyo, Tokai, Ibaraki 319-11, Japan

Characteristics of the methods of estimating vacancies in polymers from the lifetime and intensity of ortho-positronium (o-Ps) are described. From quantum mechanical consideration and from recent experimental results it can be shown that o-Ps is "a seeker and digger" of holes especially in rubbery states leading to significant overestimation of the size of vacancies. In glassy states the vacancy information will be less affected by o-Ps. Although this active nature of o-Ps might look undesirable for direct measurements of vacancies, it can be used to gain additional information about the polymer. Examples of o-Ps spectroscopy applied to the detection of glass transitions, probing sorption sites, probing vacancies in conditioned polymers, and detection of crystallization sites are presented. The o-Ps spectroscopy data are also compared with gas permeation data. In the sorption studies, the o-Ps lifetime and intensity respond in contrastingly different ways for Langmuir-type and Henry-type sorptions. In all these examples o-Ps brings forth unique information suggesting the usefulness and powerfulness of "vacancy spectroscopy" using o-Ps.

It is well recognized that the positron (e^+) is a sensitive probe of vacancy type defects in metals and semiconductors since it can be easily trapped in sites where positive ions are missing. A standardized method like the trapping model is frequently used to extract information about the size and number of the defects. Similarly positronium (Ps), a neutral particle composed of an e^+ and e^- pair, has also been noted as a unique indicator of vacancies especially in insulators like polymers and oxides (2). However due to the lack of quantitative knowledge about the behavior of Ps with regard to vacancy characteristics, it has taken some time for Ps to be adopted as a practical laboratory tool. A general reference of positron and positronium spectroscopy can be found, for example, in (1) and (2), respectively. Quite recently a simple model describing the relationship between the Ps lifetime and the size of vacancies was developed (3,4), and spectroscopy of vacancies using Ps is becoming more frequent as can be seen from the fact that the related topics occupy approximately 40% of the presentations of PPC-5 (Int. Workshop on Positron and Positronium Chemistry) (5). There are still fundamental problems to be studied before Ps can be used as a mature analytical tool, but there are already various excellent works in that direction. This article describes the information o-Ps gives concerning the vacancy characteristics in glassy polymers.

Formation and annihilation of Positronium

When positrons, either from β^+-decay radioisotopes or from nuclear reactions in accelerator facilities, are injected and stopped in polymers, they undergo a sequence of processes including formation and trapping of Ps into vacancies and eventually are annihilated emitting gamma-rays. Ps can be formed with a probability which depends on the physical and chemical conditions of the substance. There are two sub-states of Ps depending on the e^+ and e^- spin configuration: para-positronium with the spins anti-parallel (p-Ps, S=0) and ortho-positronium with the spins parallel (o-Ps, S=1). The former annihilates rapidly with the mean lifetime $\tau_1 \sim 0.12$ ns and composes the shortest lived component of the annihilation spectrum (see Figure 1). The lifetime of o-Ps is 140 ns in vacuum, but in condensed media it is substantially reduced to several ns due to the overlap with electrons from the surrounding molecules. This annihilation mode, called the "pick-off" annihilation, constitutes the main part of the o-Ps lifetime and is important with regard to the measurement of the vacancy size. The lifetime, τ_3, and its intensity, I_3, of this component are the quantities that can be measured with high accuracy, and it is customary to use these quantities to study behaviors of o-Ps. The intermediate component with $\tau_2 \sim 0.5$ ns is attributed to the annihilation from bare e^+ and from compounds containing e^+ or Ps. In addition to the lifetime measurement there are methods to measure the momentum of annihilating pairs, but for the sake of simplicity we will not deal with them in this article.

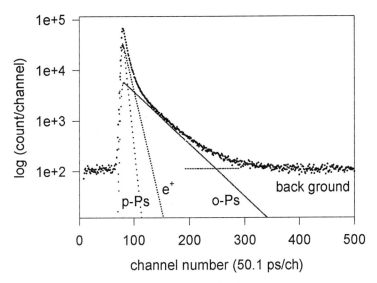

Figure 1. An example of the positron lifetime spectrum in polymers; a polycarbonate. The abscissa is the channel number of the pulse height analyzer and is proportional to time, and the ordinate is the logarithm of the count at each channel. The spectrum is composed of three exponential decay component as shown by the dotted lines, and the longest lived one is the o-Ps annihilating in vacancies.

Ps formation in polymers is not completely understood. Basically Ps can be formed in two ways. In the course of thermalization (energy loss) in polymers e^+ passes through an energy region, called the Ore gap, where e^+ can efficiently pick up an electron from the molecule M to form Ps as,

$$e^{+*} + M \quad --\rightarrow Ps^* + M^+ \tag{1}$$

Since the Ps binding energy (I_{Ps}=6.8 eV in vacuum) is smaller than the ionization energy of the molecule, I_M, this is an endothermic reaction and there is an energy threshold (E_{th}=I_M-I_{Ps}) for e^+ below which Ps formation is not possible. In this epithermal process the nascent Ps has an energy ranging up to several eV, *i.e.* the Ps is a hot atom. This takes place in a special space region, called the "short track" *(6)*, where the e^+ energy loss process is almost ending accompanied by an exceptionally high density of energy deposition. It can be shown by a crude estimation that approximately 1 mole/dm^3 of ion pairs are formed in the short track. In such an environment it is difficult for the hot Ps to survive reaction with the positive ions, and Ps is most probably oxidized to bare e+ (or electron transfer) as;

$$Ps^* + M^+ \quad --\rightarrow \quad e^+ + M \tag{2}$$

The next possible way of Ps formation is the recombination of e^+ with one of the excess electrons produced in the short track (Ps formation by the spur process).

$$e^+ + e^- \rightarrow \quad Ps \tag{3}$$

This process is quite akin to geminate recombination in radiation chemistry and indeed many experimental results show excellent parallelism between the data of Ps formation and radiation chemistry *(7,8)*. The main part of Ps we observe in polymers appears to come from reaction (3), but if the epithermal Ps *via* reaction (1) survive the oxidation reaction (2) it will also show up. The energetics of Ps formation *via* the spur process are given by,

$$-E_{e+} - E_{e-} + E_{Ps} \leq I_{Ps} \tag{4}$$

where E's are the energy levels, measured from the vacuum level, of the precursors and the product. In polymers e^+ and e^- may be sitting on certain energy levels corresponding to their work functions. It is difficult to estimate E_{Ps} since we do not know to which state in the vacancy Ps is formed at first. We suggested before *(9,10)* that in polymers where there are many different holes Ps may be resonantly formed into virtual levels in them. The level is higher the smaller the vacancy size, and hence there can be a critical hole size below which Ps formation is not possible. This critical size is dependent on the energy levels E_{e+} and E_{e}, and hence there can be an interplay between the energy levels and the hole size. When the energy level of e^- becomes lower (E_{e-} takes a larger negative value), the critical hole size must be shifted to a larger one and eventually for a fixed hole population the number of holes available for Ps formation becomes smaller. Indeed it is a well recognized fact that molecules with large electron affinity lead to a smaller Ps formation probability *(11)*. This effect can sometimes lead to a negligibly small Ps formation probability as in Kapton *(12)*.

How Positronium Sees Polymers

Ps is a neutral light particle composed of e^+ and e^-. Since it is a free radical it may undergo reactions with paramagnetic species and electron acceptors *via* oxidation, compound formation, and spin exchange reactions.

$$o\text{-Ps} + X \rightarrow e^+, PsX, p\text{-Ps} \tag{5}$$

The rate of this reaction is expressed by a pseudo-first order reaction as $\lambda = k[X]$. In diamagnetic substances, however, Ps feels strong repulsive force based on the Pauli's exclusion principle. Hence in crystals Ps will exist in interstitial sites as a Bloch wave, but in substances having density fluctuation it is trapped in open space. Describing the Ps state in the interstitial or in the trapped states by a spherical potential well with infinite height, its zero-point energy is given as,

$$E_0 = \pi^2 h^2/4m_e R^2 = 0.188 / R^2 \quad (E_0 : eV, R : nm) \tag{6}$$

where R is the radius of the well and m_e is the static mass of the electron. It must be noted that, due to the small mass ($=2 m_e$) of Ps, E_0 is as large as several eV for R normally found in polymers. An important consequence of the high zero-point energy is that Ps can expand the size R with the energy required to do so compensated by the decreased E_0. A well known example is the "Ps bubble" in liquids which Ps creates by pushing aside the surrounding molecules and is self-trapped by virtue of the lowered E_0.

It is not an established understanding that Ps does the same in solids, but, as will be made clear later in this article, there is a good reason to expect that Ps can "dig" holes in rubbery polymers. However in glassy polymers Ps will not be able to dig with the same ease. Another consequence of the high zero-point energy is that Ps may "seek" larger holes. In polymers the size of holes is normally of the order of $R \sim 0.1$ nm and, according to equation 6, a slight change in R of about 0.2% leads to an energy shift larger than thermal. Thus Ps is not able to move into smaller holes but is preferentially transported to larger ones, if any. Good evidence that Ps is "digging and seeking" holes is found in our recent work on the solids of low molecular weight compounds, where o-Ps was found to be in a large vacancy as if the solid is in a super-cooled liquid state *(13)*.

The lifetime of o-Ps is several ns in polymers. This time scale is faster than most of the segmental motions and Ps will see the polymer chains as almost static even in the rubbery polymers. Ps will probe this static vacancy distribution with its "digging and seeking" flaw character. The vacancy information brought forth by Ps must be received with a precaution, but we still expect the information is valuable. As will be discussed later it is also possible to use the "digging and seeking" nature of Ps as an active probe.

Once o-Ps is confined in a hole it stays there colliding many times on the wall until an electron, having anti-parallel spin to the e^+ spin, in the wall meets e^+ and is annihilated (pick-off annihilation). Theoretically the rate of this pick-off annihilation is proportional to the overlap integral of the e^+ wave function with those of external electrons. In a simple but useful model a spherical potential well is assumed for the hole and the external electrons are dealt with as an electron layer pasted over the wall with a thickness ΔR. The o-Ps lifetime is then given as *(3)*;

$$\tau_{pick\text{-}off} = 0.5 \left[1 - R/(R + \Delta R) + \sin(2 \pi R/(R + \Delta R))/2 \pi \right] \quad ns \qquad (7)$$

where $\Delta R = 0.166$ nm was found to give $\tau_{pick\text{-}off}$ agreeing well with the experimental τ_3 values *(4)*. Thus equation 7 gives a simple means to calculate the size of the hole, v_{Ps}, from the measured τ_3 value through $v_{Ps} = 4 \pi R^3/3$. Due to the "digging and seeking" nature it is not guaranteed that v_{Ps} thus obtained represents the intrinsic hole size. This issue must be examined separately.

The o-Ps lifetime is the sum of the chemical reaction term and the pick-off term.

$$1/\tau_3 = \lambda_3 \sim \lambda_{pick\text{-}off} + k[X] \qquad (8)$$

It must be noted that the experimental τ_3 value can be equated to the pick-off term *via* equation 7 only when one is certain that the chemical reaction term is not playing an important role. Quite often the chemical reaction term can not be neglected. Examples are found in the dependence of τ_3 on electron affinities of the acid anhydride moieties of polyimides *(12)*, reduction of o-Ps lifetime and intensity in baked and carbonized polymers (Tanaka, K., Yamaguchi Univ., unpublished data.). In these cases τ_3 no longer represents the hole size through equation 7.

The o-Ps intensity I_3 is the fraction of positrons that formed Ps and are trapped in the holes. From many evidences it appears to be correlated with the number of the holes, but no quantitative relationship between them is derived yet. One will easily understand the difficulty considering the complicated processes of Ps formation and trapping into holes. In a most crude case, however, it is assumed that I_3 is proportional to the number of holes. In such a case the free volume fraction v_f is equated to a product of the size of the o-Ps hole, v_{Ps}, and I_3 as $v_f = a \cdot v_{Ps} \cdot I_3$, where a is the proportionality factor *(14)*. In some reports it is claimed that this simple treatment works well. This is probably because all the complicated factors related to Ps formation and trapping processes and the "digging and seeking" nature are rounded off in the proportionality factor. However it is not mature to generalize this kind of treatment since we do not know the details of the processes leading to I_3.

Applications of Positronium as the Probe of Vacancies

Glass Transition   One of the direct and simplest applications of Ps is the detection of glass transition as illustrated in Figure 2 for poly(vinyl alcohol) *(15)*. PALS was measured at various temperatures and the measured o-Ps lifetime τ_3 was converted through equation 7 to the mean size of the holes in which o-Ps is trapped. Evidently there are two slopes, and the crossing point agrees very well with the known value of Tg. The slopes correspond to thermal expansion of v_{Ps} and the expansion coefficients $\alpha(v_{Ps})$, which are 6×10^{-4} and 4×10^{-3} deg^{-1} for glassy and rubbery states, respectively, are about one order of magnitude larger than the macroscopic expansion coefficients $\alpha(v_B)$. The larger coefficient for v_{Ps} than for v_B is not surprising because o-Ps is measuring the expansion of the open spaces itself. If we make a crude assumption that the bulk specific volume v_B is the sum of occupied volume and free volume ($v_B = v_0 + v_f$), and that the size of the Ps hole is proportional to the free volume fraction ($v_{Ps} \propto v_f$), it is straightforward to deduce $\alpha(v_{Ps})/\alpha(v_B) = 1/v_F$. This may explain as a first approximation why the expansion coefficient of the o-Ps hole is larger than the bulk

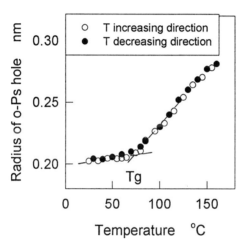

Figure 2.   Temperature dependence of the size of o-Ps holes obtained from the PALS measurements for polyvinyl alcohol.   The crossing point corresponds to the glass transition, Tg.

expansion, but the actual situation is a little more complicated. To see this we refer to the data of poly(propylene) *(16)*, for which it is reported that below Tg the values of α (v_{Ps}) and α (v_B) are 1.7×10^{-3} and 2.2×10^{-4}, respectively. The ratio is 8, which may properly be compared with $1/v_F$. However above Tg α (vPs) and α (vB) are 1.2×10^{-2} and 8.1×10^{-4}, and the ratio is 15. This larger ratio for the rubbery state is opposite to the expectation from the α (v_{Ps})/ α (v_B)=$1/v_F$ relationship, since v_F is larger in the rubbery state. This large α (v_{Ps})/ α (v_B) ratio for the rubbery state provides an evidence that o-Ps is digging holes with more ease in the rubbery states than in the glassy states.

Many data of Tg determined by the o-Ps method agree with those from the conventional methods. A compilation of Tg data obtained from PAL is found in *(17)*. The beauty of the o-Ps method lies in the ease of experiments. Since only the high energy annihilation gamma-rays have to be measured from outside, it is possible to introduce various experimental conditions to the sample polymer. For example τ_3 can be measured in vacuum or by introducing guest molecules as sorption gases, by imposing external pressure, and so on.

Sorption   Probing sorption states by o-Ps is also promising. In our first report of this research polyimide (6FDA-TMPD, Tg=377℃) and LDPE (Tg=-27℃) were measured by introducing vapors like hexane, cyclohexane, benzene, *etc.* *(18).*The experiment was simple as shown in Figure 3. A glassware having two arms connected with a stopcock was used. In one of the arms was placed a positron source (about 0.5MBq of ^{23}NaCl sealed in a Kapton foil) sandwiched by two identical pieces of sample polymer ($10 \times 10 \times 1$ mm^3). In order to ensure efficient diffusion of vapor molecules thin foils of the polymer were used and were stacked together to make thickness of about 0.5 mm, which is necessary to stop all the positrons in the sample. The liquid to be sorbed was put in the other arm. After evacuating the whole system, the stopcock connecting the two arms was opened and PALS measurements were performed.

Figure 4 shows an example of the changes of τ_3 and I_3 after introducing benzene as the sorbed molecules, but the basic tendency did not depend on the kind of the vapors used. In polyimide both τ_3 and I_3 dropped spontaneously, and then followed their gradual and delayed rise. The spontaneous drop is due to the Langmuir-type sorption in which the vapor molecules fill in the preexisting vacancies. In such conditions o-Ps finds less number and smaller size of vacancies. In LDPE, on the other hand, both τ_3 and I_3 simply increased. This result is attributed to the typical Henry-type sorption where the sorbed molecules dissolve into the chains and participate in the micro-

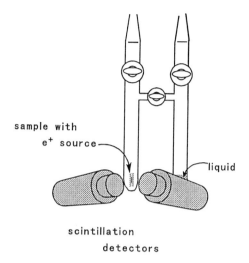

Figure 3. The experimental setup for the PALS measurements of sorption of liquid vapor in polymers. The positron source together with the sample polymer is contained in one arm of the glassware and the liquid to be sorbed is contained in the other arm. The positron lifetime measurement is performed by detecting the gamma-rays emitted from the source and from positron annihilation in the sample.

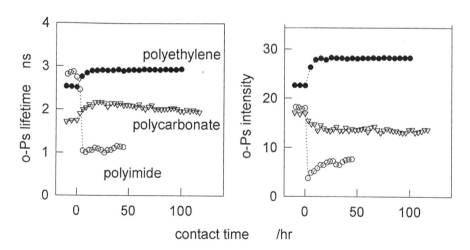

Figure 4. Changes of o-Ps lifetime and intensity during the course of sorption of benzene vapor by polymers.

Brownian motions. As a result of the broadened free volume distribution it becomes possible for o-Ps to find more and larger vacancies. The delayed rise of τ_3 and I_3 in polyimide indicates that the Henry-type sorption is also taking place after the Langmuir-type sorption process is almost terminated. It is important to note that, unlike all the other conventional methods of sorption studies, the o-Ps method can discriminate the Langmuir-type and the Henry-type sorption processes as contrastingly different effects.

The result for PET (poly(ethylene terephthalate)) is also shown in Figure 4. PET is a glassy polymer at room temperature and we would expect that τ_3 and I_3 behave in a similar way as in the polyimide. In fact I_3 decreased, but the rise of τ_3 is not what was expected. Apparently there is something more than the simple picture present thus far of the Langmuir- and Henry-type processes

Similar sorption experiments were performed using CO_2 as the sorbed molecules *(19)*. Here, in addition to the measurements of τ_3 and I_3, the sorption-related quantities, the amount of sorbed CO_2 and expansion of bulk volume, were measured and the free volume fraction under the sorbing conditions was estimated. Typical results are shown in Figure 5. In LDPE the o-Ps lifetime τ_3 and hence the size of the Ps holes rose with CO_2 pressure. The amount of the sorbed CO_2 and the bulk volume were increased, too, and the free volume fraction became larger gradually.

The results can be compared to those for polyimide (PI). The o-Ps hole size dropped at first and rose thereafter. This is similar to the results for benzene sorption in polyimide (Figure 4). The amount of sorbed CO_2 and the bulk volume were increased with the CO_2 pressure, and the free volume fraction became smaller gradually. The results for polycarbonate (PC) are similar except that the free volume fraction remained almost constant. It is puzzling why the size of the o-Ps hole can become larger at the later stage of sorption for PI and PC. Particularly for PC the size of the o-Ps hole became larger than before the sorption. It is reported that PC is still glassy as bulk when sorbing 50 atm of CO_2 *(20)*. The puzzle may be solved by remembering the "digging" nature of o-Ps. CO_2 molecules may at first be deposited at possible vacancy sites, but they will not just fill in the pores. They gradually dissolve into the chains nearby causing local plasticization, and this local softening of the chains is probed by o-Ps through its digging nature. Ps appears to be by no means a gentle probe. It must rather be regarded as a wild and active probe of holes. At a glance this active nature might seem to diminish the usefulness of Ps as a probe, but it is possible to make the best use of this particular nature. Study of the local plasticization as mentioned above is one of the promising applications. In ideally glassy polymers, *i.e.* when the chains are frozen nearly completely, o-Ps would bring forth less modified information of the holes.

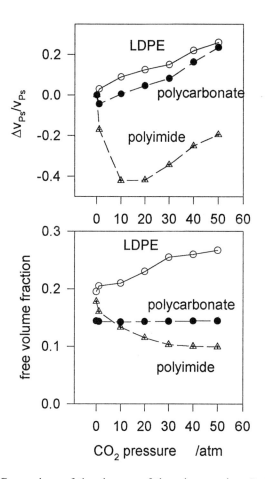

Figure 5. Comparison of the changes of the microscopic o-Ps hole size (upper figure) and the macroscopic free volume fraction (lower figure) as a function of the pressure of the sorbed CO_2 gas. Note that the o-Ps hole size always rises at the later stage irrespective of the change in the macroscopic free volume fraction.

Conditioning of Polymers by CO_2 Gas   When a polymer is exposed to high pressure CO_2 gas for sorption and then the gas is evacuated (CO_2 conditioning), the vacancy structure may be changed.   The structural change can be observed as a change in the specific volume or in the gas permeation properties, but it would be more straightforward to use o-Ps from the microscopic viewpoint.   Figure 6 shows the variation of τ_3 and I_3 with time elapsed after the CO_2 conditioning for several polymers *(21)*.   For the polyimides both τ_3 and I_3 became larger than before the conditioning. When the polymer is sorbing CO_2, the region around the sorption sites should be strongly plasticized as explained in the previous section.   On evacuation of the gas, the polymer cannot resume the initial structure immediately because it is still glassy as bulk and the sorption sites are left with substantial open space, and as such are probed by o-Ps.   In all the polyimides I_3 relaxes back to smaller values, but not completely to the initial one. τ_3 also seems to decrease in 6FDA-6FAN but it rather rises in other polyimides.   It is not clear at present whether the latter slight rise is due to delayed evacuation of residual CO_2 gas or to some real structural change.   In summary in the CO_2-conditioned polyimides the structural relaxation appears to occur mainly by the decrease in the number, and not in the size, of the holes.   For polyethylene no difference is seen before and after the conditioning in both τ_3 and I_3.

It is possible to obtain information about the size distribution of the o-Ps holes for the CO_2-conditioned polymers.   Generally the PALS data are analyzed as the sum of several discrete exponential decay functions.   But for PALS data having high statistics each component can be developed further into a set of continuous lifetime distribution. A computer program CONTIN *(22)* allows us to extract the lifetime distribution, which is then converted to the size distribution using equation 7.   Figure 7 shows the size distribution thus obtained for some CO_2-conditioned polymers.   Clearly for the polyimides the size distribution has become broader after the conditioning, but for polyethylene, which is rubbery, the size distribution is not changed.

There have been several examples where physical aging were followed by PALS for polyethylene *(23)*, polycarbonate *(24)*, and polyvinyl acetate *(25)*.   Also Suzuki *et al.* measured PALS of various polyethylenes by giving them stepwise temperature freezing *(26)*, and observed slow relaxation composing at least of two components.   Due to lack of detailed knowledge of the factors that determine I_3 the interpretation was not straightforward.   It is anticipated that more effort is directed toward detailed understanding of I_3.

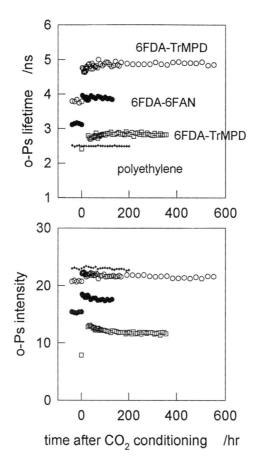

Figure 6.   The o-Ps lifetime and intensity before (time<0) and after (time>0) CO_2 conditioning of the polyimides and polyethylene. In the conditioning the sample polymer was immersed in 50 atm CO_2 gas overnight and then the gas was evacuated.   In the polyimides both o-Ps lifetime and intensity rise by the conditioning and show gradual change, while for polyethylene essentially no change is induced.

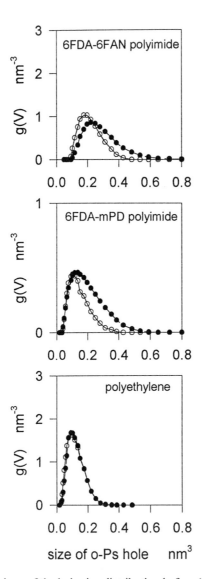

Figure 7. Comparison of the hole size distribution before (○) and after (●) CO_2 conditioning of polyimides and polyethylene. The experiments are the same as in Fig.6, but the data are analyzed to extract the distribution using the CONTIN program. (Reproduced with permission from ref.21, copyright The Chemical Society of Japan)

Crystallization of Glassy Polymer   Semicrystalline polymers are composed of amorphous and crystalline parts.   When there is no large space in the crystalline parts for o-Ps to live long enough, τ_3 and I_3 are related to o-Ps trapped in the amorphous parts.   This is the case we frequently encounter, and in such a case I_3 is excellently proportional to the amount of the amorphous parts; *i.e.* I_3 is inversely correlated with the degree of crystallinity.   An example is found for PEEK *(27)*.   There is certainly a case like polyethylene *(28)* where o-Ps appears to exist also in the crystalline parts with somewhat a smaller lifetime and different intensity than those in the amorphous part. In such a case τ_3 and I_3 are composite values containing contributions from different phase regions.   In either case, however, I_3 is reflecting the crystallinity.   But if only for the detection of the degree of crystallinity, o-Ps would not be a very valuable tool.   In this section we will show that information from o-Ps contains more than the degree of crystallinity.

Amorphous PET, prepared by quenching from the molten state, was annealed at 130°C for crystallization and PALS was measured as a function of the time of annealing *(29)*.   Together with the PALS measurements X-ray diffraction and specific density were measured. I_3 in the amorphous PET was 18.5% and after crystallization it decreased by 24% to become $I_3=14$%.   The value 24% was close to the crystallinity 25.8% and 25% obtained from the X-ray and specific density, respectively, and from this fact it is confirmed that there is no o-Ps in the crystalline parts.   The mean lifetime τ_3 slightly shifted to a larger value but it was not very significant.   To get more detailed information the PALS spectrum was analyzed using the computer program CONTIN as described in the previous section.   The results are shown in Figure 8.   Before crystallization the distribution of the o-Ps holes was broad ranging from 0.020 to 0.13 nm^3.   After crystallization the distribution became narrower and the main change is seen in the small size holes; i.e. the hole size ranging from 0.020 to 0.040 nm^3 has disappeared.   As has been discussed this distribution may not be exactly the same as the intrinsic free volume distribution, but we may conclude in a relative manner that the small size holes have been consumed by the crystallization.   It is probable that the small hole sites were the sites for nucleation of crystallization, since such regions would provide appropriate space for molecular displacement without requiring substantial energy as is claimed by the existing "solidification model" *(30)* of crystallization.

Comparison with the Gas Permeation Data   Since both τ_3 and I_3 represent quantities related to free space in polymers, they should be correlated with the gas permeation characteristics.   We have been systematically comparing PALS data and

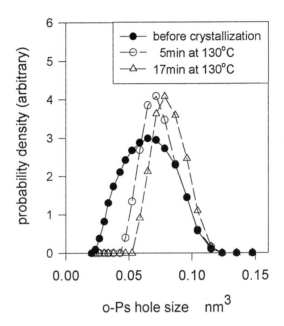

Figure 8. Change of the o-Ps hole distribution before (●) and after (○, △) crystallization of poly(ethylene terephthalate) by heat treatment. It is clearly shown that the small size holes have disappeared due to the crystallization. (Adopoted from ref.29)

gas diffusion constant D for both rubbery and glassy polymers (see Figure 9). For rubbery polymers τ_3 was found to be excellently correlated with $\log(D/T)$.[31-32] According to the free volume model for the diffusion $\log(D/T)$ is inversely proportional to the free volume per unit molecule, and the correlation between $\log(D/T)$ and τ_3 as in Figure 9 suggests that o-Ps is exactly probing the vacancies used for the gas diffusion. On the other hand a trial to find correlation for I_3 with vacancy related quantities was not successful for the rubbery polymers.

For glassy polymers at the temperature region between Tg and Tg-90K, the plot of τ_3 vs. $\log(D/T)$ was found on the same plot as for the rubbery polymers [33] (see Figure 9). This fact suggests that the free volume model for the diffusion, originally developed for rubbery polymers, applies to glassy polymers at temperatures not very far from Tg. This leads us to draw a picture that even in glassy polymers evolution of free volume holes are taking place and diffusional jumping of the penetrant molecules are allowed as in the rubbery polymers. Supported by this we may even extend our imagination that Ps is also able to dig holes in such glassy states. At temperatures substantially below Tg, however, the experimental points (the open symbols with + inside) are located a little deviated downward from the common correlation. This means that evolution of free volume is becoming less efficient. Probably the digging of holes by Ps will also be becoming less easy, but we are coming too far with this question.

Conclusion

PALS is an excellent probe of molecular size vacancies in glassy polymers and the o-Ps lifetime appears to represent the size of free volume holes fairly well at least in a qualitative manner. In rubbery polymers positronium appears to do some work on the vacancies and make the size larger, but the size information brought forth by o-Ps is still valuable provided we receive it well aware of the active nature of the o-Ps probe.

The o-Ps intensity contains information on the number of free volume holes but, since o-Ps intensity can be affected by many physico-chemical factors, it cannot be uniquely associated with the former. If we know that only the number of free volume holes has been changed, we may correlate a change in the o-Ps intensity with the hole number with some confidence. The opposite does not hold; we cannot *a priori* attribute a change in I_3 to changes in the number of free volume holes unless we are sure that no other factors are playing an important role. This imposes some limitation on the applicability of o-Ps intensity with regard to the number of holes, but still it is of great importance. In the near future when we know much better about the mechanism of Ps

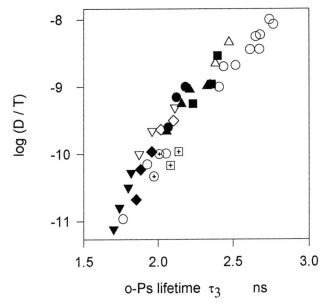

Figure 9. Plot of the effective diffusion constant of CO_2, D, through polymer films vs. o-Ps lifetime τ_3 for various polymers in rubbery (open symbols) and glassy (closed symbols). The open symbols with "+" are for glassy states very far from the glass transition point ($T<Tg-90$ degrees).

○●: poly(sulfone), ▽▼: poly(ethylene terephthalate), △▲: poly(styrene), ◇◆: polyimide (6GDA-DAN), □■: poly(carbonate)

formation and trapping into vacancies, "vacancy spectroscopy using o-Ps" will improve. It is also important always to bear in mind that the o-Ps lifetime τ_3 can contain chemical reaction terms as mentioned in section 3. Careless neglect of the chemical reaction terms leads to mistakes in interpretation of the o-Ps data. Understanding the pros and cons of Ps ensures the best use of it.

Acknowledgments   This article has been made possible with the keen and stimulating collaborations with Drs. Ken-ichi Okamoto, Kazuhiro Tanaka, Takenori Suzuki and Hamdy F. M. Mohamed.

Literature Cited

(1) *Positron Spectroscopy of Solids*, Dupasquier,A and Mills Jr. A.P., Eds.; IOS Press Ohmsha: Amsterdam-Oxford-Tokyo-Washington DC, 1995

(2) *Positron and Positronium Chemistry*, Schrader,D.M.; Jean, Y.C.; Eds., Elsevier: Amsterdam-Oxford-New York-Tokyo,1988

(3) Eldrup, M.; Lightbody,D.; Sherwood, J.N.; *Chem. Phys.* **1981,** 63, 51.

(4) Nakanishi, H.; Jean, Y.C.; Chapter 5 of ref.2

(5) Proc. PPC-5 was published as *J. Radioanal. Nucl. Chem.* **1996,** 210/211

(6) *Radiation Chemistry: Principles and Applications,* Farhatazis; Rodgers, M.A.J.; Eds.; VCH Publishers: New York, 1987

(7) Ito, Y.; Chapter 4 of Ref.2

(8) Mogensen, O.E.; *Positron Annihilation in Chemistry;* Springer Ser. Chem. Phys. 58; Springer-Verlag: Berlin-Heidelberg-New York, 1995

(9) Ito, Y.; Okamoto, K.; Tanaka, K.; *J. de Physique IV, Colloque C4, suppl. J. Physique II.* **1993,**3, 241.

(10) Ito, Y.; *J. Radioanal. Nucl. Chem.* **1996,** 210, 327.

(11) Goworek, T.; Rybka, C.; Wawryszuk. J.; Wasisewicz, R.; *Chem. Phys. Lett.* **1984,** 106, 482.

(12) Okamoto, K.; Tanaka, K.; Katsube, M.; Sueoka, O.; Ito, Y.; *Radiat. Phys. Chem.* **1993,** 41, 497.

(13) Ito, Y.; Mohamed, H.F.M.; Shiotani, M.; *J. Phys. Chem.* **1996,** 100, 14161.

(14) Deng, Q.; Sunder, C.S.; Jean, Y.C.; *J. Phys. Chem.* **1992,** 96, 492.

(15) Mohamed, H.F.M.; Ito, Y.; El-Sayed, A.M.A.; Abdel-Hady,E.E.; *Polymer,* **1996,** 37, 1529.

(16) Kristiak, J.; Sausa, O.; Bandzuch. P.; Bartos, J.; *J. Radioanal. Nucl. Chem.* **1996,** 210, 563.

354

(17) Ito. Y.; Okamoto, K.; *Kobunsi Ronbunshu* (in Japanese) **1996,** 53, 592.

(18) Ito, Y.; Sanchez, V.; Lopez, R.; Fucugauchi, L.A.; Tanaka, K.; Okamoto, K.; *Bull. Chem. Soc. Jpn.* **1993,** 66, 727.

(19) Ito, Y.; Mohamed, H.F.M.; Tanaka, K.; Okamoto, K.; Lee, K.H.; *J. Radioanal. Nucl. Chem.* **1996,** 211, 211.

(20) Hathisuka, H.; Sato, T.; Imai, T.; Tsujiya, Y.; Takizawa,A.; Kinoshita, T.; *Polym. J.* **1990,** 2,77.

(21) Tanaka, K.; Ito, M.; Kita, H.; Okamoto, K.; Ito, Y.; *Bull. Chem. Soc. Jpn.* **1995,** 68, 3011.

(22) Gregory, R.B.; Zhu, Y.; *Nucl. Instr. Methods* **1990,** A172, 290.

(23) McGervey, D.M.; Panigraph, N.; *Proc. 7-th ICPA.* **1985,** 690.

(24) Hill, A.J.; Jones, P.J.; Pearsall, G.W.; *J. Polym. Sci.* **1988,** A26, 1541.

(25) Kobayashi, Y.; Zeng, W.; McGervey, J.D.; *Proc. 8-th ICPA.* **1988,** 812.

(26) Suzuki, T.; Miura, T.; *J. de Physique IV, Colloque V4, suppl. J. Physique II,* **1993,** 3, 283.

(27) Nakanishi, H.; Jean, Y.C.; Smith, E.G.; Sandarezki, T.C.; *J. Polym Sci.* **1989,** B27, 1419.

(28) Suzuki, T.; Oki, Y.; Numajiri, N.; Miura, T.; Kondo K.; Ito, Y.; *Radiat. Phys. Chem.* **1995,** 45, 797.

(29) Mohamed, H.F.M.; Ito, Y.; Imai, M.; *J. Chem. Phys.* **1996,** 105, 4841.

(30) Stamm, M.; Fischer, E.W.; Dettenmaier, M.; Convert, P.; *Faraday Discuss. Chem. Soc.* **1979,** 68, 263.

(31) Okamoto, K.; Tanaka, K.; Katsube, M.; Kita, H.; Ito, Y.; *Bull. Chem. Soc. Jpn.* **1993,** 66, 61.

(32) Tanaka, K.; Okamoto, K.; Kita, H.; Ito, Y.; *Polymer Journal,* **1993,** 25, 275.

(33) Tanaka, K.; Katsube, M.; Okamoto, K.; Kita, H.; Sueoka, O.; Ito, Y.; *Bull. Chem. Soc. Jpn.* **1992,** 65, 1891.

Chapter 24

Subnanometer Hole Properties of Cellulose Studied by Positron Annihilation Lifetime Spectroscopy

H. Cao[1], J.-P. Yuan[1], Y. C. Jean[1,3], A. Pekarovicova[2], and R. A. Venditti[2]

[1]Department of Chemistry, University of Missouri at Kansas City, Kansas City, MO 64110
[2]Department of Wood and Paper Science, North Carolina State University, Box 8005, Biltmore Hall, Raleigh, NC 27695

Two series of cellulose samples, Avicel and Whatman CF11 cellulose ball-milled powders with different crystallinity are studied below T_g temperature by using positron annihilation lifetime spectroscopy. A good correlation is found between ortho-positronium formation probability and crystallinity as measured by Fourier transform - infrared spectroscopy. Sub-nanometer hole distributions are found to be narrowed as a function of milling time. These are interpreted in terms of microstructural changes of cellulose.

Cellulose has been known to have highly ordered morphology. It does not exist as an entirely crystalline material, but forms in different phases with different degrees of order. Irregular amorphous regions intersperse between regular crystalline phases. Crystallinity is an important structural feature for cellulose which can influence some properties critical for technological applications, for example, tensile strength and water sorption ability. Many different techniques have been applied to measure the crystallinity, including physical, chemical, and sorption methods. X-ray diffraction is commonly used as the reference evaluation. Some new techniques, such as solid-state ^{13}C nuclear magnetic resonance (NMR) and Fourier transform infrared spectroscopy (FTIR) have been shown to be very promising to determine the crystallinity of cellulose (1-3).

In recent years, positron annihilation lifetime (PAL) spectroscopy has been demonstrated to be a special sub-nanometer probe to determine the free-volume hole size, fraction and distribution in a variety of polymers (4-9). In this technique, measured lifetimes and relative intensities of the positron and positronium, Ps (a bound atom which consists of an electron and a positron), are related to the size and fraction of sub-nanometer holes in polymeric materials. Because of the positive-charge nature, the positron and Ps are repelled by the ion core of polymer molecules and trapped in open spaces, such as holes, free volumes, and voids. The observed

[3]Corresponding author.

lifetime of o-Ps (the triplet state of Ps) is found to be directly correlated to hole size and the corresponding intensity could be a measure of relative number of holes.

For semi-crystalline materials, most work has been to correlate o-Ps intensity with crystallinity. While o-Ps lifetime is found to be nearly independent of crystallinity, o-Ps intensity decreases as the degree of crystallinity increases (4). These results suggest that o-Ps is preferentially formed in the free-volume holes of the amorphous phase. The Ps formation in defects or low electron density trapping sites in crystalline phase is another possible interpretation made by others (9). In this study, PAL measurements are performed on cellulose samples with degrees of crystallinity varied by controlled ball-milling. Our objective is to correlate the microstructural changes with hole properties investigated by PAL.

Experimental

Sample Preparation. Cellulose powders, Whatman CF11, short fibrous cellulose powder with mean particle size 50-350 µm (Whatman International, Ltd.) and microcrystalline cellulose Avicel, mean particle size 27.6-102 µm, crystallinity index 0.60 (Avicel PH101, obtained from FMC, Ireland), were milled separately in a vibratory mill filled to 80% of volume with steel balls for 0 to 60 min. After milling, the samples were rubbed gently through a 50 µm sieve. All the treatments were performed at room temperature which is far below T_g temperature. The PAL temperature dependence data up to 200 °C do not show T_g onset. Detailed information about sample preparation was described elsewhere(10).

Positron Annihilation Lifetime Spectroscopy. The positron annihilation lifetime spectra were acquired by detecting the prompt γ-ray (1.28 MeV) from the nuclear decay that accompanies the emission of a positron from the ^{22}Na source and the annihilation γ-rays (0.511 MeV). A fast-fast coincidence circuit of a PAL spectrometer with the time resolution of 0.27 ns was used for PAL measurements. The positron source was ^{22}NaCl sandwiched between Kapton foils. The sample powders without any packing or pressing were placed into the sample cell with the positron source sitting in the center. All the samples were measured under vacuum at 25 °C. Detailed description of PAL spectroscopy can be found elsewhere(4).

Mean Free-Volume Hole Size and Fraction. The obtained PAL data were analyzed to finite term lifetimes using the PATFIT program (11). In these cellulose samples, it was found that three lifetime results give the best χ^2 (<1.1) and most reasonable standard deviations. The shortest lifetime τ_1 was fixed to 0.125 ns which attributes to the self-annihilation of p-Ps (singlet Ps), and the intermediate lifetime ($\tau_2 \approx 0.4$ ns) is the lifetime of the positrons. The longest lifetime ($\tau_3 \approx 1$-2 ns) is due to the pick-off annihilation of o-Ps (triplet Ps). In the current PAL method, o-Ps is regarded only formed in the free-volume holes and τ_3 is directly correlated to the free-volume hole size by the following equation(12):

$$\tau_3 = \frac{1}{\lambda_3} = \frac{1}{2}\left[1 - \frac{R}{R_0} + \frac{1}{2\pi}\sin\left(\frac{2\pi R}{R_0}\right)\right]^{-1} \qquad (1)$$

where τ_3 is the o-Ps lifetime (ns), λ_3 is the o-Ps annihilation rate (ns^{-1}), R is the mean free-volume hole radius (Å), $R_0 = R + \Delta R$, and ΔR is the electron layer thickness, semiempirically determined to be 1.66 Å.

The fractional free volume f_v (%) is expressed as an empirically fitted equation (8):

$$f_v = A V_f I_3 \qquad (2)$$

where V_f (Å^3) is the mean volume of free-volume holes calculated by using spherical radius R from equation1, I_3 (%) is o-Ps intensity, and A is an empirical constant 0.0018 determined from the specific volume data.

Free-Volume Hole Distributions. Since the free-volume holes in polymers have a distribution, a PAL spectrum can be expressed in a continuous form:

$$N(t) = \int_0^\infty \lambda\alpha(\lambda)e^{-\lambda t}d\lambda + B \qquad (3)$$

where $\lambda\alpha(\lambda)$ is the probability density function (PDF) of the annihilation with annihilation rate λ. The computer program CONTIN has been widely used to obtain PDF(λ) vs. λ. It has been tested that CONTIN results strongly depend on the statistic factor (13). Usually 10 million total counts are required to get the reliable results. In this work, we used another popular program MELT (14,15) to perform the continuous lifetime analysis to the spectra of cellulose samples with the total counts of 1 million. It appears that MELT can give reliable distribution results even at relatively low total counts. Comparison of MELT and CONTIN will be reported in the future. In this paper we only report the lifetime distributions with MELT analysis.

From PDF(λ) vs. λ, one can easily obtain PDF(τ) vs. τ considering $\tau = 1/\lambda$. The long lifetime range $\tau \geq 0.6$ ns is defined as the o-Ps lifetime and from the correlation between o-Ps lifetime and free-volume hole radius R (equation 1), PDF(R) vs. R and PDF(V) vs. V can also be computed (16).

Results and Discussion

Crystallinity of each Avicel and Whatman CF11 sample was determined by deuteration FTIR. Detailed information was described in the previous paper (10). Figure 1 shows that crystallinity decreases with the milling time. The crystallinity of Whatman CF11 is 64.7% for the unmilled sample and decreases to 33.2% after 60 min of milling. The microcrystalline cellulose Avicel shows 58.8% crystallinity for the unmilled sample and 17.0% after 60 min milling.

Moisture sorption was measured for each cellulose sample as described in the previous paper (10). It is known that moisture regain of cellulose is proportional to the amorphous fraction, or the readily accessible portion of cellulose. Figure 2 shows that moisture regain increases with the milling time which indicates the increase of the amorphous fraction.

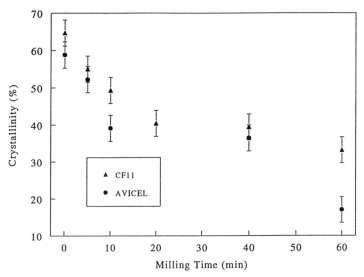

Figure 1. Crystallinity vs. milling time.

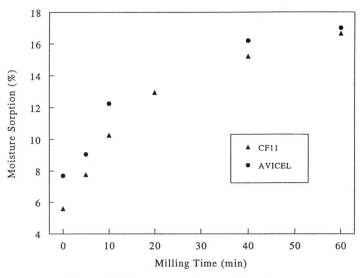

Figure 2. Moisture sorption vs. milling time.

The PAL results for the Avicel and Whatman CF11 cellulose were obtained for samples of varying ball milling time. The spectra were analyzed by PATFIT into three components. τ_1 was fixed at 0.125 ns, the lifetime of p-Ps. Part of the results of this work has been reported in previous paper (*10*). It has been found that τ_3 decreases and I_3 increases with the ball milling time. Figure 3 shows τ_3 vs. milling time for Avicel and Whatman CF11 cellulose powders. For both series, the values of τ_3 decrease significantly in the initial stage of milling. Further ball milling decreases the value of τ_3 to approximately the same value (1.25 - 1.30 ns) which then remains almost constant with longer milling time. By using equation 1, mean free-volume hole radii were calculated from τ_3 values and plotted in Figure 4. For both samples, the mean radii decrease significantly within the first 10 min of milling and then gradually decrease to a constant value of approximately 2.1 Å. This result suggests that ball milling makes the amorphous phase of cellulose powders reach a characteristic final state, which is similar for both Avicel and Whatman CF11 cellulose samples. The decrease in τ_3 or mean hole size with decreasing crystallinity differs from the previous observation that τ_3 does not change with the degree of crystallinity(*4*). It is possible that milling does not only reduce the crystallinity, but also changes the free-volume properties of the amorphous phase by generation of new free-volume holes of smaller size, and/or modification of the initial holes.

Figure 5 shows that I_3 dramatically increases for both samples in the early stage of milling and then approaches a constant value of about 30%. This result is consistent with the decrease of crystallinity and the increase of moisture sorption with the milling time and all the results indicate that the ball milling may increase the amorphous fraction of cellulose by disrupting the crystalline phase and restructuring it to amorphous phase. The similar I_3 at long milling times for both cellulose powders again suggests a similar final state for the two kinds of cellulose samples after extensive milling.

In order to further understand the microstructural changes during the milling, the Avicel data were analyzed using the program MELT to determine continuous lifetime distributions. Figure 6 shows the lifetime probability density function vs. τ for the Avicel sample with 10 min milling. The three peaks are consistent with the three lifetime components calculated by PATFIT. Since only the long lifetime component corresponds to the o-Ps annihilation in free-volume holes, PDF(τ) for the third peak is plotted in Figure 7 for each Avicel sample with different milling time. From the correlation between τ_3 and hole radius R, PDF(R) and PDF(V) were calculated and plotted in Figure 8 and Figure 9 respectively. It can be observed that the original Avicel cellulose has the broadest peak which means there exists a large variety of holes with different hole sizes in the amorphous phase of Avicel. With the first milling of 5 min, the peak shifts to smaller V and becomes narrower. This result suggests that the fraction of small holes dramatically increases and some original large size holes disappear after the milling. It can be explained by the two concurrent effects of milling, creation of new small size holes during the process of disrupting the original crystalline phase into an amorphous phase by ball-milling and modification of the initial large size holes into smaller size holes in the amorphous region by rearrangement of the polymer matrix during the milling. Those two effects contribute to the average τ_3 decrease and I_3 increase with the milling time. The last

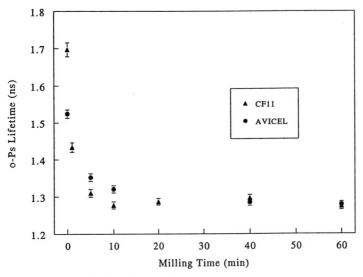

Figure 3. o-Ps lifetime vs. milling time.

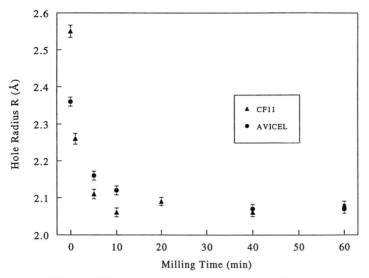

Figure 4. Mean free-volume hole radius vs. milling time.

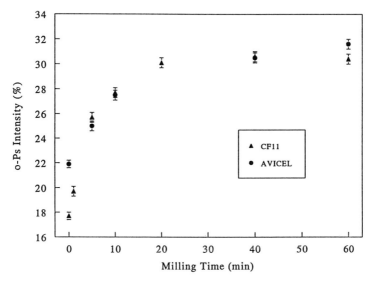

Figure 5. o-Ps intensity vs. milling time.

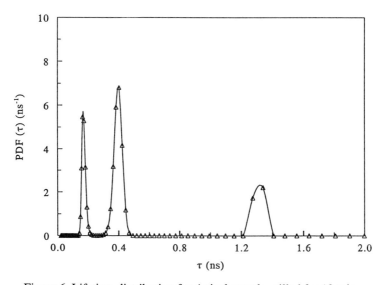

Figure 6. Lifetime distribution for Avicel sample milled for 10 min.

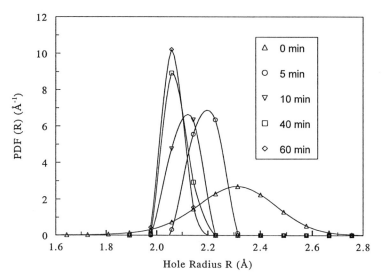

Figure 7. o-Ps lifetime distributions for Avicel samples.

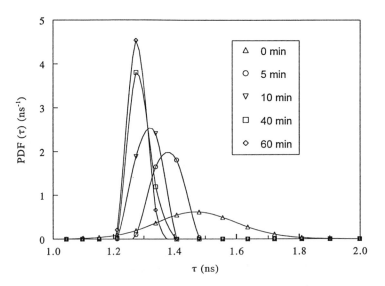

Figure 8. Hole radius distributions for Avicel samples.

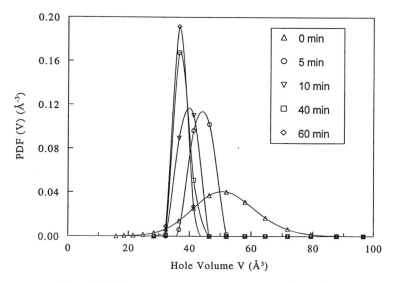

Figure 9. Hole volume distributions for Avicel samples.

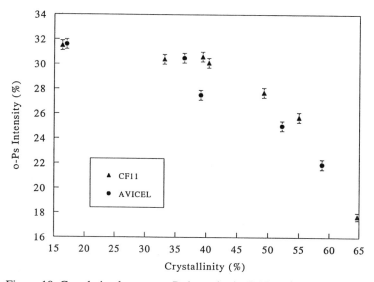

Figure 10. Correlation between o-Ps intensity by PAL and crystallinity
determined by FTIR.

two samples with 40 and 60 min of milling show similar peaks both in shape and position, which supports the claim that the microstructure of these cellulose powders approaches a final state after extended milling.

Correlation between I_3 and crystallinity is shown in Figure 10. The parameter I_3 increases as crystallinity decreases because of the increase of amorphous phase fraction. The data appear to be linear in the range from 35% to 65% crystallinity, but the data in the low crystallinity region do not follow this linear relationship. The reason for this deviation at low degrees of crystallinity is not clearly known yet. The two techniques (PAL and FTIR) are based on different measurements of indirect evidence of crystallinity. The deuteration FTIR measurement, which uses accessible OH group as an indirect measure of amorphous content, may underestimate the crystallinity due to accessible OH groups on the surfaces of crystalline regions. On the other hand, the PAL technique may overestimate the crystallinity for a number of reasons, one being that o-Ps is not sensitive to holes with a radius larger than 20 Å (17).

Extrapolation of the I_3 value at 100% crystallinity can be used to determine whether o-Ps is also located in crystalline regions. Ignoring the two data in the low crystallinity region, and fitting the other data by straight line, the extrapolated value at 100% crystallinity is $I_3 = 6\%$. But due to the above reasons concerning the differences of the two techniques, it may not be appropriate to fit the data by a straight line. It is still not clear where the o-Ps is located, in the free volume and holes solely in the amorphous phase or possibly in some trapping sites in crystalline phase. It appears that the o-Ps is annihilated mainly in amorphous phase. The sites in the crystalline phase available for Ps trapping are most likely smaller than those in the amorphous phase. The decrease of τ_3 with milling time indicates that o-Ps is not likely in the crystalline phase because the crystallinity decreases due to the milling. The newly created amorphous phase may have holes with smaller size and the holes in the original amorphous phase may also be rearranged into smaller size holes during the milling. A narrower distribution of hole size due to milling is also consistent with this suggestion.

Fractional free volume f_v for Avicel and Whatman CF11 samples are calculated by using equation 2 and plotted in Figure 11. It shows that f_v first drops in the beginning of milling and then slightly increases until approaching a constant value of 2%. This behavior can be understood by the two opposite trends with the milling time, i.e., decrease of mean free-volume hole size and increase of the free-volume hole number.

Conclusion

From PAL results for ball-milled Avicel and Whatman CF11 cellulose samples, it is found that the mean free-volume hole size decreases with milling time. This result is interpreted as being due to the generation of new holes with smaller size in the process of disrupting part of the crystalline phase transforming crystalline to amorphous phase, and due to the modification of holes in the original amorphous phase. The increase in o-Ps intensity with milling time is interpreted to be due to the increase in the amount of amorphous phase. From the o-Ps lifetime distribution analysis, the PDF peak shifts to the small size region and the distribution is narrowed

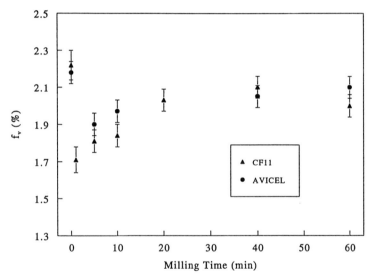

Figure 11. Fractional free volume vs. milling time.

after milling. A good correlation is found between I_3 and crystallinity measured by deuteration FTIR. These results show that PAL spectroscopy is a useful tool to probe sub-nanometer hole properties of cellulose.

Acknowledgment

This work was supported by the TAPPI Foundation under grant No. 95-1666 and the AFOSR under grant No. F 49620-97-1-0162.

Literature Cited

1. Vander Hart, D. L.; Attala, R. H. *Macromolecules.* **1984**, *17*, 1465.
2. Roberts, G. A. F. In *Paper Chemistry;* Roberts, G. A. F., Ed.; Blackie, Chapman and Hall Inc.: New York, NY, 1991, pp 9-24.
3. Hullermann, S. H. D.; Van Hazendonk, J. M.; Van Dam, J. E. G. *Carbohyd. Res.* **1994**, *261*, 163.
4. Jean, Y. C. *Microchem. J.* **1990**, *42*, 72.
5. Kluin, J.-E.; Yu, Z.; Vleeshouwers, S.; McGervey, J. D.; Jamieson, A. M.; Simha, R.; Sommer, K. *Macromolecules* **1993**, *26*, 1853.
6. Zipper, M. D.; Simon, G. P.; Cherry, P.; Hill, A. J. *J. Polym. Sci., B: Polym. Phys.* **1992**, *32*, 1237.
7. Xie, L.; Gidley, D. W.; Hristov, H. A.; Yee, A. F. *J. Polym. Sci., B: Polym. Phys.* **1995**, *33*, 77.
8. Wang, Y. Y.; Nakanishi, H.; Jean, Y. C.; Sandreczki, T. C. *J. Polym. Sci., B: Polym. Phys.* **1990**, *28*, 1431.
9. Xie, L.; Gidley, D. W.; Hristov, H. A.; Yee, A. F. *Polymer* **1994**, *32*, 3861.
10. Pekarovicova, A.; Venditti, R. A.; Cao, H.; Lou, Y. M.; Jean, Y. C. *J. Pulp and Paper Sci.* **1997**, *23(3)*, J101.
11. PATFIT package, 1989; purchased from Risø National Laboratory, Risø, Denmark.
12. Nakanishi, H.; Wang, S. J.; Jean, Y. C. In *Positron Annihilation Studies of Fluids*; Sharma, S. C., Ed.; World Scientific: Singapore, 1988; p 292.
13. Jean, Y. C.; Dai, G. H. *Nucl. Instrum. and Meth. in Phys. Res.*, **1993**, *B79*, 356.
14. Shukla, A.; Peter, M.; Hoffmann, L. *Nucl. Instrum. and Meth. in Phys. Res.* **1993**, *A 335*, 310.
15. Hoffmann, L.; Shukla, A.; Peter, M.; Barbiellini, B.; Manuel, A. A. *Nucl. Instrum. and Meth. in Phys. Res.* **1993**, *A 335*, 276.
16. Dai, G. H. 1994. M.S. Thesis. University of Missouri-Kansas City.
17. Liu, J.; Deng, Q.; Jean, Y. C. *Macromolecules* **1993**, *26*, 7149.

Chapter 25

Ionic Conductivity in Glassy PVOH–Lithium Salt Systems

M. Forsyth[1], H. A. Every[1], F. Zhou[2], and D. R. MacFarlane[2]

[1]Departments of Materials Engineering and [2]Chemistry, Monash University, Wellington Road, Clayton, Victoria 3168, Australia

Ionic conductivity (10^{-4} S/cm at 25°C) has been observed in glassy poly(vinyl alcohol)/lithium salt complexes. X-ray diffraction patterns indicate that, while the pure polymer is semi-crystalline, the addition of salt suppresses the extent of crystallinity. The glass transition temperatures of these systems are typically in the region of 50-70°C. The conductivity is dependent on the concentration and anion in the lithium salt; $LiClO_4$ producing the highest conductivities in this work. The presence of unhydrolyzed acetate groups in the polymer backbone causes T_g to be decreased, as compared to the homopolymer, and produces an order of magnitude increase in conductivity. ^7Li solid state NMR spectroscopy suggests that lithium ion motion is present below T_g indicating that, at least in part, lithium ion motion is responsible for the ionic conductivity and that this motion is decoupled from the polymer segmental motions. ^{19}F T_1 and T_2 NMR relaxation times were also measured for the anion, and the lack of correlation with the conductivity data suggests anion motion is unlikely to be a major contributor to the conductivity.

Polymer electrolytes have been under intense investigation (*1*) for the past two decades owing to their potential applications in a variety of new electrochemical devices, in particular lithium batteries. The most successful of these (ie. having high ionic conductivities with suitable mechanical properties) have been based on, or have contained segments of, the polyether unit. These have served to act as a good solvent, in particular for alkali metal cations which coordinate to the ether oxygen. In work carried out in our laboratories (*2*) and others (*3*) it has been shown that optimum conductivity in these electrolytes occurs at salt concentrations of the order of 1mol/kg, with higher salt concentrations resulting in a decreasing conductivity. This fall off in conductivity occurs as a result of (i) an increasing degree of ion aggregation and therefore the availability of fewer small ions, and (ii) an increasing glass transition temperature, T_g, and hence a decrease in ionic mobility. These systems are characterized by a T_g well below room temperature and are therefore soft elastomeric materials (4) or semi-crystalline materials at ambient temperatures.

Recently (5,6,7) solid electrolyte systems based on poly(vinyl alcohol), PVOH, (I)

$$-\left(CH_2-\underset{\underset{OH}{|}}{CH}\right)_{\overline{x}}$$

(I)

have been investigated which contain a very substantial amount of salt (up to 75wt% salt). These electrolytes have been shown to have room temperature conductivities up to 10^{-4}S/cm (5), almost an order of magnitude higher than other known solvent-free polymer electrolytes. One of the unusual and interesting characteristics of these new PVOH systems is that higher salt concentrations result in higher conductivities, up to the current limit of miscibility. Further, the glass transition temperature is above room temperature (ca. 70°C) and shows little dependence on salt concentration, in contrast to the systems based on polyethers (8).

These solid electrolyte systems are of considerable scientific and technological significance since ionic conductivity appears to be decoupled from the polymer motions, as manifest in the DSC glass transition temperature. This characteristic probably qualifies these systems as the first polymeric members of the group of fast ion conductive materials (5). Fast ion conduction occurs when the diffusive/conductive modes of motion of one or more species in the material become decoupled from the main structural modes which determine T_g. Ion motion then takes place against a static background of sites between which the ions hop. Systems having conductivities in the range of 10^{-5}- 10^{-4} S/cm at room temperature, and having T_g above room temperature, are of interest in the electrochromic window device (9) being developed by a number of groups. Further improvement of the conductivity may open other applications, for example in photoelectrochemical solar cells and low power capacitors.

Poly(vinyl alcohol) is well known to be a semi-crystalline material (T_m=220°C, T_g=85°C) although it often degrades before melting (10). The polymer is made by a hydrolysis reaction of poly(vinyl acetate) which is typically 80-99% complete. The 99% hydrolyzed polymer can be viewed as a pure homopolymer, whereas the materials having a higher fraction of remaining acetate groups should be viewed as a random copolymer. The degree of hydrolysis influences the glass transition temperature of the amorphous fraction (T_g (99%) ~ 68°C, T_g (88%) ~ 55°C) in PVOH. The glass transition temperature of PVOH is relatively high for a vinyl polymer of this structure. This is the result of the strong hydrogen bonding interactions, both intra- and interchain. The effect of acetate groups is to disrupt this to some extent as seen in the decreased T_g.

One possible interpretation of the high ionic conductivities observed in these PVOH based electrolytes is that the conduction involves protons liberated from the hydroxy group. The mobility of such protons can be high as a result of small size and mass and may even be assisted by a mechanism similar to the Grotthius mechanism observed in aqueous systems. In this model, the alkali metal ions and the corresponding anion which are dissolved in the polymer may be relatively immobile and hence may not contribute to the conductivity. In our recent work (5) we have used ^{7}Li NMR experiments to probe cation mobility. These have shown significant lithium ion mobility below T_g and this is enhanced at higher salt concentrations. Hence we hypothesize that the dissolved ions *are* a major conductive species.

Since the pure polymer is partly crystalline, the question arises as to the role of the crystal domains in the conduction process. This question also arose in the interpretation of conductivity in the polyether electrolytes (*1*) and it was shown that conduction occurs predominantly in the amorphous regions. In polyether systems containing a substantial degree of crystallinity, the conductivity is observed to increase by as much as an order of magnitude as the temperature is increased through the melting point. It has been concluded therefore that the presence of a mobile amorphous phase is necessary for conduction in these polyether systems. Ion mobility also rapidly diminishes as the temperature is decreased towards T_g (*1*). As with these polyether systems it is hypothesized that significant conduction occurs in the amorphous regions of the PVOH/salt mixtures.

In this paper we have investigated the influence of the degree of hydrolysis, salt content and the nature of the salt on conductivity in PVOH/lithium salt mixtures. These results, along with ^{7}Li and ^{19}F NMR relaxation measurements and X-ray diffraction data, are used to examine the hypotheses outlined above regarding the nature of the conduction mechanism.

Experimental method

Sample preparation. $LiCF_3SO_3$ (Aldrich), $LiBF_4$, and $LiClO_4$ salts were dissolved in poly(vinyl alcohol), PVOH. Two polymers were studied with the molecular weight ranges of 31,000-50,000 g/mol (Aldrich) and 105,600-110,000 g/mol (Aldrich), with degrees of hydrolysis being 88% and 99.8% respectively. The polymer films were prepared by dissolving the PVOH and salt in dimethyl sulfoxide (DMSO), (17g DMSO/1g PVOH) at 70°C for one hour. The solution was then cast onto a glass plate and the DMSO removed using a high vacuum pump over a period of 15 hours. The film was finally dried between two pieces of teflon in a vacuum oven at 50 to 70°C for approximately 20 hours. The DMSO removal was monitored gravimetrically. The polymer systems studied consisted of salt to polymer weight ratios between 0.25 and 1.5.

Conductivity measurements. The conductivity of the electrolytes was measured using a Hewlett Packard 4284A LCR meter in the range 20Hz-1MHz. Disc shaped samples ~1.5cm in diameter and 0.2mm thick were sandwiched between a pair of blocking electrodes after coating each side with a circular gold electrode by sputtering. The brass conductivity cell was loaded with the sample in a nitrogen drybox and sealed to prevent moisture ingress during the measurements.

Thermal analysis. Differential scanning calorimeter thermograms were obtained using a Perkin Elmer DSC7. Samples were encased in Al sample pans and quenched to liquid nitrogen temperatures prior to heating in the DSC at rates between 10 and 30°C/min. Annealing experiments were carried out by holding the sample in the DSC at temperatures just below T_g for various periods of time. After this period of annealing the sample was cooled and rewarmed to observe the resultant shape of the glass transition region.

X-ray diffraction. Wide-angle x-ray diffraction (WAXD) patterns were obtained on a Scintag PAD5 instrument in reflection mode with filtered CuKα radiation. The data was obtained from 10° through to 70° with a 0.05° step size at a scanning rate of 2°/min.

NMR Linewidth measurements. A home built permanent magnet spectrometer operating at 13.69 MHz was used to measure ^{19}F relaxation times for the PVOH electrolytes. T_1 and T_2 were obtained using the Inversion-Recovery and Spin-echo methods respectively at temperatures between -70°C to 20°C.

^{7}Li NMR linewidth measurements for the PVOH electrolytes were made using a modified Bruker CXP 300 NMR spectrometer operating at 116.6 MHz. A simple one pulse experiment was used with a pulse length of 1.5μs. The linewidths were taken at full width at half maximum (FWHM) of the peaks betweem -40°C to 80°C. The onset of narrowing was determined from plotting the magnitude of the linewidth as a function of temperature.

Results and Discussion

Thermal Analysis and XRD. Figure 1(a) shows a typical thermogram for pure PVOH (99.8%). A weak step in the heat capacity is observed between 60°C and 70°C. To prove that this event is in fact a glass transition, the polymer was annealed at 60°C for 3 hours. Such annealing produces a characteristic enthalpy overshoot at the glass transition if some non-reversible relaxation process has taken place. This is often taken as clear indication that a thermal transition is in fact a glass transition. The second thermogram confirms this hypothesis by showing a substantial enthalpy overshoot peak after the annealing. In Figure 1(b) similar behaviour in a salt/copolymer (88%) system is seen, indicating a glass transition temperature around 55°C.

Heating to higher temperatures showed a broad endotherm around 190°C which may be the melting transition in the semi-crystalline pure polymer. This transition was not distinctly observed in the polymer/salt systems, indicating suppresion of crystallinity by the salt. This behaviour is confirmed in Figure 2, which shows the wide angle X-ray diffraction pattern for the 88% hydrolyzed systems. The pure co-polymer, which contains no salt but has otherwise been prepared in the same manner, by casting from DMSO solution, exhibits a clear crystalline diffraction peak at 2θ = 20°. This peak corresponds to the 110 reflection arising from the planar zig-zag configuration of the backbone (*11*). Addition of salt to this polymer causes a marked decrease in the intensity of this peak relative to the broad amorphous peak. The degree of crystallinity in PVOH is well known to be strongly dependent on its thermal treatment (*10,12-14*). Sakurada has found that PVOH cast from solution is isotropic with a degree of crystallinity around 15% (*10*). Peppas and Hansen however show that solution cast PVOH is 20% crystalline (*12*), as does Molyneux (*13*). Annealed (180°C) isotropic PVOH, can have crystalline fractions as high as 75% (Kenny and Willcockson) (*14*). On the basis of these results, the extent of crystallinity in the pure copolymer sample in Figure 2 is likely to be of the order of 20%. The extent of crystallinity in the corresponding salt containing system appears to be lower than this, implying that the addition of salt to PVOH does indeed supress crystallinity.

Conductivity. Conductivity data as a function of salt content in the PVOH/Li triflate system are presented in Figure 3. At all temperatures the conductivity increases sharply with added salt content up to the solubility limit. This behaviour is in marked contrast to the polyether based electrolyte systems where a conductivity maximum is

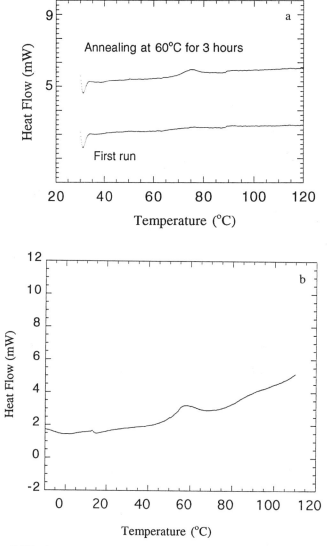

Figure 1. DSC thermograms of (a) PVOH (99.8% hydrolyzed) before and after annealing and (b) a PVOH (88% hydrolyzed) system containing 33% by weight Li triflate.

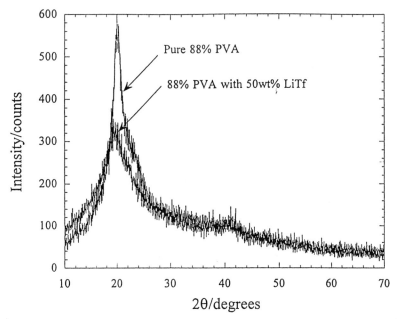

Figure 2. X-ray diffraction pattern of a PVOH (88% hydrolyzed) system containing 50 percent by weight Li triflate

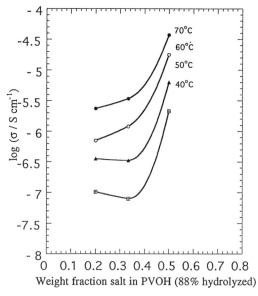

Figure 3. Conductivity as a function of salt content in the PVOH (88% hydrolyzed)/Li triflate system.

observed typically around 1mol/kg ($\simeq$ 10% (w/w) in the case of LiClO$_4$). In fact, there is a suggestion in the low concentration data in Fig. 3 that the conductivity passes through a minimum. Since pure PVOH is not known to be ion conductive, the conductivity must therefore exhibit a maximum at yet lower concentrations. This data emphasizes the very much higher salt content of these electrolytes, as compared to the traditional polyether systems, and the likelihood of a molten salt like conduction mechanism.

In Figure 4(a) conductivity data for selected systems are presented as a function of inverse temperature. Comparison of these results with those in the literature for the 88% hydrolyzed PVOH systems reveal some significant differences. The Li triflate and LiBF$_4$ data are significantly lower than the measurements of Yamamoto et al. (6) while the LiClO$_4$ measurements from the present work are quite similar to the LiBF$_4$ results of Yamamoto et al. One possible cause of such variations is the extent of residual DMSO in the sample. Yamamoto et al. quote a value of residual DMSO of 3.5% (w/w) as obtained by GLC. In our work accurate weight loss measurements have shown that typical residual DMSO content is between 5 and 10%, the higher DMSO contents being recorded for the higher salt content samples. A determination of the effect of DMSO content on conductivity shows that the conductivity changes by approximately one order of magnitude as the DMSO content is reduced from 25% to ~5%. Hence the presence of small amounts of residual DMSO cannot obviously explain the above discrepancy. The thermal analysis traces of the materials prepared in this work are also significantly different from those of Yamamoto et al., who observed melting transitions around 100°C. The lack of this transition in our samples suggests morphological differences between samples prepared in this work, which are possibly the result of i) the molecular weights used, ii) the tacticity of the PVOH samples, and iii) the precise details of the thermal history of the samples in the final stages of preparation.

Figure 4(b) compares Li triflate systems based on the homopolymer with those of the copolymer. At both salt concentrations the copolymer systems are at least an order of magnitude more conductive than the corresponding homopolymer systems. This correlates with the higher T$_g$ for the homopolymer systems and the ^{7}Li NMR data, as will be discussed further below.

The data in Figure 4 exhibit Arrhenius behaviour in the lower temperature region well below T$_g$. Near T$_g$, all data sets become curved and exhibit decreasing activation energy. Table I summarizes the activation energies obtained from linear fits to the lower temperature regions, as indicated by the lines drawn in the Figure. These activation energies are notably higher by at least a factor of 2 than polyether electrolytes of similar conductivity. However, these activation energies correspond to a sub-T$_g$ process. Under similar conditions (ie. below their T$_g$) polyether electrolytes are not conductive. A more useful comparative system for these PVOH electrolytes is the family of lithium ion conductive ceramics in which facile lithium motion takes place within a rigid inorganic oxide framework (15). The mechanism in the latter case is thought to be ion hopping between vacant isoenergetic sites. In these systems the activation energy is typically ~10^2 kJ/mol and corresponds to the energy required to break the Li$^+$ - lattice oxygen bond. In the PVOH electrolyte systems described in this work, given that the activation energies are obtained below T$_g$, there must exist a rigid framework of some sort and E$_a$ probably corresponds to the breaking of a bond between the mobile ion and this rigid framework.

374

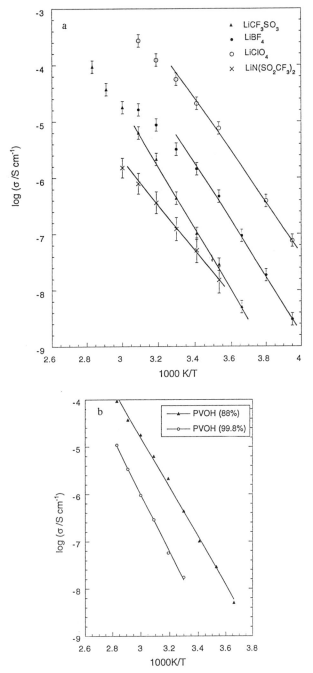

Figure 4. Conductivity as a function of inverse temperature, showing the effect of (a) different anions at fixed composition (50% salt by weight) and degree of hydrolysis (88%) and of (b) polymer degree of hydrolysis (salt=Li triflate)

Table I. Activation energies for ion conduction in PVOH based polymer electrolytes

Salt	Polymer	Weight ratio of salt/PVOH	E_a (kJ/mol)
LiCF$_3$SO$_3$	PVOH(88%)	1:2	118 ± 5
LiBF$_4$	PVOH (88%)	1:2	87 ± 2
LiClO$_4$	PVOH(88%)	1:2	72 ± 2
LiClO$_4$	PVOH(99.8%)	1:2	130 ± 6
LiClO$_4$	PVOH(88%)	1:2	72 ± 1
LiClO$_4$	PVOH(88%)	1:1	67 ± 2
LiCF$_3$SO$_3$	PVOH(88%)	1:1	99 ± 2
LiCF$_3$SO$_3$	PVOH(88%)	1:2	94 ± 3
LiCF$_3$SO$_3$	PVOH(88%)	1:4	118 ± 5

The activation energy is almost independent of salt content but is higher for the homopolymer than for the copolymer in the case of the perchlorate systems. This is consistent with the higher T_g observed for the homopolymer based electrolytes and also their lower room temperature conductivity. The activation energies decrease in the order $CF_3SO_3^- > BF_4^- > ClO_4^-$ observed (under similar conditions of hydrolysis and concentration), concomitant with the increase in conductivity as a function of the anion.

NMR.

Anion mobility by ^{19}F NMR

Figure 5 presents the ^{19}F temperature dependent T_1 relaxation times for various PVOH/triflate samples. A comparison of polymer type and salt concentration can be made here. All data pass through a minimum. The position of this minimum, in temperature, relates to the correlation time of the motion responsible for the ^{19}F relaxation, whilst the magnitude of T_1 at the minimum is indicative of the strength of the interaction governing the relaxation process. As illustrated in Figure 5 at a fixed salt concentration, ^{19}F relaxation is independent of the degree of hydrolysis of the polymer. This suggests the fluorines are unlikely to be relaxing as a result of polymer-anion interactions. The decrease in T_1 at the minimum with increasing salt confirms that the relaxation is either Li-F or F-F dominated. The shift of the T_1 minimum to lower temperatures suggests an increase in mobility (probably anion tumbling) with increasing salt concentration.

The T_2 measurements (Figure 6) support the behaviour observed in the T_1 data. Upon the addition of salt, the T_2 relaxation times increase suggesting enhanced ion mobility. For the electrolytes with the same salt concentration, but differing degrees of hydrolysis, the T_2 relaxation times are very similar. If the anion was diffusing in these systems, then it might be expected that the T_2 would approach T_1 at high temperatures. According to simple BPP theory, the relaxation times are dependent on the correlation time, τ_c, and the Larmor frequency, ω_L (16) as shown for example in the following equations:

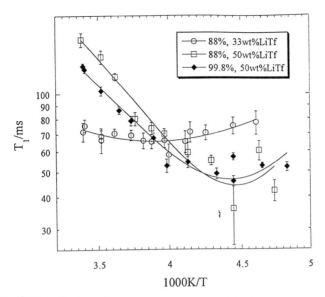

Figure 5. ^{19}F Spin-lattice relaxation measurements (T_1) for PVOH/Li triflate systems as a function of inverse temperature. Lines drawn are a guide to the eye only.

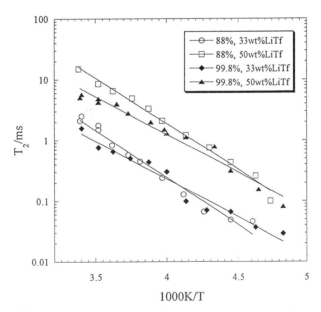

Figure 6. ^{19}F Spin-spin relaxation (T_2) measurements as a function of inverse temperature in PVOH/Li triflate systems. Lines are Arrhenius fits to the data with activation energies indicated in Table II.

$$\frac{1}{T_1} \propto \left[\frac{\tau_c}{1 + \omega_L^2 \tau_c^2} + \frac{4\tau_c}{1 + 4\omega_L^2 \tau_c^2} \right]$$

$$\frac{1}{T_2} \propto \left[3\tau_c + \frac{5\tau_c}{1 + \omega_L^2 \tau_c^2} + \frac{2\tau_c}{1 + 4\omega_L^2 \tau_c^2} \right]$$

These equations show that field components which contribute to the relaxation in T_1 also affect T_2, but T_2 is also sensitive to static fields. It is this static field component which dominates when T_2 is considerably smaller than T_1. The T_2 relaxation times for these samples are approximately an order of magnitude lower than the T_1 relaxation times at corresponding temperatures. Combining this observation with the lack of any correlation between the ^{19}F T_1 and T_2 trends and conductivity trends with salt and polymer type, it is suggested that the anion is unlikely to be diffusing in these systems. Future NMR diffusion measurements will test this hypothesis.

This conclusion is not inconsistent with the dependence of conductivity on anion type as seen in Figure 4. The influence of an immobile anion on conductivity may be due to the strength of its interaction with the Li^+ ion. This is also observed in the conductivity activation energy data in Table I. The lack of contribution of the anion to conductivity in these systems is also indicated by the very low conductivity observed by Yamamoto et al. for PVOH systems containing Bu_4N^+ ClO_4^- (Bu = butyl); the Bu_4N^+ is known to be a large weakly interacting cation. Therefore the conductivity in these systems does not appear to be due to anion motion. Comparison with the Bu_4N^+ based system suggests that Li^+ ion mobility is chiefly responsible for the observed conductivity in the Li^+ based systems of this work. Further evidence for this is found in the ^7Li NMR linewidth measurements discussed below.

Cation mobility by ^7Li NMR

The ^7Li NMR linewidth measurements as a function of temperature for the 99.8% hydrolysed PVOH:LiCF$_3$SO$_3$ = 1 sample are shown in Figure 7. At the lowest temperatures, the linewidths are broad (~5500 Hz) and the shape of the curve appears to be approaching a Gaussian line shape. As the temperature increases, the linewidth decreases and, at the highest temperature, the lineshape becomes increasingly Lorentzian.

In Figure 8 the linewidth (FWHM) data are plotted for several systems as a function of temperature. The two systems based on 99.8% hydrolyzed PVOH show little dependence of the ^7Li linewidth on salt content. The shape of the curve as a function of temperature is characteristic of a system passing out of its low temperature state in which the Li^+ ions are effectively immobile. This low temperature behaviour is termed the rigid lattice limit. With increasing temperature the line begins to narrow as a result of increased ion mobility. The high temperature (motionally narrowed) limit is set by the resolution of the spectrometer. In the case of the 88% hydrolyzed sample with salt to polymer ratio of 0.5, a rigid lattice limit is not reached over the temperature range studied. In addition, over this entire temperature range, the lithium linewidth is always smaller for the copolymer; this also therefore indicating higher lithium ion mobility.

The linewidth data can be well fitted by the following equation (19,20):

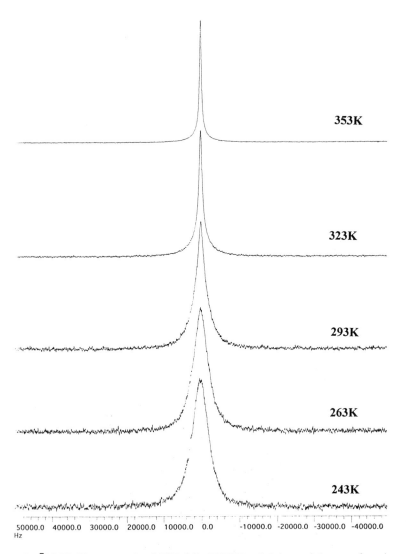

353K

323K

293K

263K

243K

50000.0 40000.0 30000.0 20000.0 10000.0 0.0 -10000.0 -20000.0 -30000.0 -40000.0
Hz

Figure 7. ^{7}Li NMR spectra for LiCF$_3$SO$_3$:PVOH = 1:1 by weight as a function of temperature.

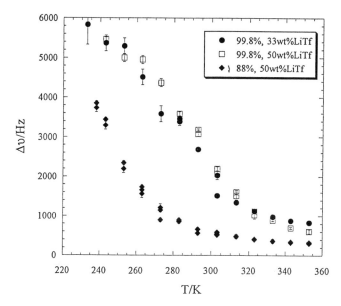

Figure 8. ^{7}Li NMR linewidths as a function of temperature for 99.8% and 88% hydrolyzed PVOH/Li triflate electrolytes. The temperature of onset of motional narrowing is estimated by extrapolating the linear portion of the curve to the rigid lattice estimate.

$$(\delta\omega)^2 = \left(\frac{2}{\pi}\right) \delta\omega_0^2 \tan^{-1}(\tau_c \delta\omega)$$

where $\delta\omega$ is the narrowed linewidth, $\delta\omega_0$ is the rigid lattice linewidth and τ_c is the correlation time related to diffusional motion. The correlation time is assumed to show an Arrhenius temperature dependence

$$\tau_c = \tau_0 \exp\left(\frac{E_a}{kT}\right)$$

The activation energies obtained from these fits are also shown in Table II.

Table II. NMR ion mobility parameters

Salt/PVOH systems (degree of hydrolysis)	Onset of Narrowing/K	$\tau_c(^7Li)$ Activation energy/kJ mol^{-1}	$^{19}F, T_2$ Activation energy /kJ mol^{-1}
1:1 (88%)	214±2	19.7±0.7	28.7 ± 0.1
1:2 (99.8%)	244±2	18.6±1.2	23.8 ± 0.1
1:1 (99.8%)	260±2	23.8±1.1	23.8 ± 0.1
1:2 (88%)	-	-	29.7 ± 0.1

For the same salt concentration, the activation energy for lithium ion mobility (as given by τ_c) is lower in the case of 88% hydrolyzed PVOH.

The onset of line narrowing can be estimated from these linewidth curves (see Table II). The higher salt content, higher degree of hydrolysis samples give the highest onset temperature. Given that the rigid lattice limit has not been reached in the case of the copolymer sample, an approximate onset temperature has been determined by assuming the rigid lattice is approximately 6000Hz. This data clearly shows the higher lithium ion mobility in the case of the copolymer. It is interesting to note that the onset of motional narrowing occurs at temperatures at least 70°C below T_g as determined from thermal analysis. This is in stark contrast to polyether and other polymer electrolytes where motional narrowing appears closely linked to T_g (17,18). It is also of interest to observe that higher conductivities are measured in samples based on the 88% hydrolyzed PVOH polymer, consistent with the higher lithium ion mobility indicated by ^7Li NMR (in contrast to the ^{19}F relaxation data). However, in comparing the temperature dependence of the linewidths for the 99.8% hydrolyzed PVOH with different salt concentrations, it appears that the higher salt content results in a slightly decreased lithium ion mobility. This is in contrast to the increased ionic conductivity. Since conductivity is related to the number of charge carriers (n_i) and their mobility (μ_i) by the equation $\sigma = \sum_i n_i q_i \mu_i$ (where q_i=charge), the increased conductivity for the higher salt content is apparently the result chiefly of an increase in the number of charge carriers. The contributions of H$^+$, Li$^+$ and anion to conductivity will be clarified in the near future by diffusion measurements.

Conclusions

The work reported here on poly(vinyl alcohol) electrolytes shows that the materials are mainly amorphous with T_g around 50-70°C and conductivity at room temperature as high as 10^{-4} S/cm. ^{19}F NMR relaxation measurements are consistent with the hypothesis that the anion, or at least the triflate anion, is not significantly mobile below T_g. ^{7}Li NMR linewidth measurements indicate, however, that the lithium ion only approaches its rigid lattice linewidth at ca. 250K. Above this temperature the ^{7}Li line is motionally narrowed, consistent with either mobile ^{7}Li ions or motion of some other species in the vicinity of the Li ion. This NMR evidence, plus the observation that tetrabutylammonium analogues of these systems show only very low conductivity, are all consistent with the hypothesis that the lithium ion is the chief conducting species in these systems below T_g. The observation of high ionic mobility below T_g is a common phenomenon in the so called fast ion conducting glasses and ceramics. To our knowledge this is the first observation of such fast ion conducting behaviour in a polymeric system. The observation of fast ion conduction in this polymer system may reflect the high solubility of the lithium salts in the systems due to the hydroxy groups in the polymer chain. This high solubility results in a medium which has more in common with a molten salt system than traditional polyether based electrolytes. The fact that the 88% PVOH has a higher conductivity than the 99% hydrolysed PVOH suggests that the polymer is not irrelevant in the conduction process. The room temperature activation energies for conduction in these systems is high compared to other polymer electrolytes and is a strong function of the anion, suggesting an anion-cation bond breaking step in the overall conduction.

Literature Cited

1. Gray, F.M. *Solid Polymer Electrolytes*, VCH Publishers Inc. New York (1991).
2. (a) Bishop, A.G.; MacFarlane, D.R.; McNaughton, D.; Forsyth, M. *J. Phys. Chem.* **1996**, *100*, 2237.
    (b) Bishop, A.G.; MacFarlane, D.R.; Forsyth, submitted to Electrochemica Acta.
3. Watanabe, M., Oohashi, S., Sanui, K., Ogata, N., Kobayashi, T. and Ohtaki, Z., *Macromolecules* **1985**, *18*, 1945.
4. Sun, J.; MacFarlane, D.R.; Forsyth, M. *Solid State Ionics* **1996**, *85*, 137.
5. Every, H.A.; Zhou, F.; Forsyth, M.; MacFarlane, D.R. submitted to Electrochemica Acta.
6. Yamamoto, T.; Inami, M.; Kanbara, T. *Chem. Mater.* **1994**, *6*, 44.
7. Kanbara, T.; Inami, M., Yamamoto, T.; Nishikata, A., Tsuru, T.; Watanabe, M.; Ogata, N. *Chem. Lett.* **1989**, 1913.
8. MacFarlane, D.R.; Sun, J.; Meakin, P.; Fasoulopoulos, P.; Hey, J.; Forsyth, M.; Rosalie, J.M. *Electrochemica Acta* **1995**, *40*, 2131-2136.
9. MacFarlane, D.R.; Sun, J.; Forsyth, M.; Bell, J.M.; Evans, L.A.; Skyrabin, I.L. *Solid State Ionics* **1996**, *86-88*, 959-964.
10. Sakurada, I. "Polyvinyl Alcohol Fibres, International fibre science and technology series, 6" New York, Marcel Dekker, Inc. (1985).
11. Hodge, R.M.; Edward, G.H.; Simon, G.P. *Polymer* **1996**, *8*, 1371.
12. Peppas, N.A.; Hansen, P.J, *J. Appl. Polym. Sci.* **1982**, *27*, 9787.
13. Molyneux, P. "Nonionic polymers - The vinyl group" Chapter 4 in "Water-soluble synthetic polymers" Vol. 1 CRC Press, Florida USA, p.119-131 (1983).
14. Kenney, J.F.; Wilcockson, G.W. *J. Polym. Sci. A-1* **1966**, *4*, 679.

15. (a) H. Aono, E. Sugimoto, Y. Sadaoka, N. Imanaka and G.-Y. Adachi, J. Electrochem. Soc. **1990**, *137*, 1023. (b) H. Aono, E. Sugimoto, Y. Sadaoka, N. Imanaka and G.-Y. Adachi, J. Electrochem. Soc. **1989**, *136*, 590.
16. Fukushima, E.; Roeder, S.B.W. "Experimental Pulse NMR: A nuts and bolts approach" Addison-Wesley Publishing Co., Inc. Reading, Massachusetts (1981).
17. Chung, S.H.; Jeffrey, K.R.; Stevens, J.R. *J. Chem. Phys.* **1991**, *97* , 1803.
18. Stallworth, P.E.; Greenbaum, S.G.; Croce, F.; Slane, S.; Solomon, M. *Electrochim. Acta* **1995**, *40*, 2137.
19. Bloembergen, N.; Purcell, E.M.; Pound, R.V. *Phys. Rev.* **1948**, *73,* 679.
20. A. Abragam, "The Principles of Nuclear Magnetism" Oxford University Press (1961), p.456.

Note added in Proof:

One of the referees points out that the degree of hydrolysis of polyvinylacetate directly influences the free volume as measured by PALS. This is consistent with a lower T_g for the 88% hydrolysed polymer and the corresponding higher conductivities measured in this copolymer.

(R.M. Hodge, T.J. Bastow and A.J. Hill, proceedings of 10th IAPRI conference on Packaging, Melbourne, March 24-27, 1997)

Chapter 26

Oxygen Transport Through Electronically Conductive Polyanilines

Yong Soo Kang[1], Hyuck Jai Lee[2], Jina Namgoong[2], Heung Cho Ko[2], Hoosung Lee[2], Bumsuk Jung[1], and Un Young Kim[1]

[1]Division of Polymer Science and Engineering, Korea Institute of Science and Technology (KIST), P.O. Box 131 Cheongryang, Seoul, Korea
[2]Department of Chemistry, Sogang University, Mapo-ku, Seoul, Korea

Emeraldine base of polyaniline was synthesized by a chemical oxidation polymerization technique. The resulting emeraldine base film was treated with 4M HCl, 1M NH_4OH, and subsequently with varying dopant HCl concentrations. The oxygen and nitrogen permeabilities through the doped polyanilines decreased with their doping level, while their oxygen selectivity over nitrogen increased up to 12.2 when doped with 0.0150 M HCl solution. When doped with 0.0175 M HCl, the membrane selectivity is expected to be higher but unmeasurable because of extremely low permeability of nitrogen. The origin of such high selectivity is explored in terms of the facilitated transport and free volume change upon doping. Because the polarons generated upon doping react with oxygen molecules *specifically and reversibly*, these polarons can act as oxygen carriers and thereby facilitate oxygen transport. It is, therefore, expected that the oxygen permeability increases with the polaron (carrier) concentration. However, permeability for oxygen in these materials decreased with the increase in polaron concentration. Instead, the permeability correlated well with the d-spacing or free volume, measured via x-ray diffraction. It is found that although facilitated oxygen transport may occur, its contribution to oxygen permeability is insignificant. The free volume change upon doping seems to play a major role in determining gas permeation.

Electronically conductive polyaniline has been paid much attention as a potential membrane material for gas separation because of high selectivity, particularly very high selectivity of oxygen over nitrogen *(1-4)*. In addition, it is thermally stable and soluble in NMP, not like common electronically conductive polymers such as polyacetylene, polypyrrole, etc. *(5,6)*. Therefore, it can be easily processed to form membranes.

The reported transport properties strongly depended upon doping conditions and doping level *(1-4)*. It has been known that the chemical and physical natures of polyaniline can be changed by doping. For example, paramagnetic polarons can be generated by doping with protonic acids such as HCl, HF, camphor sulfonic acid, etc. as illustrated in Figure 1 *(7)*. Since oxygen is also paramagnetic, specific interaction

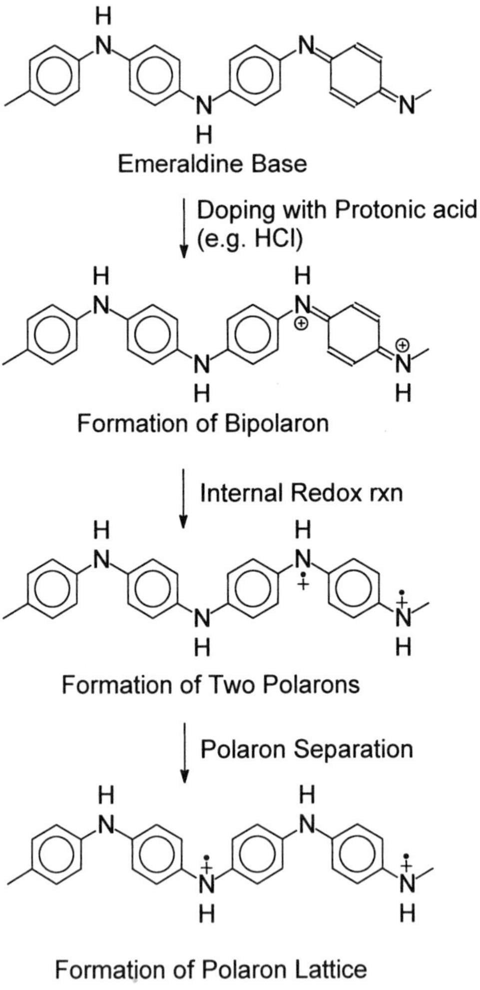

Figure 1. Schematic doping mechanism of emeraldine base polyaniline.

between polarons and oxygen molecules is expected. This interaction has been investigated by EPR spectroscopy and electronic conductivity and found to be reversible *(8)*.

When the interaction between oxygen molecules and polarons is assumed to be 1 to 1, it can be expressed as a simple reversible chemical reaction:

$$O_2 + P \Leftrightarrow [O_2 - P] \tag{1}$$

where P is the polaron, and $[O_2 - P]$ is the oxygen-polaron complex. The polaron can possibly act as an oxygen carrier because it reacts with oxygen molecules *specifically* and *reversibly (8)*. The oxygen transport could, then, be facilitated due to the presence of the oxygen carrier in addition to the normal Fickian permeation *(9,10)*. The facilitated oxygen transport might result in a high oxygen permeability as well as a high oxygen selectivity over nitrogen.

It has been reported that densification of polyaniline can occur when doped with protonic acids *(11)*. The densities of as-cast, fully doped and redoped polyanilines were reported to be 1.3, 1.4 and 1.32 g/cm^3, respectively *(11)*. The densification causes a reduction in the free volume, through which gas molecules can permeate. Therefore, it is also possible that the doping or dedoping process results in a free volume change which dictates the gas permeability.

In this study, it will be attempted to interpret the O_2 permeability and O_2/N_2 selectivity in terms of the facilitated transport due to the presence of the paramagnetic polarons and the free volume change upon doping with protonic acids.

EXPERIMENTAL

Polymerization of aniline : Polyaniline was prepared with oxidative polymerization of aniline in aqueous acidic media (1M HCl) with ammonium persulfate as an oxidant by following the method used by Mattes et al. *(1,2)*. The molar ratio of monomer/oxidant used was 4/1. The reaction was carried out at 0°C for 3 hours, and the precipitate was formed during the reaction. The precipitate was, subsequently, filtered and washed with deionized water until the filtrate was colorless. The as-synthesized polyaniline in its protonated form was treated with 1M NH_4OH for 15 hours to yield emeraldine base powder, followed by drying under vacuum for over 48 hours at room temperature.

Doping and dedoping of polyaniline membranes with aqueous HCl solution : The emeraldine base powder was dissolved in NMP (8 wt %). The emeraldine base solution in NMP was cast onto a glass plate and the solvent was removed under 120°C for 3 hours. The as-cast membrane was, then, immersed into a 4M HCl solution for 24 hours to give a fully doped membrane. The fully doped membrane was completely dedoped by immersion into 1M NH_4OH solution for 48 hours. The dedoped membrane was subsequently redoped with 0.0150, 0.0175, 0.0200 (± 0.0002), and 1 M HCl solutions. Each membrane was dried under vacuum for 48 hours at room temperature.

EPR experiment : Narrow strips of polyaniline membrane (1mm x 5mm) were put into an EPR cell. In order to study the interaction of oxygen molecules with polarons, the cell was connected to a large volume (1 L) of oxygen reservoir so that the applied pressure maintained constant during the experiment. The EPR line intensity was monitored with oxygen contact time from a Bruker EPR spectrometer (ER 200E-SRC). The spectrum obtained was doubly integrated to obtain the polaron concentration.

X-ray experiment : X-ray diffractogram was obtained by an X-ray diffractometer (Rigaku Geigerflex D/Max-B System). The d-spacing was obtained using the Bragg's law

$$d = \frac{\lambda}{2\sin\theta} \tag{2}$$

where λ is the X-ray wave length.

Gas permeation experiment : A constant volume technique was used to measure the gas permeability (12). After mounting a membrane in a permeation cell, the cell system was evacuated and its leak rate was checked, which was typically 45 mTorr/hr. The true pressure increase data were obtained by substracting the leak from the measured pressure increase. The permeability P_i was calculated from the slope of a plot of pressure vs time t at steady state,

$$P_i = \frac{\Delta Q_i L}{\Delta t A p_0} \tag{3}$$

where $\Delta Q_i = \Delta p_i V T_0 / 760(T_0 + T)$; L and A are the thickness and the surface area of the membrane, respectively; p_0 is the applied pressure; V and T_0 are the volume of the downstream side and 273 K, respectively. Note that the permeability P_i has a unit barrer where 1 barrer is $1\times10^{-10} cm^3 (STP)cm/cm^2$ sec $cmHg$. The effective diffusion coefficient D_i was also calculated from the time lag, θ_i, obtained from the pressure vs time curve (12).

$$D_i = \frac{L^2}{\theta_i} \tag{4}$$

The relationship between P_i and D_i is $P_i = D_i S_i$ where S_i is the solubility coefficient. The ideal separation factor is defined in a usual manner

$$\alpha_{ij} = \frac{P_i}{P_j} \tag{5}$$

and is termed the selectivity hereafter.

RESULTS AND DISCUSSION

Gas permeation. The oxygen and nitrogen permeabilities were measured as a function of the doping level by a conventional time-lag method. Their effective diffusion coefficients were calculated from the time-lag, and their solubility coefficients from the relationship of $S_i = P_i/D_i$. The results are listed in Table I. The permeabilities of both gases and the O_2/N_2 selectivities through the polyaniline membranes were plotted as a function of doping level in Figure 2. The oxygen permeability and selectivity for the as-cast membrane were 0.124 barrer and 6.53, respectively. When fully doped with 4M HCl, the permeabilities of both O_2 and N_2 were too low to be measured with the current permeation equipment. When dedoped, the oxygen permeability was 4.82×10^{-2} barrer and its selectivity increased to 9.03. In the case of the redoped samples, the oxygen permeability decreased with the doping level, while the selectivity increased from 9.03 to 12.2. When doped at HCl concentrations higher than 0.0175 M, the nitrogen permeability was unmeasurably small. Therefore, it is expected that the selectivity of the polyaniline doped with 0.0175 M HCl is higher than that doped with 0.0150 M HCl solution. Mattes et al. (1-3) also found the maximum selectivity when doped at 0.0175 M HCl solution.

Table I. Transport characteristics of various polyaniline membranes

	Gas	P_i	D_i	S_i	$\alpha_{i,j}$	d-spacing ($\overset{o}{A}$)
As-cast	O_2	1.24×10^{-1}	88.0	2.32	6.53	4.50
	N_2	1.90×10^{-2}	26.7	0.72		
4 M HCl	O_2	x	x	x	x	3.64
	N_2	x	x	x		
Dedoped	O_2	4.82×10^{-2}	21.5	2.24	9.03	4.50
	N_2	5.35×10^{-3}	3.21	1.57		
0.0150 M HCl redoped	O_2	2.64×10^{-2}	3.84	6.68	12.2	4.36
	N_2	2.17×10^{-3}	0.59	3.37		
0.0175 M HCl redoped	O_2	1.27×10^{-2}	2.61	6.48	x	4.30
	N_2	x	x	x		
0.0200 M HCl redoped	O_2	8.13×10^{-3}	1.42	7.20	x	4.30
	N_2	x	x			

P_i : barrer; D_i : (cm^2/sec x 10^{10}) and S_i : (cm^3(STP)/cm^3 cmHg x 10^3)
x: unmeasurable

The effective diffusion coefficient D_i, calculated from the time-lag, consistently decreases with the doping level while the solubility coefficient S_i increases. Oxygen molecules are dissolved in the polyaniline matrix by an ordinary sorption process and additionally are adsorbed on the polaron sites by forming a polaron-oxygen complex thus causing an increase in the oxygen solubility coefficient. Since oxygen molecules are momentarily adsorbed on the polaron sites due to the reversible reaction, its retention time inside the membrane will be lengthened, which causes the decreased effective diffusion coefficient. From these experimental results, it is found that both the oxygen and nitrogen permeabilities primarily depend upon the doping level and that they decrease with the doping level regardless of their sample history. The O_2/N_2 selectivity seems to increase with the doping level; however, it is not conclusive due to the limited number of experimental data available.

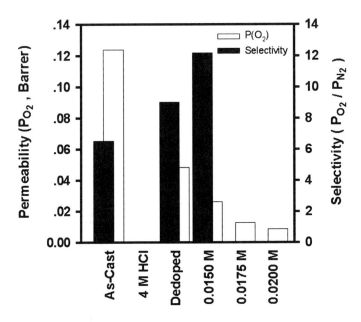

Figure 2. Oxygen permeability and oxygen/nitrogen selectivity as functions of the dopant HCl concentration.

Facilitated oxygen transport. It is well known that polarons are generated when polyanilines are doped with protonic acids as schematically illustrated in Figure 1. Because both the polaron and oxygen are paramagnetic, their specific magnetic interaction is anticipated. The interaction was investigated by EPR spectroscopy and electronic conductivity measurement and found to be *specific* and *reversible (8)*. The polaron can act as an oxygen carrier for the facilitated oxygen transport. In facilitated transport, the total effective permeability is a summation of the Fickian permeation due to a concentration gradient and the carrier-mediated permeation *(9,10,14-16)*. Therefore, the effective permeability will increase with the carrier concentration, if all the other factors such as structure are kept unchanged. In this respect, the same authors measured the polaron concentrations with varying doping levels *(8)*. The polaron concentration initially increased with the doping level, reached a maximum near pH=0, and decreased with further doping (Figure 3). Thus in our permeation experimental range for dopant concentration up to 0.02 M HCl, the polaron concentration monotonically increased with the doping level. Therefore, the oxygen transport may be facilitated owing to the presence of the polaron, an oxygen carrier, if all other factors are unchanged upon doping with HCl. If this is the case, the oxygen permeability should increase with increasing doping level up to 0.02 M HCl. However, the experimental results showed that the permeability decreased with the doping level. It is thus concluded that although the facilitated transport is very likely to take place, its contribution to the oxygen permeability is insignificant. However, this conclusion can be made only if the structure of the doped polyanilines does not change upon doping.

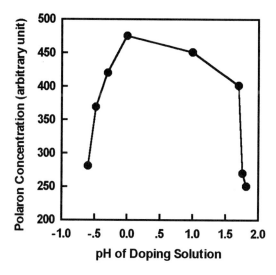

Figure 3. Polaron concentration as a function of dopant HCl concentration at ambient environment. Adapted from ref 8.

The intersegmental distance and free volume theory. It has been known that the polyaniline subchains become rigid upon doping resulting in a densification (10) and an increase in the glass transition temperature (17). In order to characterize the densification, X-ray diffraction was employed to obtain the d-spacing via Bragg's law. X-ray diffractograms in Figure 4 show two peaks for the doped polyanilines and one broad amorphous peak for the as-cast sample at ca.19°. The peak at ca. 10° increases with increasing dopant HCl concentration, which is consistent with the results by Pouget et al. (18). It is the broad amorphous peak at ca.19° that is used to calculate d-spacing. The calculated values for d-spacing are listed in Table I. It is clearly seen that the d-spacing decreased with increasing doping level regardless of the doping condition (dedoped or redoped). The d-spacing represents the intersegmental distance, through which solute molecules can pass. Therefore, the small d-spacings may be primarily responsible for low permeability.

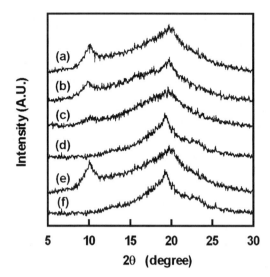

Figure 4. X-ray diffractograms of polyanilines by varying dopant HCl concentration: (a) 1.000M redoped, (b) 0.050M redoped, (c) 0.001M redoped, (d) dedoped, (e) 4M doped and (f) as-cast.

According to Fujita's free volume theory *(12)*, the diffusion coefficient is

$$D_i = D_0 \exp(-B_d / f) \tag{6}$$

where D_0 is the pre-exponential factor; B_d represents the minimum hole size for a solute to diffuse; f is the free volume fraction defined as a ratio of free volume to total volume. Here, we assume that f is linearly proportional to d^3 although a sound physical basis for the linearity has not been provided. According to the free volume theory, a plot of $\ln(D_i)$ vs $1/d^3$ should be linear. In Figure 5 is given the plot of $\ln(D_i)$ vs $1/d^3$, along with the effective diffusion coefficient data for polyimide, polysulfone and polycarbonate *(19)*. The relationship is fairly linear regardless of the chemical and physical structure of the polymers. This result suggests that the diffusion behavior is influenced primarily by the free volume. This result is also consistent with the high separation factor for gas mixture of H_2, CO_2, CH_4 and N_2 in conductive polyanilines which do not have any specific interaction with polarons *(1)*.

Oxygen diffusion in polyanilines. Two dynamic processes are taking place simultaneously in the doped polyaniline: oxygen diffusion and its reversible reaction with the polarons. Let us assume that the time scale for the reaction between the polarons and oxygen molecules is much smaller than that for the diffusion process. In other words, it is assumed here that the diffusion process is the rate determining step for mass transfer. Further, if we assume that the interaction between oxygen molecules and polarons is 1 to 1, the amount of oxygen adsorbed on the polarons at

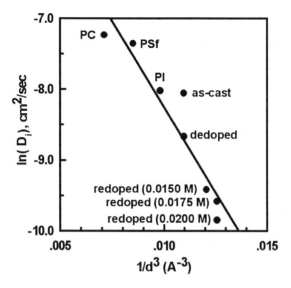

Figure 5. Plot of $\ln(D_i)$ as a function of $(1/d^3)$.

time t, M_t, can be calculated from the polaron concentration change with the oxygen exposure time. In Figure 6, M_t/M_∞ is plotted against the square root of time/thickness for a polyaniline membrane (30 μm thick) doped with 1 M HCl at an oxygen pressure of 137.9 kPa, where M_∞ is the equilibrium amount of oxygen adsorbed on the polarons at a given pressure. The data clearly show a linear relationship at the early stage. This result demonstrates that the mass transfer is Fickian and that diffusion is the rate determining step. In other words, the chemical reaction between oxygen molecules and polarons is much faster compared to the oxygen diffusion process.

When Fick's law is valid, the diffusion coefficient is readily calculated from the initial slope of a sorption curve (20).

$$D_i = \frac{\pi}{16}R^2 \tag{6}$$

where R is the initial slope of the linear portion of the reduced sorption curve defined as $R = d(M_t/M_\infty)/d(t/L^2)^{1/2}$ at the early stage. The effective diffusion coefficient was 1.88×10^{-10} cm^2/sec, which is comparable to that obtained from permeation experiments. The solid line in Figure 6 is the theoretical predictions by Fick's law with $D_i = 1.88 \times 10^{-10}$ cm^2/sec. This sorption experiment provides the diffusion coefficient which can not be easily obtained from the regular permeation experiments because of the low permeability in the doped samples.

392

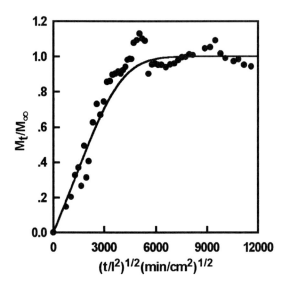

Figure 6. Reduced sorption curve of oxygen in polyaniline doped with 1 M HCl at 137.9 kPa oxygen. The solid line is from theoretical predictions by Fick's law with $D=1.88 \times 10^{-10}$ cm^2/sec.

CONCLUSIONS

The oxygen permeability of polyaniline membranes doped with HCl decreased whereas the O_2/N_2 selectivity increased with doping level regardless of their sample history. Polarons can act as oxygen carriers because of their specific and reversible reaction with oxygen molecules. It was experimentally observed that the oxygen permeability decreased with polaron concentration. The results demonstrated that facilitated oxygen transport can take place owing to the presence of the polarons, but it contributes to the oxygen permeability insignificantly. Instead, the oxygen diffusion coefficient correlates well with the free volume estimated from intersegmental distance. Therefore, the structure change upon doping seems to be a major factor in determining gas permeation behavior of these materials. In addition, the oxygen diffusion was found to be Fickian from the EPR-based sorption measurement in the doped polyaniline membranes.

Literature Cited

1. Mattes, B. R.; Anderson, M. R.; Conklin, J. A.; Reiss, H.; Kaner, R. B. *Synth. Met.* **1993**, *55-57*, 3655.
2. Anderson, M. R.; Mattes, B. R.; Conklin, J. A.; Reiss, H.; Kaner, R. B. *Synth. Met.* **1991**, *41-43*, 1151.

3. Anderson, M. R.; Mattes, B. R.; Conklin, J. A.; Reiss, H.; Kaner, R. B. *Science* **1991**, *252*, 1412.
4. Pellegrino, J.; Radebaugh, R.; Mattes, B. R. *Macromolecules* **1996**, *29*, 4985.
5. Oh, E. J.; Min, Y.; Wiesinger, T. M.; Manohar, S. K.; Scherrer, E. M.; Prest, P. J.; MacDiarmid, A. G.; Epstein, A. J. *Synth. Met.* **1993**, *55-57*, 977.
6. MacDiarmid, A. G.; Epstein, A. J. *Synth. Met.* **1994**, *65*, 103.
7. Epstein, A. J.; Ginder, J. M.; Zuo, F.; Bigelow, R. W.; Woo, H.-S.; Tanner, D. B.; Richter, A. F.; Huang W.-S.; MacDiarmid, A. G. *Synth. Met.* **1987**, *18*, 303.
8. Kang, Y. S.; Lee, H. J.; Namgoong, J.; Lee, H.; Jung, B. *Polymer* (to be submitted).
9. Kang, Y. S.; Hong, J.-M.; Jang, J.; Kim, U. Y. *J. Memb. Sci.* **1996**, *109*, 149.
10. Hong, J.-M.; Kang, Y. S.; Jang, J.; Kim, U. Y. *J. Memb. Sci.* **1996**, *109*, 159.
11. Gow, C. J.; Zukoski, C. F. *J. Colloid & Interface Sci.* **1990**, *136*, 175.
12. *Polymer Permeability* ; J. Comyn, Eds.; Elsevier, New York, **1985**.
13. Ray, A.; Richter, A. F.; MacDiarmid, A. G. *Synth. Met.* **1994**, *29*, E151.
14. Noble, R. D. *J. Memb. Sci.* **1991**, *60*, 297.
15. Cussler, E. L.; Aris, R.; Bhown, A. *J. Memb. Sci.* **1989**, *43*, 149.
16. Nishide, H.; Ohyanagi, M.; Okada, O.; Tsuchida, E. *Macromolecules* **1987**, *20*, 417.
17. Chen S.-A.; Lee, H.-T. *Macromolecules* **1995**, *28*, 2858.
18. Pouget, J. P.; Jozefowicz, M. E.; Epstein, A. J.; Tang, X.; MacDiarmid, A. G.; *Macromolecules* **1991**, *24*, 779.
19. Koros, W. J.; Fleming, G. K.; Jordan, S. M.; Kim, T. H.; Hoehn, H. H. *Prog. Polym. Sci.* **1988**, *13*, 339.
20. Crank, J. *The Mathematics of Diffusion*, 2nd ed., Oxford University Press, Oxford, 1975.

Chapter 27

Xenon-129 NMR as a Probe of Polymer Sorption Sites: A New View of Structure and Transport

J. M. Koons, W.-Y. Wen, P. T. Inglefield, and A. A. Jones

Carlson School of Chemistry, Clark University, Worcester, MA 01610

Xenon-129 NMR is used to study sorption sites in high density polyethylene, poly(4-methyl-1-pentene), and nafion. At room temperature, only a single resonance is observed for xenon dissolved in the first two polymers. However, for the poly(4-methyl-1-pentene), two lines are observed at 223 K which are interpreted as xenon dissolved in the amorphous phase and xenon dissolved in the crystalline phase. Exchange by translational diffusion collapses these two lines to a single line at room temperature. An independent measurement of domain size from proton spin diffusion experiments allows for an estimate of the xenon diffusion constant at the coalescence temperature. For comparison, at a temperature 223 K, polyethylene remains a single line consistent with xenon being dissolved in only the amorphous phase. At 203 K, the xenon resonance associated with the crystalline phase of poly(4-methyl-1-pentene) further splits into two lines, indicating two sorption sites as predicted by computer simulation. At 183 K, xenon in amorphous polyethylene also broadens the lines, indicating the heterogeneity of a polymeric glass. At room temperature, the spectrum of xenon in nafion consists of two lines corresponding to two different environments. The first environment is an amorphous fluorocarbon environment and the second is a disordered ion/polar group environment.

Xenon-129 NMR spectroscopy has been used for investigating a variety of microporous materials including zeolites(*1-3*), clathrates(*4-5*), polymers(*6-12*), and biological substances(*13-15*). The application of xenon-129 NMR has been reviewed(*16-19*) including xenon optical-pumping which leads to a dramatic enhancement of sensitivity(*20-21*).

Stengle and Williamson(*6*) measured the shift of xenon-129 in a variety of polymers as well as simple liquids(*22-22a*) and could correlate the observed dependence with refractive index and density. A better correlation was found between closely related substances such as the alkanes and polyethylene and poorer correlations were found for comparisons of more dissimilar systems.

In some cases, more than a single resonance is observed in a polymer system, indicating more than one sorption environment. Stengle and Williamson observed two overlapping resonances in linear low density polyethylene. Cheung and Chu*(11)* observed a shoulder for xenon-129 in semicrystalline poly(4-methyl-1-pentene) at low temperatures. The characterization of the sources of more than one resonance in these two cases was not completely determined, but in the cases of polymer blends, more thorough interpretations*(8,10,17)* were presented. In single phase blends, an average signal is observed while in phase separated systems two signals, one for each component of the blend, are observed. The size of the domains in the phase separated system was of the order of microns. Exchange between domains occurs as a result of translational diffusion. If the diffusion constant is known, domain size can be estimated*(8)* and if the domain size is known an average diffusion constant for the two components can be calculated*(10)*. Diffusion constants for xenon in some rubbers are available*(17)* which allowed for the estimate of domain size in a phase separated blend of two rubbers*(8)*. In a layered composite, two-dimensional NMR was used to obtain an average diffusion constant*(10)*. Measurement of a xenon diffusion constant in a glassy polymer, polystyrene, was made by observing the exchange of xenon between polystyrene microspheres and free xenon gas using xenon-129 NMR*(12)*. In all three of these studies of translational diffusion and exchange, the domain size was of the order of microns.

In this report, we wish to reduce the size scale of domains studied by xenon-129 NMR to the nanometer range. This would allow the examination of a wider variety of structural features of polymer systems. To reach this size scale, diffusion must be slowed so that exchange is in turn slowed. This can be accomplished by either lowering temperature or by studying polymeric glasses. In rubbers*(17)*, xenon diffusion constants are of the order of 10^{-7} cm^{2}s^{-1} at room temperature while in glasses*(12)* they are of the order of 10^{-9}cm^{2}s^{-1}. Another important factor affecting xenon spectra is the concentration of xenon in the polymer which is controlled by the pressure of xenon gas over the solid polymer.

The first system, high density polyethylene, is presented as a reference case. High density polyethylene is typically 90% crystalline and the sorbed xenon is only present in the amorphous phase. In this supposedly simple system, the effects of temperature and xenon pressure can be observed.

In the second system presented here, poly(4-methyl-1-pentene), (abbreviated PMP) solubility and permeability studies*(23)* indicate that gases can be sorbed into both the crystalline and the amorphous phases. In essentially all other semi-crystalline polymers, gas is not expected to be sorbed into the crystalline domains. However, computer simulations also indicate that sorption into the crystalline domains of PMP is possible*(24)*. The low temperature shoulder of xenon-129 spectrum in PMP observed earlier*(11)* was attributed to sorption into a second environment, probably the crystalline domains. Conditions of temperature and pressure will be sought where resolved signals from xenon-129 in amorphous and crystalline regions of PMP can be observed. To assign the resonances to the appropriate domains, the level of crystallinity in the PMP will be varied. The level of crystallinity can be determined from the solid state proton signal line shape. Also, to more definitely characterize gaseous diffusion in these systems, the size of the amorphous domains will be measured from proton spin diffusion measurements*(25,26)*. This gives an independent measurement of domain size so that an average diffusion constant can be determined.

Nafion is another polymer which has a complex relationship between morphological structure and transport*(27)*. This is an ionomer based on tetrafluoroethylene copolymerized with units which contain sulfonic acid groups on short fluorinated side chains. The morphology of this polymer is quite complicated and includes regions of crystalline fluorocarbon, amorphous fluorocarbon, ionic clusters and ionic groups in disordered regions. Nafion membranes are quite permeable to water and various models have been proposed to account for this behavior. Xenon-129 NMR should provide information on the various sorption environments which a xenon atom encounters in Nafion. To aid in the interpretation, the spectra of xenon-129 in Nafion, spectra of xenon in perfluorinated heptane and in polytetrafluoroethylene were also acquired.

Experimental

Xenon gas was purchased from Matheson Gas Products which contained xenon-129 in natural abundance at a level of 26.44%. Polymer samples were placed in heavy-walled, 10 mm NMR tubes (with a 7 mm i.d.) and packed to a depth of 55 to 60 mm. This corresponded to sample masses of about 1 gram. The tubes were then connected to a vacuum line with a transducer electrometer system to monitor the pressure. After removal of the sorbed air, xenon gas was introduced into the entire manifold. Placing the tubes into liquid nitrogen cryopumped xenon quantitatively into the tubes and the tubes were then flame sealed to give final pressures in the range of 3 to 15 atmospheres at ambient temperature.

High density polyethylene in the form of pellets was purchased from Scientific Polymer Products, Inc.. The molecular weight was reported as 125,000. Commercial PMP from two sources was studied. Beads from Polysciences, Inc., were measured as received and a second sample of these beads was annealed at a temperature of 140^0C for two days. A third sample of commercial PMP was provided by Professor Don Paul of the University of Texas at Austin. The preparation and characterization of this sample (labeled Q) has been reported[23] and the important parameter for the purposes of the current study is that the sample is 53.6% crystalline as determined by X-ray analysis. Nafion beads (NR50) were purchased from Aldrich Chemical Company in the hydrogen form. The equivalent weight per ionic group is 1250 g and the molecular weight is greater than 10^5. One sample of dry Nafion was studied and a second sample to which 5% by weight of water was added was also studied. A solid block of commercial poly(tetrafluoroethylene) was machined to fit into a 10 mm NMR tube. A hole was drilled down the center of the cylinder to allow the presence of free gas in the area of the receiver coil in the NMR experiment. Liquid perfluoroheptane was purchased from PCR Inc.

All xenon-129 NMR experiments were performed with a Varian Unity 500 NMR spectrometer interfaced to a Sun IPX workstation running the VNMRS 4.1A software package. A two-channel, 10 mm broadband probe was employed. The ^{129}Xe NMR spectra were acquired at 138.3 MHz with a 20 μs 90^0 pulse and a relaxation delay of about 100s, typically the number of scans required was 1000. At low temperatures, shorter pulse widths were used to avoid making the relaxation delay even longer. The spectral width was typically 250 ppm though for some samples broader widths were employed.

Solid state proton line shape experiments on PMP and Goldman-Shen proton spin diffusion experiments on PMP were performed on a Bruker MSL 300. The 90^0 pulse width was typically 2μs. The Goldman-Shen experiments were performed at a temperature of 350 K with a discrimination time of 50μs.

Results

Figures 1-3 display xenon-129 spectra as a function of temperature on polyethylene samples at pressures of 3, 5 and 10 atm, respectively. In all the xenon-129 spectra

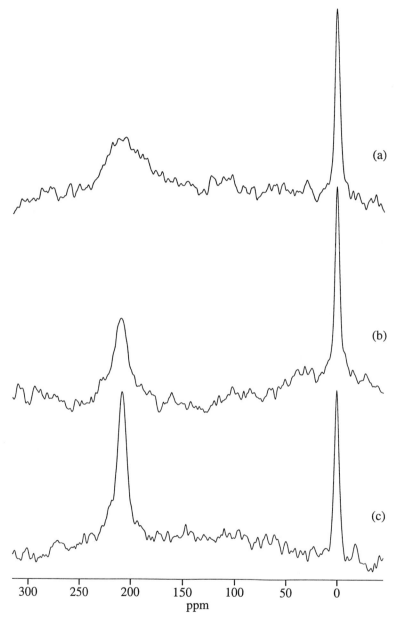

1. Spectra of Xenon-129 in High Density Polyethylene The pressure at 25^0C is 3 atm and the spectra are at temperatures of -90, -70 and -50^0C from top to bottom in the figure.

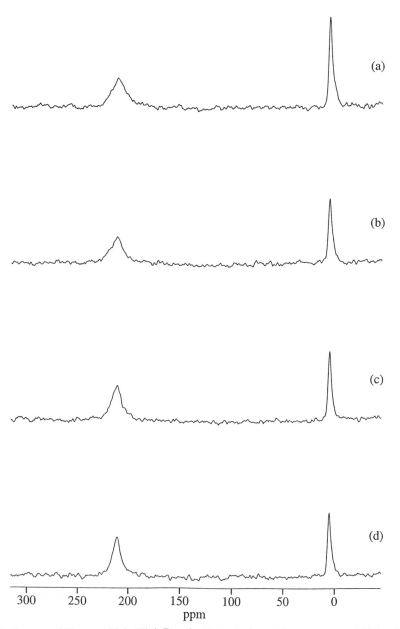

2. Spectra of Xenon-129 in High Density Polyethylene The pressure at 25⁰C is 5 atm and the spectra are at temperatures of -80, -70, -60 and -50⁰C from top to bottom in the figure.

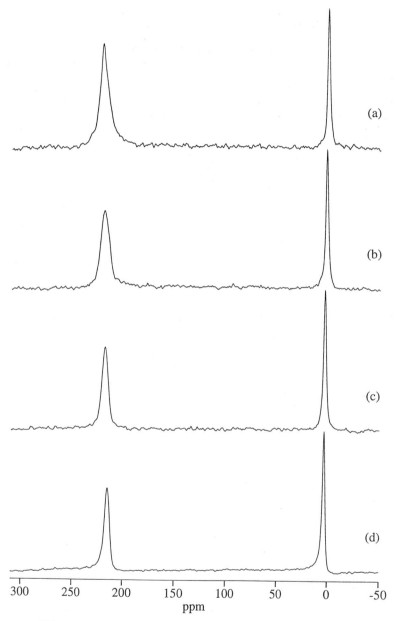

3. Spectra of Xenon-129 in High Density Polyethylene The pressure at 25⁰C is 10 atm and the spectra are at temperatures of -80, -70, -60 and -50⁰C from top to bottom in the figure.

the resonance for the gas is set to 0 ppm. Figures 4-6 display xenon-129 spectra as a function of temperature and at a pressure of 5 atm on PMP beads, annealed PMP beads and PMP film. Figure 7 shows xenon-129 spectra as a function of temperature and at a pressure of 10 atm on annealed PMP beads. Figure 8 shows the xenon-129 spectra at room temperature for Nafion beads under 10 atm and under 3 atm pressure of xenon, of wet Nafion beads under 4 atm, of polytetrafluoroethylene under 10 atm and perfluoroheptane under 5 atm.

Figure 9 is the solid state proton spectrum of the PMP film at 350 K which is used to determine the level of crystallinity. This is a typical proton spectrum for a semicrystalline polymer taken above the glass transition temperature and consists of a narrow Lorentzian line from the amorphous component superimposed on a broader Gaussian line from the crystalline component, as has been observed in related systems. Figure 10 shows the Goldman-Shen free induction decays for a series of mix times. In the Goldman-Shen experiment a discrimination time of 50µs was employed and free induction decays were acquired after mix times ranging from 0.3 to 100 ms. At short mix times, the free induction decay consists of the narrow Lorentzian component and as the mix time is lengthened, the broader Gaussian component grows back at the expense of the Lorentzian component(25,26). Figure 11 shows the recovery curve for the magnetization associated with the Gaussian component which can be fit to obtain a domain size if the spin diffusion constant is known.

Interpretation

In one sense, the polyethylene spectra shown in Figures 1-3 are straightforward. At most temperatures and pressures, the spectra consist of one resonance for the free gas and a second at about 200 ppm for xenon-129 in the amorphous regions. However, at a particular temperature, the line is narrower at higher pressures. Similarly, at a particular pressure, the line is narrower at higher temperatures. The major source of line broadening is the result of the inhomogeneous environments encountered by xenon gas in the amorphous domains. Local density fluctuations expected in any amorphous polymer would lead to fluctuations in chemical shift. If translational diffusion is sufficiently fast, the observed resonance is an average of the environments. As diffusion slows, individual environments would cause inhomogeneous broadening. At lower temperatures, diffusion obviously slows, allowing a more complete reflection of the inhomogeneity of the amorphous regions in the line width. Similarly, higher pressures produce significant narrowing. In this case, the gas is plasticizing the polymer, leading to more rapid exchange between environments. In this regard, it is important to note that the penetrant is not just a reporter of the environments but may also be affecting the polymer.

At the lowest temperature (-90°C) and at the lowest pressure (3 atm), the xenon spectrum for the sorbed line is rather broad, possibly asymmetric or bimodal. This would indicate an asymmetric or bimodal distribution of sorption environments on a fairly short length scale. More complicated descriptions of amorphous environments might consider the effects of the presence of an amorphous/crystalline interfacial region or of two sorption environments as proposed in the dual mode model. The low pressure, low temperature spectra being discussed have the poorest signal to noise and better spectra are needed to characterize this situation properly.

To interpret the xenon-129 spectra of Figures 4-7 it is required that the proton line shape and proton spin diffusion results for the three PMP samples be reported first. The per cent crystallinity was determined by fitting the proton line

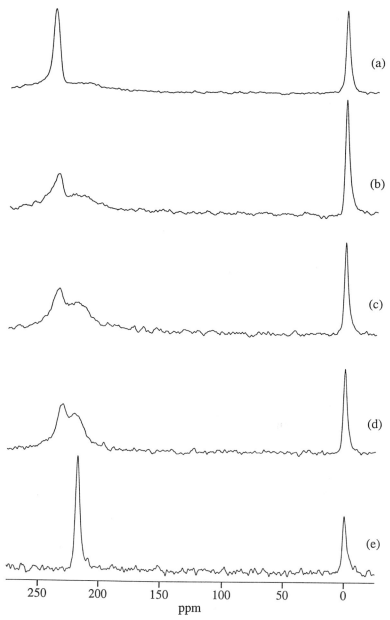

4. Spectra of Xenon-129 in Non-Annealed Beads of Poly(4-methyl-1-pentene) The pressure at 25^{0}C is 4 atm and the spectra are at temperatures of -80, -70, -60, -50 and +25^{0}C from top to bottom in the figure.

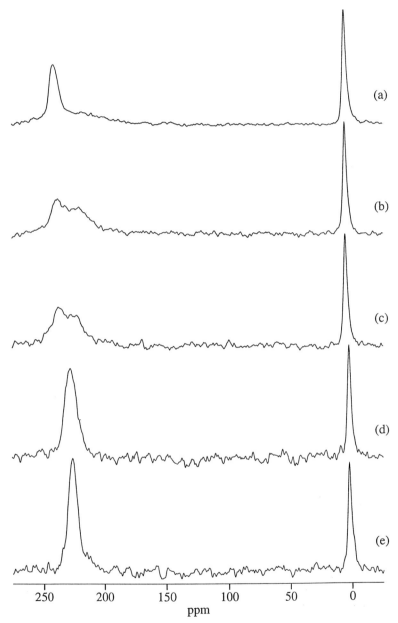

5. Spectra of Xenon-129 in Annealed Beads of Poly(4-methyl-1-pentene) The pressure at 25⁰C is 4 atm and the spectra are at temperatures of -80, -60, -50, -40 and -30⁰C from top to bottom in the figure.

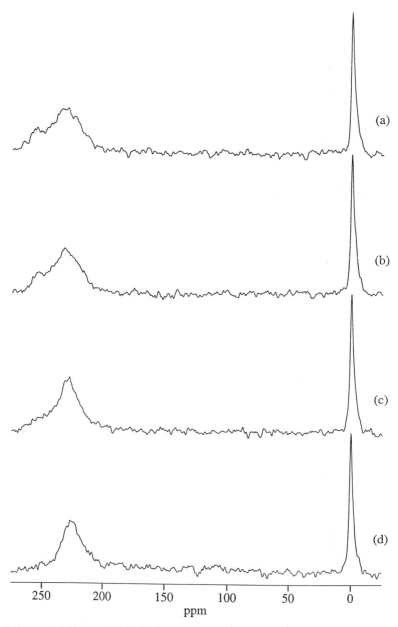

6. Spectra of Xenon-129 in Film of Poly(4-methyl-1-pentene) The pressure at 25⁰C is 4 atm and the spectra are at temperatures of -80, -70, -60 and -50⁰C from top to bottom in the figure.

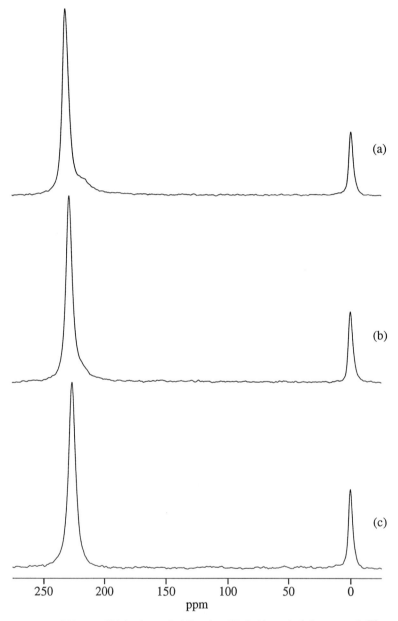

7. Spectra of Xenon-129 in Annealed Beads of Poly(4-methyl-1-pentene) The pressure at 25°C is 10 atm and the spectra are at temperatures of -70, -60 and -50°C from top to bottom in the figure.

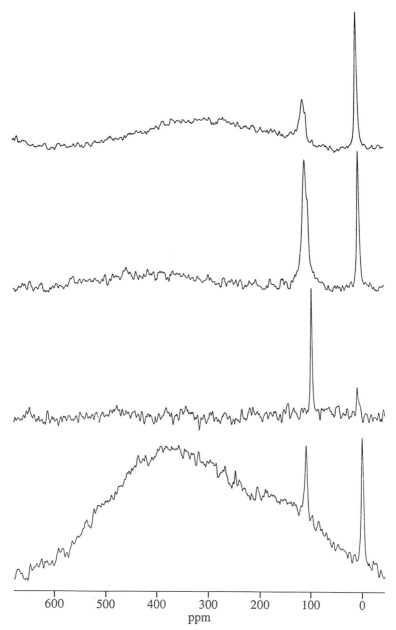

8. Spectra of Xenon-129 in Fluorinated Polymers at Ambient Temperature The top spectrum is Nafion beads at 10 atm pressure. The second spectrum is of Nafion beads at 3 atm pressure. The third spectrum is of Teflon at 10 atm. The fourth spectrum is of Nafion containing 5% water at 3 atm.

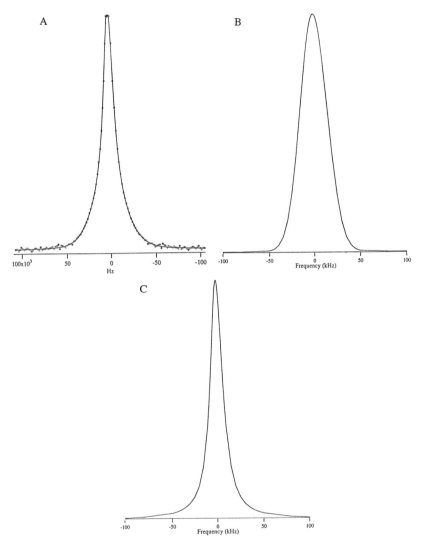

9. (a) The solid spectrum of the PMP film at 350°K which is used to determine the level of crystallinity.
(b) The Gaussian (crystalline) component of the fit. The T_2 for this component is 8.9 μs and constitutes 49% of the experimental spectrum.
(c) The Lorentzian (amorphous) component of the fit. The T_2 for this component is 19.7 μs.

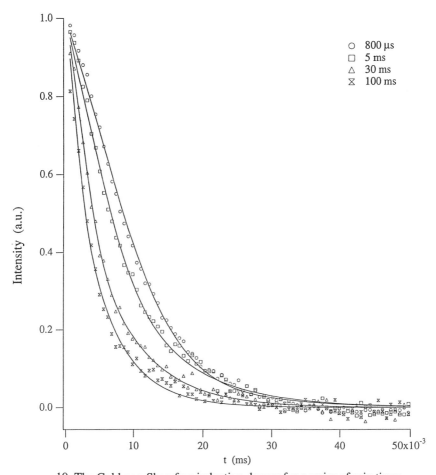

10. The Goldman-Shen free induction decays for a series of mix times.

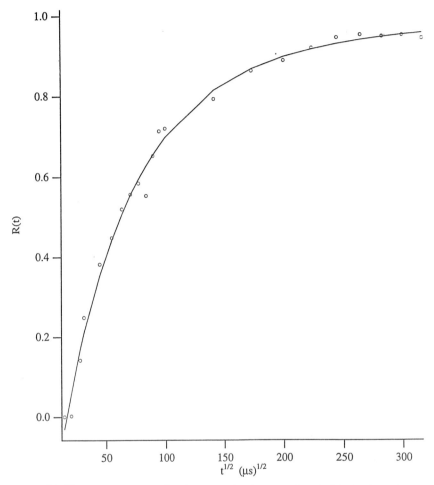

11. The recovery curve for the magnetization associated with the Gaussian component along with the fit to obtain a domain.

shapes such as Figure 9 to a Gaussian component plus a Lorentzian component where the per cent crystallinity corresponds to the per cent Gaussian in the line shape fit. The line in Figure 9 is the fit. For the unannealed PMP beads the per cent crystallinity is 28.9%; for the annealed PMP beads, 41.5%; and for the film, 49.4%. The X-ray result(23) on the film was 53.6% which is in reasonable agreement.

In all cases, the xenon-129 spectra of PMP in Figures 4-6 show structure for the sorbed peak below -50⁰C. In particular, Figures 4 and 5 show an obvious splitting of the sorbed peak at -50⁰C into two roughly equal components. The component with the lower chemical shift broadens as temperature is lowered below -50⁰C while the peak with the higher shift remains sharp. In Figure 6, the PMP sample with the highest crystallinity, the higher shift component is dominant. The broadening of the lower shift component with decreasing temperature and the reduced intensity of the lower shift component in the higher crystallinity sample both point to this component arising from xenon in the amorphous environment. The higher shift component therefore arises from xenon-129 in the crystalline environment.

The importance of pressure on the xenon-129 spectra is demonstrated in Figure 7. At most temperatures, the spectrum consists of a more intense line with a shoulder, though at -80⁰C there is a second lower intensity line. As with the spectra in polyethylene, higher pressures of xenon lead to narrower lines, which was attributed to more rapid exchange between environments. In PMP at low xenon pressures, the two approximately equal intensity lines are good evidence for two environments while exchange appears to obscure this information at higher pressures.

In Figure 6 for the most crystalline sample, another feature is noticeable in the spectrum. At -70, -80 and -90⁰C, a less intense resonance is seen at 250 ppm from the free gas resonance. In Figures 4 and 5, the two main resonances assigned to the amorphous and crystalline regions are at about 220 and 230 ppm, respectively. In the computer simulation(24) of methane gas in PMP, two different sites for gas sorption in the crystalline phase were noted. The simulation suggests that this is due to the gas being partly absorbed into the helix surface with one site being outside the helix perimeter and the other within. Both xenon and methane are of comparable diameter, so the second feature which is more apparent in the higher crystallinity spectra may reflect a second distinct environment for xenon in the crystalline phase. In retrospect, if one goes back to Figures 4 and 5, the peak at 250 ppm is detectable in some cases (Figure 5, -70⁰C) but would not justify further consideration without the information from the higher crystallinity sample.

An estimate of the average xenon diffusion constant in PMP can be made from the coalescence of peaks if independent domain size information(8,17) is available. A domain size for the amorphous region can be determined by fitting the proton magnetization recovery curve shown in Figure 11. The equation describing the recovery of the magnetization(26) is

$$R(t) = 1 - \phi_x(t)\,\phi_y(t)\,\phi_z(t) \qquad (1)$$

where

$$\phi(t) = \exp(Dt/b^2)\,\text{erfc}(Dt/b^2)^{1/2} \qquad (2)$$

To obtain a value of b, the domain size, the spin diffusion constant D is estimated from the equation

$$D = 0.13a^2 / T_{21} \qquad (3)$$

The average distance, a, between protons was set at 1.8 Angstroms and the value of spin-spin relaxation time of the amorphous component, T_{21}, determined from the proton spectra is 9.51 ms. The value of the amorphous domain, b, for the annealed sample of PMP is 57 Angstroms.

The average diffusion constant for the xenon gas in PMP, D_{Xe}, can then be estimated from the equation

$$b = \sqrt{6tD_{Xe}} \qquad (4)$$

where the time t is determined from the low temperature separation in hz, δv, of the coalescing lines.

$$t = \sqrt{2} / \delta v \qquad (5)$$

The separation of the xenon resonances from the amorphous and crystalline domains is 11.5 ppm or 1590 Hz at -70^{0}C. This leads to an average diffusion constant, D_{Xe}, of 6.4x 10^{-11}cm^2/s at the coalescence temperature of about -40^{0}C. There are a number of simplifications and approximations in this calculation which could be reduced by a more careful analysis of the line shape collapse with temperature. Nevertheless, the essence of the relationship between line shape collapse, translational diffusion and domain size is contained in this analysis. The value of D_{Xe} obtained here for PMP is close to the value for amorphous polystyrene at the same temperature. Values for D_{Xe} in glassy polystyrene were measured over a 70^0 range above room temperature(12), yielding an activation energy which can be used to estimate a value of 4.4x10^{-11}cm^2/s at -40^{0}C. The PMP D_{Xe} is an average of the value for the amorphous and crystalline domains and the glass transition temperature for PMP is 70 degrees lower than for polystyrene. At 35^{0}C Puelo, Paul and Wong(23) obtained values of D for methane in PMP of 9.1x10^{-7} and 3.2x10^{-7} cm^2/s for the crystalline and amorphous phases, respectively.

The Nafion spectrum in Figure 8 consists of two overlapping lines near 100 ppm and a very broad line centered at about 350 to 400 ppm. A similar spectrum is observed at both 10 atm and 3 atm. In the sample of Nafion containing 5 wt% water, the broad line is more intense and only one line is observed near 100 ppm. To assist in the interpretation of this spectrum, Figure 8 also contains the spectrum of perfluoroethylene. Only one line is observed near 100 ppm. Presumably this is the resonance for xenon-129 in the amorphous regions of perfluoroethylene. However, this shift is quite different from the shifts of hydrocarbon polymers which lie in the range of 180 to 250 ppm. The spectrum for xenon-129 in the simple liquid perfluoroheptane with a peak near 90 ppm is similar to the spectrum of xenon-129 in perfluoroethylene reinforcing the interpretation of this unusual shift as typical for xenon-129 in a perfluoroalkane system. Stengle, Reo and Williamson(22) found the shifts of xenon-129 in a variety of liquids to roughly correlate with the quantity $(n^2 - 1)^2/(2n^2 +1)^2$, where n is the refractive index of the liquid. The refractive index of perfluoroheptane(28) is 1.2618, while most organic compounds are about 1.5 near 20^{0}C. Thus the lower shift of xenon in perfluoroheptane is expected from the Stengle, Reo and Williamson correlation(22).

The broad peak centered at 350 to 400 ppm in Nafion is not typical of polymeric systems and is assigned to ion/polar group containing regions as distinct from the non-polar perfluoroethylene regions assigned above. The intensity of this peak is larger in the sample which contained 5 wt% water and water is only expected to be in the ion/polar group containing regions. Metal ions are found in small clusters in Nafion and other ionomers but the majority of ions are distributed in disordered regions. These interconnected, disordered regions account for the high water mobility through Nafion and xenon NMR reflects the disorder of this region as a very inhomogeneous broad line width.

Conclusions

The spectra of xenon-129 in high density polyethylene are relatively simple, with only a single resonance about 200 ppm for the free gas resonance. The sorbed xenon-129 is present in the amorphous phase and the line width increases with decreasing temperature and xenon pressure. At the lowest temperature and pressure, the line width is about 50 ppm and the line shape is asymmetric, indicating a distribution of sorption environments in the amorphous component.

In PMP at temperatures below about $-40^{0}C$, two xenon resonances are observed for sorbed xenon. One is from crystalline domains and the other is from amorphous domains. At higher temperatures, only a single resonance is observed because of exchange of xenon between the two environments by translational diffusion. Measurement of domain size by proton spin diffusion allows for a determination of the average diffusion constant. The amorphous domain size in the samples studied was near 60 Angstroms, indicating that xenon-129 NMR can be used to study sorption environments on a short length scale. Indeed, the crystalline domains appear to have two sorption environments which are distinguishable by xenon-129 NMR. Lower temperatures, about $-70^{0}C$, were required to resolve the two sorption sites in the crystal. Higher pressure increases exchange or modifies the polymer and obscures the ability to detect different sorption environments.

In Nafion, two general types of environments were noted. One near 100 ppm is characteristic of an amorphous fluoropolymer environment and the other was attributed to ionic environments. The line width of xenon in the ionic region was very large indicating a very heterogeneous environment.

Acknowledgment

This work is supported by the Army Research Office Grant #DAAG04-96-1-0094. The authors also wish to thank Professor Don Paul for suggesting xenon NMR measurements on PMP.

Literature Cited

1. Ito, T.; Fraissard, J., *J. Chem. Phys.* **1982**, *76*, 5225.
2. Fraissard, J.; Ito, T., *Zeolites* **1988**, 8, 350.
3. Bansal, N.; Dybowski, C., *J. Magn. Reson.* **1990**, *89*, 21.
4. Davidon, D.W.; Handa, Y.P.; Ripmeester, J.A., *J. Phys. Chem.* **1986**, *90*, 6549.
5. Ripmeester, J.A.; Ratcliffe, C.I.; Tse, J.S., *J. Chem. Soc. Faraday Trans. I* **1988**, *84*, 3731.
6. Stengle, T.R.; Williamson, K.L., *Macromolecules*, **1987**, *20*, 1428.

412

7. Walton, J.H.; Miller, J.B.; Roland, C.M., *J. Polym. Sci.: Part B: Polym. Phys.* **1992**, *30*, 527.
8. Walton, J.H.; Miller, J.B.; Roland, C.M.; Nagode, JB., *Macromolecules,* **1993**, *26*, 4052.
9. Kentgens, A.P.M.; van Boxtel, H.A.; Verweel, R.J.; Veeman, W.S., *Macromolecules*, **1991**, *24*, 3712.
10. Tomaselli, M.; Meier, B.H.; Robyr, P.; Suter, U.W.; Ernst, R.R., *Chem. Phys. Lett.* **1993**, *205*, 145.
11. Cheung, T.T.; Chu, P.J., *J. Phys. Chem.* **1992**, *96*, 9551.
12. Simpson, J. H.; Wen, W.-Y.; Jones, A. A.; Inglefield, P.T.; Bendler, J.T., *Macromolecules,* **1996**, *29*, 2138.
13. Miller, K.W.; Reo, N.V.;Uiterkamp, A.J.M.S.; Stengle, D.P.; Stengle, T.R.; Williamson, K.L., *Proc. Natl. Acad. Sci. USA,* **1981**, *78*, 4946.
14. Tilton, Jr., R.F.; Kuntz, I.D., *Biochemistry,* **1982**, *21*, 6850.
15. Albert, M.S.; Cates, G.D.; Driehuys, B.; Happer, W.; Saam, B.; Springer, Jr., C.S.; Wishnia, A., *Nature,* **1994**, *370*, 199.
16. Barrie, P.J.; Klinowski, J., *Progress NMR Spectroscopy,* **1992**, *24*, 91.
17. Walton, J.H., *Polym. & Polym. Composites,* **1994**, *2*, 35.
18. Raftery, D.; Chmelka, B.F., *NMR Basic Principles and Progress,* **1994**, *30*, 111.
19. Dybowski, C.; Bansal, N., *Ann. Rev. Phys. Chem.* , **1991**, *42*, 433.
20. Raftery, D.; Long,H.; Meersmann, T.; Grandinetti, P.J.; Reven, L.; Pines, A., *Phys. Rev. Lett.,* **1991,** *66*, 584.
21. Raftery, D.; Reven, L.; Long, H.; Pines, A.; Tang, P.; Reimer, J.A., *J. Phys. Chem.,* **1993,** *97*, 1649.
22. Stengle, T. R.; Reo, N. V.; Williamson, K. L., *J. Phys. Chem.* **1981**, *85*, 3772.
22a.Miller, K.W.; Reo, N.V.; Uiterkamp, A.J.M.S.; Stengle, D.P.; Stengle,T.R.; Williamson, K.L., *Proc. Natl. Acad. Sci. USA,* **1981**, *78*, 4946.
23. Puelo, A. C.; Paul, D. R.; Wong, P. K.; *Polymer,* **1989**, *30* , 1357.
24. Mueller-Plathe, F., *J. Chem. Phys.* **1995**, *103*, 4346.
25. Goldman, M.; Shen, L. *Phys. Rev.* **1966**, *144*, 321.
26. Cheung, T. T. P.; Gerstein, B. C., *J. Appl. Phys.* **1981**, *52*, 5517.
27. Eisenberg, A. and Yeager, H. Eds. *Perfluorinated Ionomer Membranes,* ACS Symposium Series 180 , American Chemical Society, Washington DC, 1982.
28. Weast, R. C. Ed., *Handbook of Chemistry and Physics,* CRC Press Inc., Cleveland, Ohio, 1975.

Chapter 28

Dissolution Mechanism of Glassy Polymers: A Molecular Analysis

B. Narasimhan[1] and N. A. Peppas[2]

[1]Department of Chemical and Biochemical Engineering, Rutgers University, 98 Brett Road, Piscataway, NJ 08854–8058
[2]School of Chemical Engineering, Purdue University, West Lafayette, IN 47907–1283

The dissolution of glassy polymers in solvents was analyzed using a combination of molecular theories and continuum mechanics arguments. The effect of molecular parameters such as the disentanglement rate of the polymer chain and the diffusion coefficient on the dissolution mechanism was established through the solution of the model equations. Approximate solution of the model equations under pseudo-steady state conditions yielded information on the temporal evolution of the gel layer thickness. Simulations established the polymer molecular weight as the most important parameter controlling the dissolution mechanism. The model predictions were compared to experimental data for poly(ethylene glycol) dissolution in water and for polystyrene dissolution in methyl ethyl ketone. The predictions agreed well with the data within experimental error.

Polymer dissolution is an important phenomenon in polymer science and engineering. For example, in microlithographic applications, selectively irradiated regions of a photosensitive polymer are dissolved in appropriate solvents to obtain desired circuit patterns (*1*). In the field of controlled drug release, zero-order drug release systems have been designed (*2*) by rendering the polymer dissolution phenomenon as the controlling step in the release process. Polymer dissolution also finds applications in membrane science (*3*), treatment of unsorted plastics for recycling (*4-5*), the semiconductor industry (*6*) and in packaging applications (*7*).

Polymer dissolution in a solvent involves two transport processes, namely, solvent diffusion and chain disentanglement. When an uncrosslinked, amorphous, glassy polymer is brought in contact with a thermodynamically compatible solvent, the latter diffuses into the polymer and when the solvent concentration in the swollen polymer reaches a critical value, chain disentanglement begins to dominate and eventually the polymer is dissolved. Ueberrieter and co-workers (*8-10*) summarized the various types of dissolution and the surface structure of glassy polymers during dissolution. Important parameters such as the polymer molecular weight, the solvent diffusion coefficient, the gel layer thickness, the rate of agitation and temperature were identified. Since then, various mathematical models have been proposed to understand polymer dissolution.

The approaches to model polymer dissolution can be broadly classified as :

i.   use of phenomenological models and Fickian equations (11-14);

ii. models with external mass transfer as the controlling resistance to dissolution (15);

iii. models based on stress relaxation (16-17); and

iv. analysis using anomalous transport models for solvent transport and scaling laws for actual polymer dissolution (18-19).

The objectives of the present research were: (i) to develop a solvent transport model accounting for diffusional and relaxational mechanisms, in addition to effects of the viscoelastic properties of the polymer on the dissolution behavior; (ii) to perform a molecular analysis of the polymer chain disentanglement mechanism, and study the influence of various molecular parameters like the reptation diffusion coefficient, the disentanglement rate and the gel layer thickness on the phenomenon; and (iii) to experimentally characterize the dissolution phenomenon by measuring the temporal evolution of the various fronts in the problem.

Mathematical Analysis

A polymer dissolution process is depicted in Figure 1. During the initial stage of the dissolution process, a glassy polymer of thickness 2l starts swelling due to the penetration of the solvent into it and the simultaneous transition from the glassy to the rubbery state. Thus, two distinct fronts are observed - a swelling interface at position R and a polymer/solvent interface at position S. Front R moves inwards while front S moves outwards. When the concentration of the penetrant in the polymer exceeds a critical value, macromolecular disentanglement begins. After the concentration exceeds the critical value, true dissolution commences. After macromolecular disentanglement is complete, the polymer is dissolved. During this time, front R continues to move towards the center of the slab, while front S moves inwards as well. After the disappearance of the glassy core, only front S exists and it continues to move inwards towards the center of the slab till all of the polymer is dissolved.

The entire concentration field is divided into three regimes. We define the swollen polymer (i.e. the region R < x < S in Figure 1) as the *concentrated regime*. We postulate the existence of a diffusion boundary layer adjacent to the rubbery - solvent interface, S, through which the disentangled chains diffuse. The diffusion boundary layer is defined as the *semi-dilute regime* and has a constant thickness δ. When the polymer is fully dissolved, the disentangled chains move freely in the solvent and exhibit Brownian motion. This region is referred to as the *dilute regime*. These regimes are depicted in Figure 2. Having characterized the entire concentration range into these regimes, our approach is to write transport equations in each of these regimes and couple them through conditions at the moving boundary.

A model for one-dimensional solvent diffusion followed by chain disentanglement in amorphous, uncrosslinked, linear polymers was developed by us (21). This model describes transport in a polymer film only in the x direction and the model assumptions are discussed in detail elsewhere (21). A two-component system is considered, with the solvent being component 1 and the polymer being component 2. The present work will focus on refining our previous model by providing more molecular insight into the mechanism of chain disentanglement. This aspect is discussed in the following two sections.

The equation for solvent transport consists of a diffusional term and a term due to osmotic pressure. The osmotic pressure term arises by using linear irreversible thermodynamics arguments (20). The osmotic pressure is related to the viscoelastic properties of the polymer through a constitutive equation. In our analysis, the Maxwell element has been used as the constitutive model. Thus, the governing equations for solvent transport in the concentrated regime are

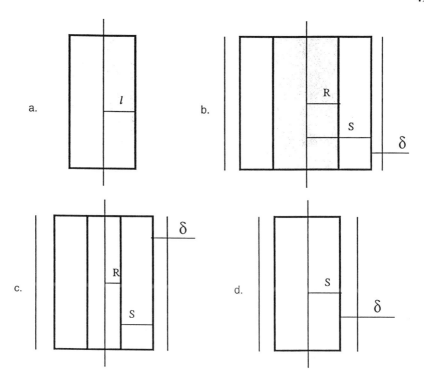

Figure 1. Schematic representation of a one-dimensional solvent diffusion and polymer dissolution process: a) initial slab thickness, $2l$; b) initial swelling step showing the increasing position of the rubbery/solvent interface, S, and the decreasing position of the glassy/rubbery interface, R; c) beginning of the dissolution step showing the decreasing position of the interface S along with the decreasing position of the interface R; and d) final dissolution step where the slab has been transformed into a rubbery material (disappearance of interface R) and the position of interface S still decreases.

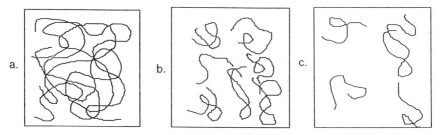

Figure 2. The solvent concentration field varies during polymer dissolution in a controlled release device. Before dissolution starts, there is no disentanglement in a swellable system (concentrated regime, a). The onset of dissolution in the diffusion boundary layer leads to the semi-dilute regime (b). Finally, the dissolution is complete and the disentangled chains exhibit Brownian motion in the solvent in the dilute regime (c).

$$\frac{\partial v_1}{\partial t} = \frac{\partial}{\partial x} [D_{12} \frac{\partial v_1}{\partial x}] + \frac{\partial}{\partial x} [\frac{D_{12} \overline{V}_1 v_1}{RT(1 - v_1)(1 - 2\chi v_1)} \frac{\partial \sigma_{xx}}{\partial x}] \qquad (1)$$

$$\frac{\partial \sigma_{xx}}{\partial t} = - \frac{\sigma_{xx}}{(\eta/E)} + \frac{E}{(1 - v_1)^2} \frac{\partial v_1}{\partial t} \qquad (2)$$

Here, v_1 is the volume fraction of the solvent in the swollen polymer, σ_{xx} is the stress developed within the polymer, D_{12} is the mutual diffusion coefficient, χ is the polymer-solvent interaction parameter, $\overline{V}_1$ is the molar volume of the solvent, T is the temperature, E is the modulus of the polymer and η is the viscosity of the polymer. Equation (1) is valid in the region between x = R and x = S.

As the polymer chains disentangle, they move out of the gel-like phase through a diffusion boundary layer. The chain transport through this boundary layer is described as

$$\frac{\partial v_2}{\partial t} = \frac{\partial}{\partial x} [D_p \frac{\partial v_2}{\partial x}] - \frac{dS}{dt} \frac{\partial v_2}{\partial x} \qquad (3)$$

Here, D_p is the polymer diffusion coefficient. The above equation is valid in a diffusion boundary layer of constant thickness δ. Appropriate initial and boundary conditions are written (20-21) for equations (1) through (3). The most important boundary condition is the one at the position S. It is postulated that the flux of disentangling polymer chains is equal to zero till a time equal to the reptation time (22-24) of the polymer elapses, and after this, the flux proceeds at a disentanglement rate, k_d. It is worthwhile to note that the positions R and S are both moving boundaries. This completes the formulation of the moving boundary problem.

Molecular Analysis of the Diffusion Coefficient. The next step was to obtain expressions for the diffusion coefficients and the disentanglement rate of the polymer. It has been shown by us that the mode of diffusion of the polymer undergoes a change (20) from a reptation-type to a classical Zimm type as the polymer dissolves. Thus, expressions were needed for the "reptation" diffusion coefficient and the Zimm diffusion coefficient.

The reptation theory (22-23) for polymer solutions predicts that the viscosity of the solution, η, varies with the polymer molecular weight, M, as

$$\eta \sim M^3 \qquad (4)$$

However, experiments (25) have indicated that the molecular weight exponent is close to 3.4. From the models that have been proposed to improve the reptation approach, it appears that in concentrated polymer solutions in good solvents, the solution viscosity is given as

$$\eta = c^{0.5} (cM)^\beta \qquad (5)$$

where c is the number of monomers per unit volume. If β equals 3.4, as is usually observed, then we obtain

$$\eta = c^{3.9} M^{3.4} \tag{6}$$

A mean field description (24) of the plateau modulus, G_N, shows that it is independent of the polymer molecular weight and that

$$G_N \sim c^2 \tag{7}$$

Also by definition,

$$\frac{\pi^2}{12} G_N t_{rep} = \eta \tag{8}$$

Hence,

$$t_{rep} \sim c^{1.9} M^{3.4} \tag{9}$$

The reptation diffusion coefficient is defined as

$$D_2 = \frac{1}{6} \frac{r_g^2}{t_{rep}} \tag{10}$$

where r_g is the radius of gyration of the polymer. The latter is given (26) as

$$r_g = 0.408248 \, b \, N^{0.5} \tag{11}$$

where b is the Kuhn length. Hence, D_2 scales to

$$D_2 \sim c^{-1.9} M^{-2.4} \tag{12}$$

Using exact expressions (24) for G_N and η, an expression for the reptation time can be derived as

$$t_{rep} = \frac{3\pi\eta_1}{kT} \left(\frac{N}{N_e}\right)^{3.4} a^3 \tag{13}$$

where η_1 is the solvent viscosity, k is the Boltzmann constant, T is the temperature, N is the number of monomers in the chain, N_e is the number of monomers per entanglement segment and a is the average length of a segment between entanglements. The average length of a segment between entanglements (23) can be expressed as

$$a = b \, N_e^{0.5} \tag{14}$$

De Gennes (23) has shown that

$$N_e \sim r_g f\left(\frac{c^*}{c}\right) \tag{15}$$

where f is a dimensionless function and c^* is the crossover concentration between concentrated and semi-dilute regimes. From the above definitions of r_g and a, it can be shown that

$$N_e = \frac{3.9219 \times 10^{29}}{c} \tag{16}$$

Using equations (14) and (16) in equation (13), we have

$$t_{rep} = \frac{0.01368 \, \eta_1}{T} \, v_2^{1.9} \, N^{3.4} \tag{17}$$

Substituting equation (14) and equation (17) in equation (10), we obtain

$$D_2 = \frac{4.8157 \times 10^{-14} \, T}{\eta_1} \, v_2^{-1.9} \, N^{-2.4} \tag{18}$$

Molecular Analysis of the Chain Disentanglement Rate. We consider a monodisperse system of chains which relaxes the anisotropy induced by an instantaneous strain imposed at time zero. This is the starting point used by Doi and Edwards (27). The effect of the neighboring chains is felt by any one chain in the form of a tube of constraints (22), which bounds the chain laterally. The diameter of this tube is a, which depends on the polymer concentration, but is independent of the polymer molecular weight. The curvilinear length of the chain, L, is called its primitive path (28). This length is proportional to the polymer molecular weight and is given by

$$L \, a = R^2 \tag{19}$$

where R is the end-to-end distance of the polymer chain. As soon as the instantaneous strain is imposed on the polymer chain, it begins to relax to equilibrium through the process of tube renewal which begins from the ends. As the chain reptates back and forth along its primitive path, segments of the original tube (at t = 0) vanish and are "renewed" or replaced by new ones. If F(t) is the fraction of polymer that has not relaxed, i.e., is still within the deformed original tube, then it can be shown (29) that

$$F(t) = \frac{8}{\pi^2} \sum_{n \, odd} \frac{1}{n^2} \exp\left(-\frac{n^2 \pi^2 D}{L^2} t \right) \tag{20}$$

Here, D is the reptation diffusion coefficient, which is inversely proportional to the chain length.

According to the viscoelastic theory of Doi and Edwards (24), the time dependent shear modulus, G(t), is related to the plateau modulus, G_N^0, as

$$G(t) = G_N^0 \, F(t) \tag{21}$$

The plateau modulus is independent of time and is given (27) by

$$G_N^0 = \frac{4 \, c \, k_B \, T \, L}{5 \, a} \tag{22}$$

where c is the number of polymer chains per unit volume, k_B is the Boltzmann constant and T is the absolute temperature. It is to be noted that in all of the above equations, it has been assumed that the tube diameter, a, is a constant.

The concept of time-dependent reptation is not a new one (*30-32*). For the problem to be self-consistent, F(t) should be related to a (t). Clearly, when this happens, equation (20) is no longer valid. During the process of chain disentanglement, the mobility of the polymer chain increases indicating that within the reptation framework, the tube diameter must increase. To obtain a relationship between F and a , we utilize the result from equation (7).

Disentanglement occurs when the solvent concentration in the polymer exceeds a critical value (*20*). Hence, "dilution" of the polymer by the solvent results in constraint release, leading to disentanglement. Figure 3 shows the test polymer chain in an entangled system. Figure 4 shows the neighboring chains represented as obstacles, which can move. When these obstacles move away from the chain, the tube diameter increases, resulting in disentanglement. Thus, the relationship between F and a can be written like a power law of the form

$$a(t) = a_0 \, F(t)^{-k} \tag{23}$$

where a_0 is the tube diameter at very short relaxation times. Since R is independent of concentration, L can be substituted from equation (19) into equation (22) to yield

$$G_N^0 = \frac{4 \, c \, k_B \, T \, R^2}{5 \, a^2} \tag{24}$$

Combining equations (7) and (24), we can obtain

$$a \sim c^{-0.5} \tag{25}$$

Since disentanglement occurs only as a result of solvent penetration, we can use the result of equation (25) in equation (23) to yield k = 0.5. In other words, we treat the mobility of the chains surrounding the test chain similar to solvent mobility and this makes the result of equation (25) analogous to the one in equation (23). Since equation (19) is always true, we can write

$$L = L_0 \, F(t)^{0.5} \tag{26}$$

where L_0 ($= R^2/a_0$) is the initial primitive path of test polymer chain. Clearly, the above arguments become invalid when the relaxation process approaches equilibrium, i.e., F = 0. Hence, for the self-consistent tube model to hold, we must have

$$a \ll R \ll L \tag{27}$$

To obtain actual solutions, we need to solve a diffusion problem, where the primitive path, L, varies with time. The expression for F(t) now becomes

$$F(t) = \frac{8}{\pi^2} \sum_{n \ \mathrm{odd}} \frac{1}{n^2} \exp \left(- n^2 \, \pi^2 \int_0^t \frac{D}{L^2} \, dt \right) \tag{28}$$

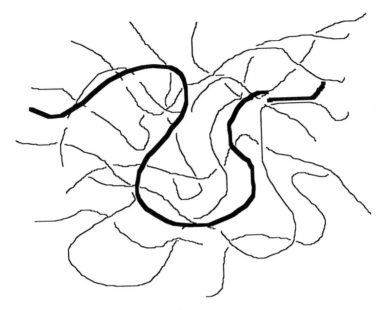

Figure 3. A test polymer chain in an entangled system.

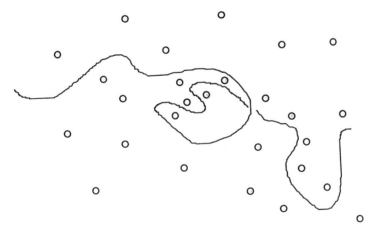

Figure 4. A fixed chain in the presence of obstacles.

As a first approximation, the diffusion coefficient, D, is kept constant. In general, it varies with concentration. Substituting equation (26) into equation (28),

$$F(t) = \frac{8}{\pi^2} \sum_{n\ odd} \frac{1}{n^2} \exp\left(\frac{-n^2\pi^2 D}{L_0^2} \int_0^t \frac{dt'}{F(t')}\right) \quad (29)$$

The term $n = 1$ predominates in the summation in equation (29). Hence, we can write

$$F(t) = \exp\left(\frac{-\pi^2 D}{L_0^2} \int_0^t \frac{dt'}{F(t')}\right) \quad (30)$$

where the front factor has been adjusted from $8/\pi^2$ to unity so as to maintain the initial condition that $F = 1$ at $t = 0$. Differentiating equation (30) with respect to time, we obtain

$$\frac{dF}{dt} = -\frac{1}{t_{rep}} \quad (31)$$

$$t_{rep} = \frac{L_0^2}{\pi^2 D} \quad (32)$$

It is noted that using the appropriate expression for D, the expression for t_{rep} can be made equivalent to the one appearing in equation (17). Integrating equation (31), we have

$$\begin{aligned} F(t) &= 1 - \frac{t}{t_{rep}} & t < t_{rep} \\ &= 0 & t > t_{rep} \end{aligned} \quad (33)$$

Using the scaling law developed for the tube diameter (equation (23)), we have

$$\begin{aligned} \frac{a(t)}{a_0} &= \left(1 - \frac{t}{t_{rep}}\right)^{-0.5} & t < t_{rep} \\ &= \infty & t > t_{rep} \end{aligned} \quad (34)$$

The above equation shows that as a time period equal to the reptation time elapses, the tube diameter goes to infinity, and hence reptation no longer becomes the mode of diffusion for that chain. This completes the self-consistent formulation for the tube model.

Having developed an expression for the tube diameter as a function of time, we now define the disentanglement rate, k_d, as

$$k_d = \frac{L(t)}{t_{rep}} \qquad (35)$$

The justification for the above definition comes from the argument that the disentanglement rate represents a characteristic length for a disentangling chain divided by a characteristic time. We have chosen this length to be the primitive path and the time to be the reptation time. Using the scaling laws developed in the previous section, we can write

$$\begin{aligned} \frac{k_d}{k_{d,0}} &= \sqrt{1 - \frac{t}{t_{rep}}} \qquad & t < t_{rep} \\ &= 0 \qquad & t > t_{rep} \end{aligned} \qquad (36)$$

This completes the prediction of all the parameters in the system. In the above expression, $k_{d,0}$ is the value of the disentanglement rate at $t = 0$ and this can be approximated by using the definition given in (21), which is given by

$$k_d = \frac{r_g}{t_{reptation}} \qquad (37)$$

Simulation Results

Equations (1), (2) and (3) lead to a system of two coupled, nonlinear partial differential equations, one of which is coupled with an ordinary differential equation. The solution of the above system of equations would also generate the temporal evolution of the two moving boundaries and hence the gel layer thickness. The concentration flux can be integrated to obtain the mass of the polymer dissolved as a function of time. The moving boundary problem was transformed into a fixed boundary problem by using "front fixing" techniques (33) that utilize a new set of space coordinates. A modified Landau transform (34) was applied to the concentrated regime, i.e. in the region $R < x < S$. This transform is given by

$$\xi_1 = \frac{x - R}{S - R} \qquad (38)$$

This fixed the moving boundary as ξ_1 varies from 0 to 1. This transform is valid till the glassy core disappears. Once the glassy core disappears, symmetry conditions for both the solvent concentration and the stress prevail at $x = 0$. The new Landau transform that fixes the rubbery solvent interface is

$$\xi_1 = \frac{x}{S} \qquad (39)$$

A similar modified Landau transform was applied to the diffusion boundary layer, i.e. in the region $S < x < S + \delta$. The transform is given by

$$\xi_2 = \frac{x - S}{\delta} \qquad (40)$$

This fixed the moving boundary as ξ_2 varies from 0 to 1 and is valid even after the polymer becomes completely rubbery. These transforms suitably modified the model equations. A fully implicit backward time centered space finite difference technique was then utilized to transform the set of differential equations to a set of non-linear algebraic equations at each time step. The details of the numerical algorithm are presented elsewhere (Narasimhan and Peppas, in preparation). The resulting system was solved by using the Thomas algorithm (35).

To perform simulations with the model, the system methyl ethyl ketone (MEK)/polystyrene was used as a model system. The effect of the polymer molecular weight on the dissolution mechanism was investigated. Figure 5 shows the solvent concentration profile in the polymer ($\overline{M}_n$ = 52,000) as a function of normalized position based on the undeformed coordinate system. The center of the slab is at $\xi_1 = 0$ and the rubbery-solvent interface is at $\xi_1 = 1$. The steep profiles are indicative of a relaxation-controlled dissolution mechanism thus leading to Case II type behavior. The flat profiles in the rubbery region have been attributed to very small diffusional resistance. The glassy core essentially behaves like an impervious wall and as diffusional resistance increases, smoother concentration profiles are observed. Figure 6 shows the temporal evolution of the glassy-rubbery interface, R , the rubbery-solvent interface, S and the gel layer thickness, defined as (S - R). The concentration flux was integrated to obtain the fraction of polymer dissolved and this is shown in Figure 7. The profile is linear, once again providing evidence of Case II transport.

As the polymer molecular weight is increased to 520,000, the reptation time increases and hence the diffusion coefficient decreases. The velocity of the moving interfaces decreases as shown in Figure 8. The mass fraction of the polymer dissolved decreases with increase in the molecular weight as expected (Figure 9). Also, with increase in the molecular weight, the dissolution starts shifting towards a disentanglement-controlled mechanism as is seen from the thicker gel layers. It is interesting to observe that as the mechanism shifts to a disentanglement-controlled one, evidence of Case II transport still persists.

Approximate Solution. Approximate solutions of the model equations (1) - (3) were obtained under pseudo steady state conditions and it can be shown that the normalized gel layer thickness, α (α = (S - R)/l), is given as

$$\alpha = \sqrt{\frac{2\,(\upsilon_2^* - \upsilon_{2,eq})}{l^2}\left[\frac{D_{12}\,(2 - \upsilon_2^*)}{1 - \upsilon_2^*} - D_d\right]t} \qquad (41)$$

Here, υ_2^* is the polymer volume fraction at the glassy-rubbery interface and $\upsilon_{2,eq}$ is the polymer volume fraction at the rubbery-solvent interface.

Experimental verification of the approximate solution of the dissolution model is presented. An example of polystyrene ($\overline{M}_n$ = 179,000) dissolution (11) in methyl ethyl ketone that presents the variation of the gel layer thickness with time is compared with the model predictions using equation (39). Figure 10 shows the gel layer thickness, α, plotted as a function of the square root of time. It is observed that there is good agreement between the model predictions and the experimental data over a range of about 80 minutes.

Experimental Studies

The system chosen for experimental characterization of dissolution was poly(ethylene glycol) (PEG) - water. PEG, with a molecular weight range between 10,000 and 35,000, was obtained from Fluka AG, Switzerland. In a typical experiment, a 75%

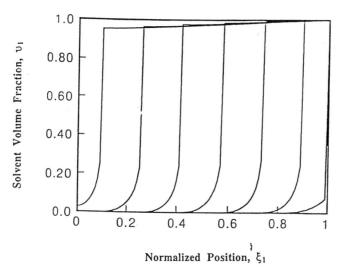

Figure 5. MEK volume fraction, υ_1, as a function of normalized position, ξ_1. The polystyrene molecular weight was $\overline{M}_n = 52,000$. The position $\xi_1 = 0$ is the glassy/rubbery interface. The time increment starting from the first curve on the right is $\Delta t = 3600$ s.

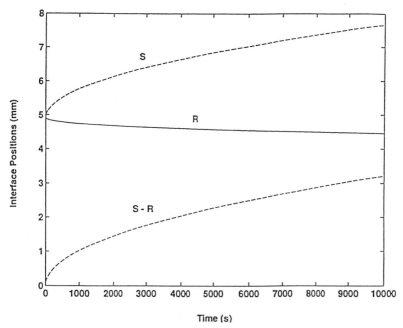

Figure 6. The rubbery-solvent interface (S), the glassy-rubbery interface (R) and the gel layer thickness (S - R) as a function of dissolution time. The polystyrene molecular weight was $\overline{M}_n = 52,000$.

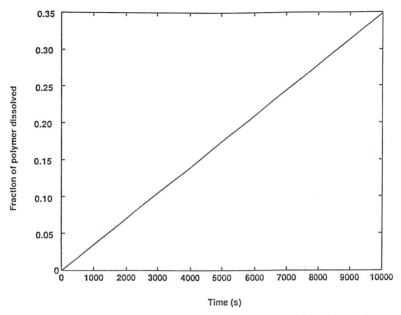

Figure 7. The fraction of polystyrene dissolved as a function of time. The polymer molecular weight was $\overline{M}_n = 52,000$.

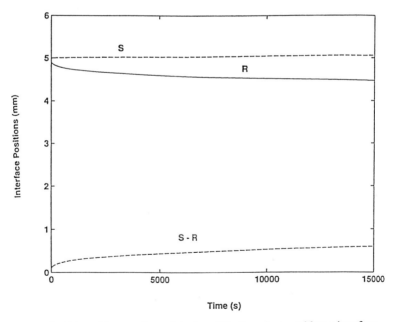

Figure 8. The rubbery-solvent interface (S), the glassy-rubbery interface (R) and the gel layer thickness (S - R) as a function of dissolution time. The polystyrene molecular weight was $\overline{M}_n = 520,000$.

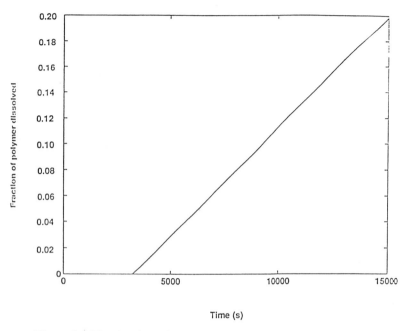

Figure 9. The fraction of polystyrene dissolved as a function of time. The polymer molecular weight was $\overline{M}_n$ = 520,000.

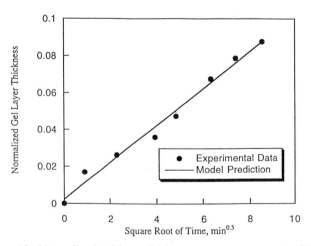

Figure 10. Normalized gel layer thickness versus square root of time for dissolution of polystyrene in methyl ethyl ketone (data of Tu and Ouano (11)).

w/v solution of PEG was prepared in deionized water at 21⁰C. The solution was poured into Petri dishes which were transferred to a vacuum oven maintained at 21⁰C and left to dry for 48 hours. The polymer solution was then poured into cylindrical vials and placed in a freezer for a period of 12 hours. This freeze/thaw process melted all the crystals and a cylinder of nearly amorphous polymer was obtained. Disks of the polymer (aspect ratio = 13) were then cut using a Buehler Isomet low speed saw (Buehler Ltd., Lake Bluff, IL).

A dissolution apparatus (Hanson Research, Northridge, CA) consisting of a dissolution cell equipped with a stirrer placed in a constant temperature bath, was used to conduct front measurements (Figure 11). Plexiglas® disks of diameter 38 mm and height 5 mm were made. The PEG sample was placed in between two such disks (Figure 12) held together with the aid of three screws. This arrangement enabled only radial transport of the solvent. The sample holder was then placed into the dissolution cell and the agitating paddle was lowered into the cell containing the solvent. At regular intervals, the sample holder was withdrawn from the cell, thoroughly dried and photographed with a camera (Reichert Scientific Instruments, Model No. JE 1010, Javelin Electronics Inc.). The position of the rubbery-solvent interface was recorded with the aid of a video recorder (Magnavox, Model No. VR9640AT01, Japan). In a typical experiment, a 1.3 cm diameter and 0.1 cm height sample would be placed in 900 ml of deionized water and the positions of the two fronts would be observed every 10 minutes. It is worthwhile mentioning at this point that the high aspect ratio of the polymer sample and the constraining walls of the plexiglas® disks make one-dimensional measurements relevant.

Dissolution experiments were conducted on the PEG-water system by investigating the effect of polymer molecular weight on the gel layer thickness. Experiments were conducted with PEG samples of molecular weight 10,000, 20,000 and 35,000. All the experiments were conducted at 21⁰C. Figure 13 shows the position of the normalized gel layer thickness as a function of time for various polymer molecular weights during PEG dissolution in deionized water. The positions were normalized with respect to the initial position, which is the initial radius of the disk (6.5 mm). It is instructive to note that since PEG is rubbery at 21⁰C, the whole section essentially represents the gel layer.

The behavior of the gel layer thickness can be explained by considering two different mechanisms for polymer dissolution, chain disentanglement by reptation, and chain diffusion in the boundary layer. In the high molecular weight region, chain entanglement is significant and hence the diffusion through the boundary layer becomes insignificant. This indicates a thicker gel layer as is shown by the experimental data

The experimental data were compared to the model predictions for all three molecular weights studied. The model presented earlier was modified (since PEG is rubbery) and used to predict the temporal evolution of the gel layer thickness for PEG dissolution in water at 21⁰C. Figures 14-16 show the comparison between the model predictions and the experimental data for PEG ($\overline{M}_n$ = 10,000; 20,000; and 35,000) dissolution in deionized water. It is seen that there is good agreement between the predictions and the data within experimental error. This indicated that the model for rubbery polymer dissolution was able to correctly predict dissolution behavior of such polymers.

Conclusions

This work has attempted to provide a molecular understanding of the dynamics of the macromolecular chains during dissolution in thermodynamically compatible solvents.

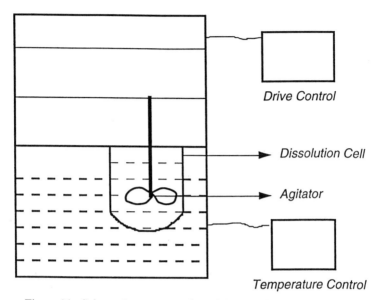

Figure 11. Schematic representation of the dissolution apparatus.

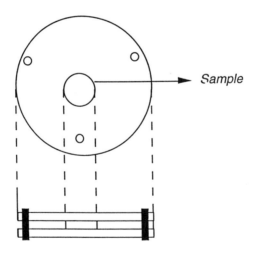

Figure 12. Schematic of the plexiglas® sample holder.

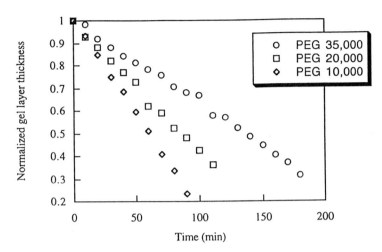

Figure 13. Effect of entanglements on the gel layer thickness during PEG dissolution in water at 21°C. The gel layer thickness was normalized with respect to the initial radius of the polymer sample (6.5 mm).

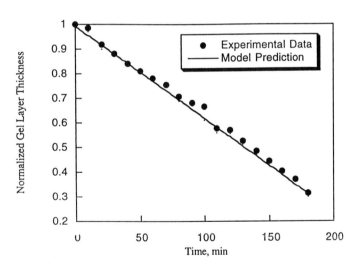

Figure 14. Normalized gel layer thickness as a function of time during PEG ($\overline{M}_n$ = 35,000) dissolution in water at 21°C. Comparison with model predictions.

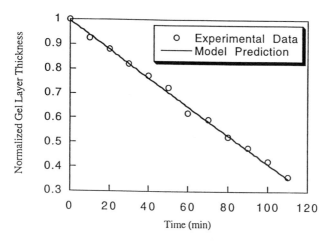

Figure 15. Normalized gel layer thickness as a function of time during PEG ($\overline{M}_n$ = 20,000) dissolution in water at 21⁰C. Comparison with model predictions.

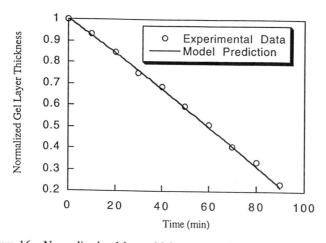

Figure 16. Normalized gel layer thickness as a function of time during PEG ($\overline{M}_n$ = 10,000) dissolution in water at 21⁰C. Comparison with model predictions.

Experimental studies involved measurement of the various moving fronts in the polymer-solvent system. In addition, mathematical models were developed to understand the dissolution mechanism of both rubbery as well as glassy polymers and the model predictions were compared to the experimental results.

The dissolution of poly(ethylene glycol) (PEG) in water was studied by measuring the temporal evolution of the rubbery-solvent interface with the aid of a photographic technique. The effect of the polymer molecular weight on the dissolution mechanism was studied. It was observed that as the PEG molecular weight was increased, the gel layer thickness increased, indicative of disentanglement-controlled dissolution.

The role of chain reptation and disentanglement on the dissolution mechanism of glassy polymers was studied by developing appropriate models. The effect of the solvent diffusional behavior as well as the viscoelastic properties of the polymer on the dissolution mechanism was investigated. The simulations demonstrated that the dissolution mechanism was disentanglement-controlled for higher molecular weight polymers and shifted to diffusion-controlled behavior on increasing the diffusion boundary layer thickness. This was observed by the difference in the profiles of the fraction of polymer dissolved with time. Approximate model solutions were obtained using a pseudo steady state argument. The model predictions were verified with available experimental data on gel layer thickness as a function of time and showed good conformity. The mechanism polystyrene dissolution in methyl ethyl ketone was established to be of the Case II type, which agrees with the predictions available in the literature.

Acknowledgment

This work was supported by grant no. CTS 92-12482 from the National Science Foundation.

References

1. O'Brien, M. J.; Soane, D. S. In *Microelectronics Processing: Chemical Engineering Aspects*; Hess, D. W.; Jensen, K. F., Eds.; ACS Adv. Chem. Series; ACS: Washington, DC, 1989, Vol. 221; pp. 325-376.
2. Colombo, P.; Catellani, P. L.; Peppas, N. A.; Maggi, L.; Conte, U. *Int. J. Pharm.* **1992**, *88*, 99.
3. Tsay, C. S.; McHugh, A. J. *J. Polym. Sci., Polym. Phys. Ed.* **1990**, *28*, 1327.
4. Nauman, E. B.; Lynch, J. C. *US Patent* 5,198,471 **1993**.
5. Nauman, E. B.; Lynch, J. C.; *US Patent* 5,278,282 **1994**.
6. Yeh, T. F.; Reiser, A.; Dammel, R. R.; Pawlowski, G.; Roeschert, H. *Macromolecules* **1993**, *26*, 3862.
7. Wielgolinski, L. *Polym. Prepr. (Am. Chem. Soc., Div. Polym. Chem.)* **1991**, *32* (2), 15.
8. Ueberreiter, K. In *Diffusion in Polymers*; Crank, J; Park, G. S., Eds.; Academic Press: New York, N. Y., 1968.
9. Ueberreiter, K.; Asmussen, F. *J. Polym. Sci.* **1962**, *57*, 187.
10. Asmussen, F.; Ueberreiter, K. *J. Polym. Sci.* **1962**, *57*, 199.
11. Tu, Y. O.; Ouano, A. C. *IBM J. Res. Develop.* **1977**, *21*, 131.
12. Devotta, I.; Ambeskar, V. D.; Mandhare, A. B.; Mashelkar, R. A. *Chem. Eng. Sci.* **1994**, *49*, 645.
13. Ranade, V. V.; Mashelkar, R. A. *AIChE J.* **1995**, *41*, 666.
14. Devotta, I.; Badiger, M. V.; Rajamohanan, P. R.; Ganapathy, S.; Mashelkar, R. A. *Chem. Eng. Sci.* **1995**, *50*, 2557.

432

15. Lee, P. I.; Peppas, N. A. *J. Contr. Rel.* **1987**, *6*, 201.
16. Brochard, F.; de Gennes, P. G. *Physico Chemical Hydrodynamics* **1983**, *4*, 313.
17. Herman, M. F.; Edwards, S. F. *Macromolecules* **1990**, *23*, 3662.
18. Papanu, J. S.; Soane, D. S.; Bell, A. T.; Hess, D. W. *J. Appl. Polym. Sci.* **1989**, *38*, 859.
19. Peppas, N. A.; Wu, J. C.; von Meerwall, E. D. *Macromolecules* **1994**, *27*, 5626.
20. Narasimhan, B.; Peppas, N. A. *J. Polym. Sci. Polym. Phys.* **1996**, *34*, 947.
21. Narasimhan, B.; Peppas, N. A. *Macromolecules* **1996**, *29*, 3283.
22. De Gennes, P. G. *J. Chem. Phys.* **1971**, *55*, 571.
23. De Gennes, P. G. In *Scaling Concepts in Polymer Physics*; Cornell University Press: Ithaca, New York, 1979.
24. Doi, M.; Edwards, S. F. *The Theory of Polymer Dynamics;* Oxford University Press: New York, N. Y., 1986.
25. Raju, V. R.; Menezes, E. V.; Marin, G.; Graessley, W. W.; Fetters, L. J. *Macromolecules* **1981**, *14*, 1668.
26. Flory, P. J., *Principles of Polymer Chemistry*; Cornell University Press: Ithaca, N. Y., 1953.
27. Doi, M.; Edwards, S. F. *J. Chem. Soc. Faraday Trans. II* **1978**, *74*, 1802.
28. Edwards, S. F. *Proc. R. Soc. Lond.* **1982**, *A385*, 267.
29. Crank, J. *The Mathematics of Diffusion*; Oxford University Press: New York, N. Y., 1975.
30. Marrucci, G. *J. Polym. Sci., Polym. Phys. Ed.*, **1985**, *23*, 159.
31. Des Cloizeaux, J. *Macromolecules* **1990**, *23*, 3992.
32. Des Cloizeaux, J. *Macromolecules* **1990**, *23*, 4678.
33. Crank, J. *Free and Moving Boundary Problems*; Oxford University Press: New York, N. Y., 1984.
34. Landau, H. G. *Q. Appl. Math.* **1950**, *8*, 81.
35. Hoffman, J. D. *Numerical Methods for Engineers and Scientists*; McGraw Hill: New York, N. Y., 1992.

Chapter 29

Gas Barrier and Thermal Properties
of (Alkylsulfonyl)methyl-Substituted Poly(oxyalkylene)s

Jong-Chan Lee, Morton H. Litt, and Charles E. Rogers

Department of Macromolecular Science, Case Western Reserve University,
Cleveland, OH 44106

Glass transition temperatures, densities, and oxygen permeability coefficients of (alkylsulfonyl)methyl-substituted poly(oxyalkylene)s have been measured to determine the relationship between the barrier property and the polymer structure. In general, polymers with shorter side chains and higher glass transition temperatures showed better barrier properties. Some of the pairs of (methylsulfonyl)methyl-substituted poly(oxyalkylene)s were found to be miscible. The oxygen permeability coefficients of these 50/50 miscible blends were lower than that of either homopolymer. The oxygen permeabilities of some homopolymers and the miscible blends were comparable to those of EVAL-F and lower than those of any other commercial high barrier polymers.

Polymers exhibiting low permeability to oxygen and other gases are referred to as barrier polymers. They have become increasingly important in the packaging industry over the past three decades (1-4). To develop new high barrier polymers, the structural factors that give low permeability must be understood. Much effort has been devoted to correlating chemical structure with permeability coefficients, and various empirical relations have been suggested (3-8). In general, polymers with highly polar groups show low permeability. Polar groups in polymer give inter- and intra-chain interactions that increase T_g and reduce segmental motion and gas solubility. Another factor that generates high barrier properties is a flexible main chain (8). Ethylene and oxyethylene repeat units are flexible because they can change their conformation very easily; polymers with only these repeats such as polyethylene and poly(ethylene oxide) show very low T_g. This requirement is normally misunderstood because polymers with flexible backbones are usually rubbery at room temperature. These rubbers have high permeability because they have large free volume and segmental motion. However, flexible backbone polymers with non-bulky polar side groups are generally glassy at room temperature and always show very low permeability. In the glassy state, the segmental mobility is greatly decreased and the permeability largely depends upon the residual free volume (9-11). The lower the free volume, the lower the permeability. A flexible backbone allows polymer chains to relax to a small free volume (8). So the polymer with a flexible backbone and non-bulky side groups such as poly(vinyl alcohol) and polyacrylonitrile in the glassy state has very low permeability. Polymers with flexible backbones and bulky side groups (e.g. polystyrene and PMMA) and polymers with rigid backbones (e.g. PPO, and polyimide (Kapton)) are also glassy. However, as their chain packing cannot be very efficient (bulky side groups or a rigid backbone cannot allow the polymer chain to relax to a small free volume), they have large free volumes and relatively high permeability. Their permeability coefficients are normally low in comparison to those of more rubbery polymers, e.g. polyethylene, polybutadiene or poly(dimethyl siloxane) (PDMS) (Table I) (12).

From the above considerations, a high barrier polymer should have a flexible backbone and polar substituents. Polymers such as poly(vinyl alcohol), polyacrylonitrile, poly(vinylidene chloride), poly(ethylene terephthalate) or Nylon 6 are barrier polymers that meet these requirements. Still, the permeability coefficients of poly(ethylene terephthalate) or Nylon 6, where the polar group is in the backbone, are higher than those of poly(vinyl alcohol), polyacrylonitrile, or poly(vinylidene chloride), where the polar group is attached to the backbone. This is true even though the polarity of the amide or ester group is similar to or even higher than the polarity of the hydroxyl, nitrile or chloro group. A polar group in the main chain tends to make the backbone less flexible, but a polar group in the side chain may have only a small effect on backbone flexibility. (Alkylsulfonyl)methyl-substituted poly(oxyalkylene)s composed of a flexible alkylene oxide backbone and very polar sulfone side groups (13) were synthesized previously (14). Scheme 1 shows the structures and the acronyms for the polymers. In this report, the structure-barrier property relationship of these polymers and blends will be discussed in terms of the glass transition temperature, density, and oxygen permeability coefficient.

Experimental

Materials. (Alkylsulfonyl)methyl-substituted poly(oxyalkylene)s were synthesized previously. Detailed synthetic procedures are given elsewhere (14).

Film preparation. 10 wt% solutions of (see Scheme 1 for designation of sample names) MSE, MSEE, MST, ESE, MSE/MST blend, MSEE/MST blend and MSE/MSEE blend were prepared using formic acid as solvent. Chloroform solutions (10 wt%) of n-PrSE, i-PrSE, BSE or PeSE were also prepared. The polymer solutions were filtered using a pressure filter funnel with 10-15 micron pore size to remove any solid particles.

For density measurements, free-standing films were prepared by casting filtered solutions on a trimethylchlorosilane-treated glass plate at room temperature in a glove box under nitrogen atmosphere. The films were dried at room temperature

Scheme 1

Chemical structures, full names, and acronyms of alkylsulfonylmethyl sustituted poly(oxyalkylene)s

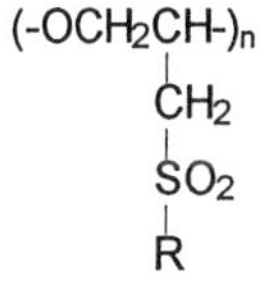

$(-OCH_2CH-)_n$ where R= -CH$_3$;   poly[oxy(<u>m</u>ethyl<u>s</u>ulfonylmethyl)<u>e</u>thylene], MSE
      | -CH$_2$CH$_3$;   poly[oxy(<u>e</u>thyl<u>s</u>ulfonylmethyl)<u>e</u>thylene], ESE
   CH$_2$ -(CH$_2$)$_2$CH$_3$; poly[oxy(<u>n-p</u>ropyl<u>s</u>ulfonylmethyl)<u>e</u>thylene], *n*-PrSE
      | -CH(CH$_3$)$_2$; poly[oxy(<u>i-p</u>ropyl<u>s</u>ulfonylmethyl)<u>e</u>thylene], *i*-PrSE
   SO$_2$ -(CH$_2$)$_3$CH$_3$; poly[oxy(<u>n-b</u>utyl<u>s</u>ulfonylmethyl)<u>e</u>thylene], BSE
      | -(CH$_2$)$_4$CH$_3$; poly[oxy(<u>n-p</u>entyl<u>s</u>ulfonylmethyl)<u>e</u>thylene], PeSE 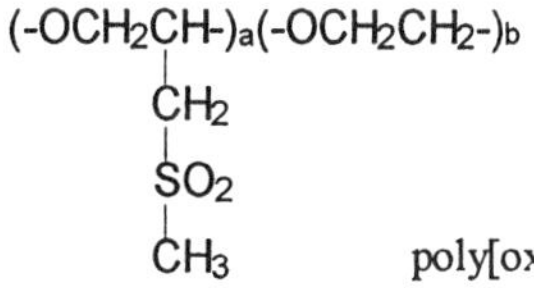
   R

$(-OCH_2CH-)_a(-OCH_2CH_2-)_b$
      |
    CH$_2$
      |
    SO$_2$
      |
    CH$_3$ poly[oxy(<u>m</u>ethyl<u>s</u>ulfonylmethyl)<u>e</u>thylene-*co*-oxy<u>e</u>thylene], MSEE

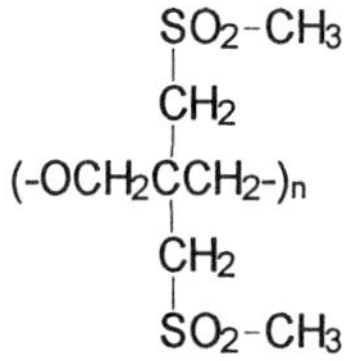

    SO$_2$-CH$_3$
       |
    CH$_2$
$(-OCH_2\overset{|}{C}CH_2-)_n$ poly[oxy-2,2-bis(<u>m</u>ethyl<u>s</u>ulfonylmethyl)<u>t</u>rimethylene], MST
    CH$_2$
       |
    SO$_2$-CH$_3$

under nitrogen atmosphere for 24 h, under vacuum at room temperature for 24 h, and then at incrementally higher temperatures until 130°C was reached after about 4 days. The films were held at 130°C under vacuum for at least 3 days.

For permeability measurements, composite films were prepared because of the brittleness of the free standing film. The composite films were obtained by casting polymer solutions on 2 mil thick Kapton film using a Gardner doctor blade. They were dried using the same procedure described above for the free standing film.

Table I. Oxygen Permeability Coefficients of Known Polymers.

Polymer	Permeability coefficient × 10^{13} [cm^3(STP)·cm/cm^2·s·Pa]	
	Obtained	Literature[b]
LDPE	2.8[a]	2.0[c]
Kapton	0.10[a]	0.13[c]
Barex	0.0037[a]	0.0041[c]
PDMS		367[d]
Poly(butadiene)		14.3[c]
Poly(styrene)		1.9[c]
PMMA		0.116[e]
PET (amorphous)		0.044[c]
Nylon 6		0.028[a]
Poly(vinylidene chloride) (Saran)		0.00383[a]
Lopac		0.0026[c]
EVAL-F[f]		0.001-0.0001[c]
Poly(vinyl alcohol)[f]		0.00665-0.00005[c]
Poly(acrylonitrile)		0.00015[c]

[a]Measured at 30°C. [b]Ref. 12. [c] Measured at 25°C. [d]Measured at 0 °C. [e]Measured at 34°C. [f]Permeability coefficients of poly(vinyl alcohol) and EVAL-F from different laboratories are very different according to the measuring conditions and film preparation methods because poly(vinyl alcohol) and EVAL-F are very sensitive to humidity.

Density measurements. Polymer densities were measured using the neutral buoyancy method using a pyconometer (*15*). The neutral buoyancy medium was a mixture of tetrachloromethane and hexane.

Permeability measurements. The oxygen permeability measurements were conducted using the ASTM D1434 volumetric method at 0% relative humidity (*16*). The volume of the atmospheric pressure side of the permeability cell is very sensitive to changes in temperature and barometric pressure. When the permeability of a polymer is very low and measurements must be made over several days, this

under nitrogen atmosphere for 24 h, under vacuum at room temperature for 24 h, and then at incrementally higher temperatures until 130°C was reached after about 4 days. The films were held at 130°C under vacuum for at least 3 days.

For permeability measurements, composite films were prepared because of the brittleness of the free standing film. The composite films were obtained by casting polymer solutions on 2 mil thick Kapton film using a Gardner doctor blade. They were dried using the same procedure described above for the free standing film.

Table I. Oxygen Permeability Coefficients of Known Polymers.

Polymer	Permeability coefficient × 10^13 [cm^3(STP)·cm/cm^2·s·Pa]	
	Obtained	Literature[b]
LDPE	2.8[a]	2.0[c]
Kapton	0.10[a]	0.13[c]
Barex	0.0037[a]	0.0041[c]
PDMS		367[d]
Poly(butadiene)		14.3[c]
Poly(styrene)		1.9[c]
PMMA		0.116[e]
PET (amorphous)		0.044[c]
Nylon 6		0.028[a]
Poly(vinylidene chloride) (Saran)		0.00383[a]
Lopac		0.0026[c]
EVAL-F[g]		0.001-0.0001[c]
Poly(vinyl alcohol)[g]		0.00665-0.00005[c]
Poly(acrylonitrile)		0.00015[c]

[a]Measured at 30°C. [b]Ref. 12. [c] Measured at 25°C. [d]Measured at 0 °C. [e]Measured at 34°C. [f]Measured at 23 C. [g]Permeability coefficients of poly(vinyl alcohol) and EVAL-F from different laboratories are very different according to the measuring conditions and film preparation methods because poly(vinyl alcohol) and EVAL-F are very sensitive to humidity.

Density measurements. Polymer densities were measured using the neutral buoyancy method using a pyconometer (*15*). The neutral buoyancy medium was a mixture of tetrachloromethane and hexane.

Permeability measurements. The oxygen permeability measurements were conducted using the ASTM D1434 volumetric method at 0% relative humidity (*16*). The volume of the atmospheric pressure side of the permeability cell is very sensitive to changes in temperature and barometric pressure. When the permeability of a polymer is very low and measurements must be made over several days, this becomes very important. The permeability cell with a connected graduated capillary column was immersed in a constant temperature water bath at 30°C. The temperature variation of the water bath was less than 0.002°C. A graduated capillary tube with 0.00205 cm^2 internal cross sectional area was used to measure the volume change in the downstream side. The indicating liquid was Victoria Blue B in 4-methyl-2-pentanone. The downstream side was closed except to the capillary tube.

The volume changes due to changing barometric pressure were then measured. When the barometric pressure decreased from 29.910 to 29.900 inch Hg, the indicator in the capillary tube moved up by 1.85 cm. The length of indicator column was fixed at 1 cm and the volume of the downstream chamber was 11.3 cm^3. The volume change per barometric pressure change was confirmed using the ideal gas equation; the indicator height change in the capillary tube due to a 0.0100 inch Hg pressure change was 1.84 cm.

From the above calibration, when the permeation of a polymer film was measured by measuring the change in indicator column height, the volume change was recalculated for a constant barometric pressure. The permeability cell was first calibrated with known polymers and their permeability coefficients were calculated according to the ASTM D1434 volumetric method. The pressure difference across the film was 330 kPa. A comparison of oxygen permeability obtained from the literature with that measured in the present work is shown in Table I. Considering the different experimental conditions and techniques used, quite a good agreement was found.

The permeability coefficients of (alkylsulfonyl)methyl-substituted poly(oxy-alkylene)s and their blends were calculated by measuring the permeability coefficients of the composite films and using the following equation (17):

$$l/P = l_1/P_1 + l_2/P_2$$

where $l = l_1 + l_2$; l, l_1 and l_2 are the thicknesses of the composite film, substrate film (Kapton) and coated film respectively; and P, P_1 and P_2 are the permeability coefficients of the composite film, the substrate film and the coated film. The thickness of the substrate film (Kapton) was 2.0 mils. The oxygen permeability coefficient of Kapton film was measured as 1.0×10^{-14} cm^3(STP)·cm/cm^2·sec·Pa, Table I.

Oxygen permeability coefficients for each polymer were measured at least 3 times and the deviations were 5-15%. The permeability coefficients listed in Tables I and II are the average values.

Thermal Properties. Differential Scanning Calorimeter (DSC) measurements of polymers (5-10 mg) were carried out under nitrogen at a heating rate of 20°C/min using a Du Pont 921 DSC. The glass transition temperatures of dry samples were obtained from the second scan after quenching the melted samples from above their glass transition temperatures (200°C). The second scan DSC thermograms are shown in Figure 1. TGA curves were obtained using a Du Pont 2000 analyzer at a 50°C/min high-resolution heating rate in an atmosphere of nitrogen gas flowing at 50

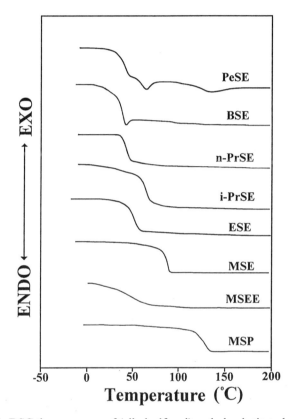

Figure 1. DSC thermograms of (alkylsulfonyl)methyl-substituted poly(oxyalkylene)s obtained from the second run at 20 °C/min after quenching the melted sample at 200 °C.

ml/min. It should be noted that during times when there is a weight change, the rate decrease in proportion to the weight change rate (minimum heating rate is 0.1°C/min), returning to 50°C/min when there is no weight change. So the initial decomposition temperatures obtained from this procedure are closer to the values obtained from the isothermal experiments. As the heating rate changes in proportion to the weight change rate, comparison of the decomposition temperature between the samples are also possible.

Table II. Physical Properties of (Alkylsulfonyl)methyl-Substituted Poly(oxyalkylene)s and Their Blends

Polymers and Blends (weight ratio)	Tg (°C)	Experimental Density (g/cc)	Permeability coefficient[a] $\times 10^{13}$ [cm^3(STP)·cm/cm^2·s·Pa]	[η] (dL/g)
MSE	85	1.442	0.0014	1.63[b]
ESE	57	1.361	0.0088	1.31[b]
n-PrSE	43	1.292	0.22	1.17[b]
i-PrSE	64	1.291	0.12	0.83[b]
BSE	37	1.265	0.40	0.90[b]
PeSE	38	1.249	0.42	1.02[c]
MSEE	52	1.340	0.0036	1.68[b]
MST	126	1.480	0.013	0.84[b]
MSE/MST=1/9	120	1.482		
MSE/MST=3/7	113	1.480	0.0015	
MSE/MST=5/5	103	1.473	0.0007	
MSE/MST=7/3	96	1.458	0.0009	
MSE/MST=9/1	90	1.446		
MSEE/MST=1/9	117	1.479		
MSEE/MST=3/7	105	1.473	0.0013	
MSEE/MST=5/5	92	1.458	0.0008	
MSEE/MST=7/3	70	1.426	0.0009	
MSEE/MST=9/1	57	1.375		

[a]Permeability coefficient of Oxygen at 30 °C measured in this lab. [b]Formic acid at 30 °C. [c]THF at 30 °C.

Results and Discussions

Thermal properties of homopolymers. MSE, MSEE, and MST have a common methylsulfonylmethyl side group. MSE and MSEE have the same oxyethylene backbone structure, but MSE has one sulfone group per repeat unit while MSEE has on average one sulfone group per two backbone units. The backbone of MST is oxytrimethylene, with two sulfone side groups on the middle carbon of each backbone unit (Scheme 1). The more sulfone groups the polymer has the more polar

ml/min. It should be noted that during times when there is a weight change, the rate decrease in proportion to the weight change rate (minimum heating rate is 0.1°C/min), returning to 50°C/min when there is no weight change. So the initial decomposition temperatures obtained from this procedure are closer to the values obtained from the isothermal experiments. As the heating rate changes in proportion to the weight change rate, comparison of the decomposition temperature between the samples are also possible.

Table II. Physical Properties of (Alkylsulfonyl)methyl-Substituted Poly(oxyalkylene)s and Their Blends

Polymers and Blends (weight ratio)	Tg (°C)	Experimental Density (g/cc)	Permeability coefficient[b] × 10^13 [cm^3(STP)·cm/cm^2·s·Pa]	[η] (dL/g)
MSE	85	1.442	0.0014	1.63[c]
ESE	57	1.361	0.0088	1.31[c]
n-PrSE	43	1.292	0.22	1.17[c]
i-PrSE	64	1.291	0.12	0.83[c]
BSE	37	1.265	0.40	0.90[c]
PeSE	38	1.249	0.42	1.02[d]
MSEE	52	1.340	0.0036	1.68[c]
MST	126	1.480	0.013	0.84[c]
MSE/MST=1/9	120	1.482		
MSE/MST=3/7	113	1.480	0.0015	
MSE/MST=5/5	103	1.473	0.0007	
MSE/MST=7/3	96	1.458	0.0009	
MSE/MST=9/1	90	1.446		
MSEE/MST=1/9	117	1.479		
MSEE/MST=3/7	105	1.473	0.0013	
MSEE/MST=5/5	92	1.458	0.0008	
MSEE/MST=7/3	70	1.426	0.0009	
MSEE/MST=9/1	57	1.375		

[a]From eq 1. [b]Permeability coefficient of Oxygen at 30 °C measured in this lab. [c]Formic acid at 30 °C. [d]THF at 30 °C.

Results and Discussions

Thermal properties of homopolymers. MSE, MSEE, and MST have a common methylsulfonylmethyl side group. MSE and MSEE have the same oxyethylene backbone structure, but MSE has one sulfone group per repeat unit while MSEE has on average one sulfone group per two backbone units. The backbone of MST is oxytrimethylene, with two sulfone side groups on the middle carbon of each backbone unit (Scheme 1). The more sulfone groups the polymer has the more polar the polymer and the higher its T_g (Table II). T_g's of MST, MSE, and MSEE are 126, 85, and 52 °C respectively.

MSE, ESE, i-PrSE, n-PrSE, BSE, and PeSE have a common backbone, but have different side chain alkyl groups (Scheme 1). As the number of carbon atoms in the alkyl side chain increases from 1 to 4, T_g decreases, Figure 2. It is known that as the length of side chain increases, neighboring chains are pushed apart, decreasing the hindrance to chain backbone motions, so T_g decreases (18-20). When an isopropyl group is in the side chain, hindrance to backbone motion increases due to its bulkiness, so T_g increases. For example, the T_g of i-PrSE (64°C) is higher than those of ESE (57°C) and n-PrSE (43°C). For PeSE (five carbon atoms in the n-alkyl side chain), the T_g is about the same or slightly higher than that of BSE (four carbon atoms in the n-alkyl side chain). The increase in T_g with increase in side chain length beyond a critical number (N_c) has been ascribed to the side chain crystallization (21); the crystallites acts as fillers in the amorphous phase and raise the T_g (22,23). The N_c values of poly(alkyl vinyl ether)s, poly(alkyl acrylate)s, poly(alkyl styrene)s and poly(alkyl methacrylate)s are 8, 9, 10 and 12 respectively (21). The N_c value of the ASE series is 4, which is comparatively small. The sulfone group in the side chain, which is the strongest polar group among all the simple functional groups, increases the interaction between the side chains as well as their tendency to form ordered phases, such as liquid crystals, even though the side chains are short. PeSE shows two endothermic peaks (Figure 1), at 60°C ($\Delta H_m = 0.38$ J/g) and 128°C ($\Delta H_m = 0.69$ J/g) which might come from the side chain organization. Liquid crystallinty of PeSE has been discussed previously in detail (14). Liquid crystalline phase formation due to strong interchain dipole-dipole interaction between sulfone groups was also reported by several authors (24,25). Recent results on the (n-octylsulfonyl)methyl-substituted poly(oxyethylene) confirm the liquid crystal behavior in this series (26).

Table III. Thermal Stability of Polymers by TGA

	$T_{d,0}$ [a] (°C)	$T_{d,1/2}$ [b] (°C)	Y [c] (%)
MSE	308	333	5.4
ESE	309	330	2.9
i-PrSE	326	342	2.3
n-PrSE	305	331	1.7
BSE	310	337	2.4
PeSE	320	335	0.8
MSEE	330	360	14.7
MST	312	348	6.9

[a] Initial decompositiom temperature. [b] Half decomposition temperature. [c] Char yield at 600°C.

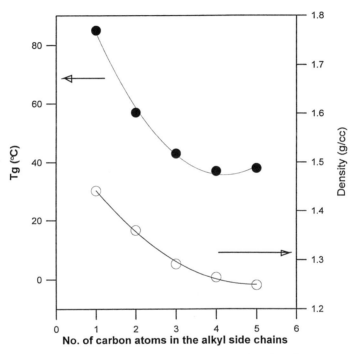

Figure 2. Glass transition temperatures (●) and densities (O) of (alkylsulfonyl)methyl-substituted poly(oxyethylene) as a function of the number of carbon in the alkyl side chains.

The thermal stability of the (alkylsulfonyl)methyl-substituted polymers was evaluated from dynamic thermogravimetric analysis (TGA) under N_2 (Table II). All the (alkylsulfonyl)methyl-substituted polymers were stable to 300°C. Figure 3 shows that all of the (alkylsulfonyl)methyl-substituted polymers have similar decomposition behavior; their initial decomposition and half decomposition temperatures were very close. Table III shows that the side chain length does not affect the thermal decomposition behavior. Backbone structure does affect the decomposition behavior slightly. MST shows a two-step decomposition and MSEE shows the highest thermal stability (Figure 4) of the three (methylsulfonyl)methyl-substituted polymers.

Gas barrier properties of homopolymers. The permeability of a polymer is strongly related to its free volume (or chain packing) and segmental motion. The less free volume (or the more effective chain packing), the lower the permeability. To have low free volume at room temperature, a polymer should have a flexible backbone and non-bulky polar side groups as mentioned in the introduction part; these conditions allow the polymer chains to relax to a small free volume before they become glassy. From this point of view, we might expect that MSEE should have very low permeability because it has the most flexible backbone (on average two ethylene oxide backbone units per one sulfone group, so the conformation change of the backbone is easiest) among the three (methylsulfonyl)methyl-substituted polymers. However, the permeability coefficient of MSEE (0.0036×10^{-13} cm³(STP)·cm/cm²·s·Pa) is slightly higher than that of MSE (0.0014×10^{-13} cm³(STP)·cm/cm²·s·Pa). As the Tg of MSEE is lower than that of MSE and only 22 °C higher than the measuring temperature (30 °C), so MSEE should have more chain segmental motion than MSE. This might increase the permeability of MSEE.

MST has the highest Tg among the three (methylsulfonyl)methyl-substituted poly(oxyalkylene)s, so this polymer should have the lowest chain segment mobility among the (methylsulfonyl)methyl-substituted polymers. The permeability coefficient of MST (0.013×10^{-13} cm³(STP)·cm/cm²·s·Pa) is 5 to 10 times higher than those of MSE and MSEE. MST has two methylsulfonylmethyl side groups which are attached to the same carbon in the backbone; this apparently gives rise to steric hindrance, and the backbone can no longer relax to a very small free volume. MSE has only one methylsulfonylmethyl group per repeat unit; that adds little or no steric hindrance to the main chain so the backbone flexibility is high enough to allow the polymer to relax to very small free volume.

For MSE, ESE, n-PrSE, BSE and PeSE, as the side chain length increases, the permeability coefficient increases (Table II and Figure 5). The permeability coefficient of i-PrSE is lower than that of n-PrSE but higher than that of ESE (Table II). Because the T_g of i-PrSE (64°C) is much higher than 30°C, much less segmental mobility is expected at 30 °C. However, the Tg of n-PrSE is 43°C, only slightly higher than 30°C; there should be much more segmental motion and its permeability coefficient is found to be higher. The T_g of ESE is 57°C, and its segmental motion is therefore expected to be much less than that of n-PrSE. Also the ethyl group is not very bulky, the main chain remains flexible and can relax to a relatively small free

445

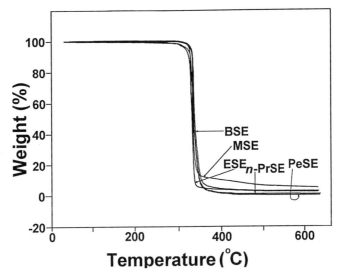

Figure 3. High resolution TGA thermograms of (alkylsulfonyl)methyl-substituted poly(oxyethylene)s under N_2.

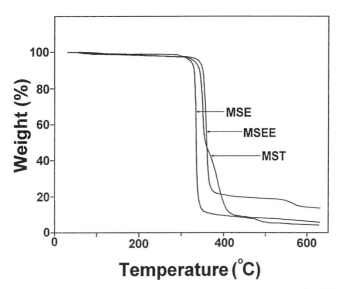

Figure 4. High resolution TGA thermograms of MSE, MSEE and MST under N_2.

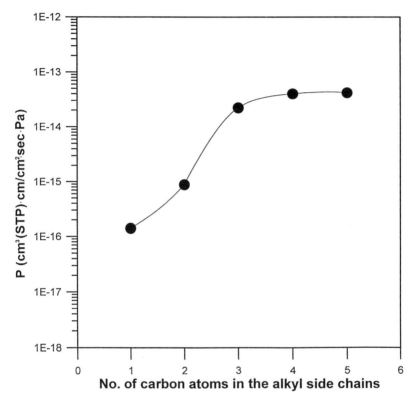

Figure 5. Oxygen permeability coefficients of (alkylsulfonyl)methyl-substituted poly(oxyethylene) as a function of the number of carbon in the alkyl side chains.

volume. The permeability coefficient of ESE is lower than those of i-PrSE and n-PrSE. The T_g's of BSE (37°C) and PeSE (38°C) are slightly lower than that of n-PrSE and the permeability coefficients of these polymers are slightly higher than that of n-PrSE. The plot of permeability coefficient vs. number of carbon atoms in the alkyl side chain, Figure 5, rises sharply as T_g drops, and reaches a plateau when T_g and density plateau.

The oxygen barrier properties of dry MSE, MSEE and ESE are much better than many good barrier polymers, e.g. PET and Nylon 6, and are better than or comparable with the better barrier polymers, such as Barex, Lopac and Saran. The best high barrier polymer, EVAL-F, has the lowest permeability of all commercial barrier polymers with a permeability coefficient in the range of 0.001×10^{-13} to 0.0001×10^{-13} cm³(STP)·cm/cm²·s·Pa (8,17). The highest quoted permeability of EVAL-F, 0.001×10^{-13} cm³(STP)·cm/cm²·s·Pa, is close to the permeability coefficient of MSE.

Gas barrier properties of blends. Several papers have examined the relationship between composition and gas permeability properties for several miscible polymer blends (27-33). When the blends show zero volume change on mixing, the permeability coefficients of these systems show zero or positive deviation from the semilogarithmic mixing rule. However the permeability coefficients of polymer blends that have significant volume contraction upon mixing show negative deviations from the semilogarithmic mixing rule. Cellulose/poly(vinyl alcohol) (33), poly(methyl methacrylate)/bisphenol chloral polycarbonate (29), tetramethyl bisphenol A polycarbonate/styrene-acrylonitrile copolymer (31) and poly(phenylene oxide)/polystyrene (27) blends are of this type. In some cases, such as cellulose/poly(vinyl alcohol) and tetramethyl bisphenol A polycarbonate/styrene-acrylonitrile copolymer blends, the permeability coefficient of the blend at a certain composition (normally 50/50) is lower than that of either homopolymer.

If two barrier polymers are miscible and the blend shows volume contraction upon mixing, the blend could be a better barrier than either polymer separately, because the free volume in the blend may be less than in either polymer alone. Among the (alkylsulfonyl)methyl-substituted poly(oxyalkylene)s, MSE/MST and MSEE/MST blends were miscible over their entire composition range. Figures 6 and 7, show DSC curves of MSE/MST and MSEE/MST miscible blends.

Figure 8 shows the densities and oxygen permeability coefficients of the MSEE/MST blends as a function of composition. The density versus composition curve shows density increase upon mixing. As might be expected the oxygen permeability coefficients show negative deviations. When the weight composition of the MSEE/MST blend is 50/50, a significant density increase, about 3.4 %, from that calculated using the linear additivity rule, was observed. Its oxygen permeability was very low, 0.0008×10^{-13} cm³(STP)·cm/cm²·s·Pa. In Figure 9 the density vs. composition curve for MSE/MST blends also shows a positive deviation, but smaller than in the MSEE/MST blends (0.8% at 50/50 composition). Similarly, the permeability coefficient vs. composition curve for the MSE/MST blend shows less negative deviation than that of MSEE/MST blends. Still the oxygen permeability

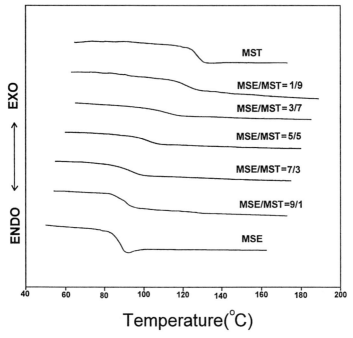

Figure 6. DSC thermograms of MSE/MST blends obtained from the second run at 20 °C/min after quenching the melted sample at 200 °C.

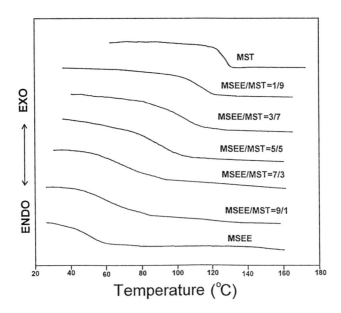

Figure 7. DSC thermogram of MSEE/MST blends obtained from the second run at 20 °C/min after quenching the melted sample at 200 °C.

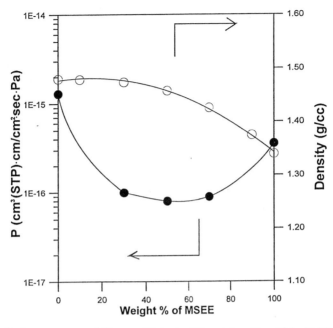

Figure 8. Oxygen permeability coefficients (●) and densities (O) of MSEE/MST blends.

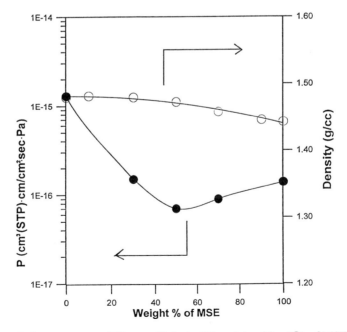

Figure 9. Oxygen permeability coefficients (●) and densities (O) of MSE/MST blends

coefficient for the 50/50 MSE/MST blend, 0.0007×10^{-13} cm^3(STP)·cm/cm^2·s·Pa, is the lowest for the systems studied in this report. The oxygen permeabilities of these blends are comparable to those of EVAL-F and are much lower than those of any other commercial high barrier polymers.

Conclusions

The thermal and barrier properties of (alkylsulfonyl)methyl-substituted poly(oxy-alkylene)s were investigated. For (alkylsulfonyl)methyl-substituted poly(oxy-ethylene)s (MSE, ESE, n-PrSE, BSE, PeSE), as the number of methylene groups in the side chains increased from 1 to 3, the T_g and density decreased rapidly and the oxygen permeability coefficient increased rapidly. As the number of methylene groups increased from 3 to 5, the properties did not change much. For (methyl-sulfonyl)methyl-substituted poly(oxyalkylene)s (MSE, MSEE, and MST), the permeability is strongly related to their structures. The permeability coefficient of MSEE which has the most flexible backbone among the three polymers plus a polar side group was slightly higher than that of MSE because its T_g was near room temperature. The permeability coefficient of MST was much higher than those of MSEE and MSE because the two bulky substituents on one center atom prevented relaxation to a low free volume. MSE showed the best gas barrier properties among the alkyl(sulfonylmethyl)-substituted poly(oxyalkylene)s; its oxygen permeability coefficient is lower than that of any commercial barrier polymer except EVAL-F.

MSE/MST and MSEE/MST blends were miscible by DSC measurements. Their density vs. composition curves showed positive deviations from the linear additivity rule; volume contraction upon mixing two miscible polymer pairs was confirmed. Permeability vs. composition plot also showed large negative deviations from the semilogarithmic mixing rule. The permeability coefficient of the 50/50 MSE/MST blend was the lowest of all the materials tested; it is comparable to EVAL-F and much lower than any other commercial barrier polymers. The volume shrinkage in the compatible blends show that the free volume is reduced, thus reducing the permeability.

All the alkyl(sulfonylmethyl)-substituted polymers were thermally stable until 300°C, which is well above their glass transition temperatures. These polymers should be melt processable, allowing production of barrier films.

Acknowledgment

Financial support of this work by Edison Polymer Innovation Corporation (EPIC) and Allied-Signal Corporation is gratefully acknowledged.

Literature Cited

1. Koros, W. J. *Barrier Polymers and Structures*, American Chemical Society, Washington DC, 1990.
2. Gomez, I. L. *High Nitrile Polymers for Beverage Container Application*, Technomic Publishing Company, Inc., Lancaster, 1990.
3. Steinger, S; Nemphos, S. P; Salame, M. In *Encyclopedia of Polymer Science and Technology*, Vol. 3, Interscience Publisher, New York, 1965, p. 480.
4. Combellick, W. A. In *Encyclopedia of Polymer Science and Engineering*, Vol. 2, John Wiley & Sons, Inc., New York, 1985, p. 176.
5. Jia, L.; Xu, J. *Polym. J.* **1991**, *23*, 417.
6. Salame, M. *Polym Eng. Sci.* **1986**, *26,* 1543.
7. Lee, W. M. *Polym Eng. Sci.* **1980**, *20,* 65.
8. Zhang, T.; Litt, M. H.; Rogers, C. E. *J. Polym. Sci., Polym. Phys. Ed.* **1994**, *32,* 1671.
9. Litt, M. H.; Tobolsky, A. V. *J. Macromol. Sci., Phys. B*, **1967**, *3*, 433.
10. Plate, N. A.; Durgarjan, S. G.; Khotimskii, V. S.; Teplyakov, V. V., Yampolskii, Y. P. *J. Membr. Sci.* **1990**, *52,* 289.
11. Litt, M. H. *J. Rheol.* **1986**, *30,* 853.
12. Pauly, S. Permeability and Diffusion Data, In *Polymer Handbook*, 3rd ed., J. Brandrup and E. H. Immergut, Eds., John Wiley & Sons, Inc., New York, 1989, p. VI 435.
13. *CRC Handbook of Chemistry and Physics*, 60th ed., Weast, R. C. Ed. CRC. Press, Cleveland, 1979, p. E64.
14. Lee, J.-C.; Litt, M. H.; Rogers, C. E. *Macromolecules* **1997**, *30,* 3774.
15. Runt. J. P. In *Encyclopedia of Polymer Science and Engineering*; Mark, H.; Bikales, N. M.; Overberger, C. E.; Menges, G., Eds.; John Wiley & Sons, Inc.: New York, 1986; Vol. 4; p 458.
16. Annual Book of ASTM Standard, D 1434, Sec. 15, Vol. 15.09, 1991.
17. Ashley, R. J. In *Polymer Permeability*, J. Comym, Ed., Elsevier Applied Science Publishers, London, 1985, p. 269.
18. Rogers, S. S.; Mandelkern, L. *J. Phys. Chem.* **1957**, *61,* 985.
19. Hayes, R. A. *J. of Appi. Polym. Sci.* **1961**, *15,* 318.
20. Wesslin, B.; Lenz, R. W.; MacKnight, W. J.; Karasz, F. E. *Macromolecules* **1971**, *4,* 24.
21. Shalaby W. S. In *Thermal Characterization of Polymeric Materials*; Turi, A Ed; Academic Press, Inc,: New York, 1981; p344.
22. Van Krevelen, D. W. *Properties of Polymers*, 2nd ed.; Elsevier Science Publishers: New York, 1990.
23. Jordan, Jr. E. F. *J. of Polym. Sci.* **1971**, *9*, 3367.
24. Zhang, T.; Litt, M. H.; Rogers, C. E. *J. Polym. Sci., Polym. Chem. Ed.* **1994**, *32,* 2291.
25. Dass, N. N.; Date, R. W.; Fawcett, A. H.; McLaughlin, J. D.; Sosanwo, O. A. *Macromolecules* **1993**, *26,* 4192.
26. Lee, J.-C.; Litt, M. H.; Rogers, C. E. *Polym. Prepr.* **1997**, *38*(2), 169.

27. Paul, D. R. *J. Membrane Sci.* **1984,** *18,* 75.
28. Chiou J. S.; Paul, D. R. *J. Appl. Polym. Sci.* **1986,** *32,* 2897.
29. Chiou J. S.; Paul, D. R. *J. Appl. Polym. Sci.* **1987,** *33,* 2935.
30. Chiou J. S.; Paul, D. R. *J. Appl. Polym. Sci.* **1987,** *34,* 1037.
31. Chiou J. S.; Paul, D. R. *J. Appl. Polym. Sci.* **1987,** *34,* 1503.
32. Ping, Z; Nguyen, Q. T.; Néel, J. *Macromol. Chem.* **1994, 195,** 2107.
33. Patel, K. Manley, R. St. J. *Macromolecules,* **1995,** 28, 5793.

INDEXES

Author Index

Subject Index